ALERT !

We have arrived at the Prophet Daniel's

LAST predicted 'DAY'

Daniel 12:12 'Blessed is the one who waits for and reaches the end of the 1,335 days'.

Hold on to the end!

SEVEN TIMES
-EGYPT TO ISTANBUL-

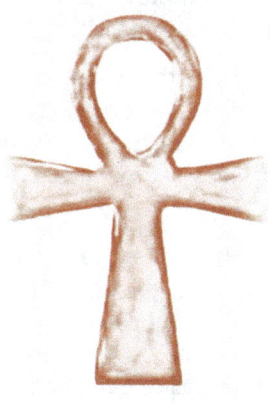

'I will go before you'

'I will go before you and will level the mountains; I will break
down gates of bronze and cut through bars of iron.
I will give you the treasures of darkness, riches stored in secret places, so that you may know that I am the Lord, the God
of Israel, who summons you by name'.

Isaiah 45:2

This first edition

Copyright © 2019 by Scope 'n Compass

ISBN 978-0-6487168-0-8
ISBN 0-6487168-0-5

Rev 1.4
(2022-2024)

All rights reserved. No part of this publication may be reproduced, distributed or transmitted in any form or by any means, without the permission of the publisher.

Notwithstanding the above, the content may be quoted for the purposes of dissemination and review, and other noncommercial uses permitted by copyright law. The object is to spread the key findings of the message worldwide.

Printed and bound by IngramSpark (Ingram Book Group) and Associates in Australia, the United States and United Kingdom.

All Scripture quotations, unless otherwise indicated, are taken from the Holy Bible, New International Version®, NIV®. Copyright ©1973, 1978, 1984, 2011 by Biblica, Inc.™ Used by permission of Zondervan. All rights reserved worldwide. www.zondervan.com
The "NIV" and "New International Version" are trademarks registered in the United States Patent and Trademark Office by Biblica, Inc.™

Bible texts designated (RSV) are quotations taken from the Common Bible: Revised Standard Version, copyright © 1973 National Council of the Churches of Christ in the United States of America. Used by permission. All rights reserved worldwide.

Scripture quotations tagged as (YLT) are from the 1898 Young's Literal Translation by Robert Young who also compiled Young's Analytical Concordance.

Front piece:
The Sign of Aquarius - Sign of a Man. Of the end of our current age Jesus said 'Then will appear the sign of the Son of Man in heaven.'

The Ankh – life; here representing the Son of Man. Jesus said 'I am the resurrection and the life' and 'No one has ever gone into heaven except the one who came from heaven – the Son of Man.'

    David and Solomon 231
VISIONS OF ELOHIM 244

  A Prophet or Two 244

    To whom it may concern 244
    Conventional wisdom 253
  Rules of Engagement 273

    A day for a year 273
    The year is how long? 276
    Time, times and half a time 286
    A new millennium 293
  Age of the Gentiles 303

    Good time, bad times 303
    Diaspora and the Brit thing 314
    Tribulation 330
    The goat and the prophet 338
  Animal Kingdoms 345

    The first 'three and half times' 345
    Messianic prophecy 367
    Forty years of fear 384
  Babylon the Great 387

    The half time beast 387
    Caliphates and crusades 407
    The end of the age 426
SIGN OF A MAN 434

  The Water Carrier 434

    The 'fig tree' buds 434
    The sixth seal 438
    Woes and wonders 446
    Celebration in the sky 458
  Whore on the Water 463

Back to the future	463
Abraham and Senusret	471
Mystery Babylon	477
Epilogue	495
Appendix 1	506
Lyrics to a rock tune	506
Appendix 2	507
The false prophet	507
Appendix 3	514
Unravelling Ezra	514
Appendix 4	521
A dating dilemma	521
Run, hop, step and jump	523
Calendar confusion	533
Appendix 5	542
Meton and Callippus	542
Heavenly visitors	547
Appendix 6	552
Exodus revisited	552
Appendix 7	554
Apocalypse vow	554
Bibliography	560

SEVEN TIMES

–EGYPT TO ISTANBUL–

For my parents who nurtured me and guided my youthful thoughts, to my family for watching out for me in trials of later years, and to fellow travellers who have walked the path and sought the way

"...if my people, who are called by my name,

will humble themselves and pray

and seek my face

and turn from their wicked ways,

then I will hear from heaven

and will forgive their sin

and heal their land."

(RSV)

Table of Contents

- Seventh of the Pharaohs 11
- IN THOSE DAYS 22
  - Mists of Time 22
    - Beginnings 22
    - The long-legged Aramean 36
    - Scripture and science 44
  - On Creation 49
    - The changing universe 49
    - Six day confusion 60
  - Ice and Flood 85
    - Giza 85
    - Reminiscing 91
    - Melting of the ice sheets 92
    - Back to 2350 BC 99
  - Hieroglyphs and Pictograms 111
    - Language and writing 111
    - Towers and tables 115
    - On the road 122
  - Cuneiform and Clay 131
    - Mari where are you? 131
    - The numbers game 155
  - Exodus Cataclysm 165
    - See you in Santorini 165
    - Avaris and Goshen 183
    - The Moses mystery 188
  - Israel 199
    - The Exodus narrative 199
    - Return to the land 210
    - King for a while 223

Illustrations

Figure 1. A map of the Eastern Mediterranean, the Near East, Egypt and surrounding regions
Plate 1. The Giza pyramids
Plate 2. The Sphinx
Plate 3. Khufu's funerary river boat
Plate 4. The natural harbour of Santorini (Thera)
Plate 5. Jebel Katrine (Mount Saint Catherine) Sinai Peninsula
Plate 6. Saint Catherine's Monastery
Plate 7. The Siq, Petra, Jordan
Plate 8. Queen Hatshepsut's temple at Deir el Bahari, Luxor
Plate 9. Ramses II temple at Abu Simbel, Upper Egypt
Plate 10. Aswan and the Nile
Plate 11. The Aswan Coptic Cathedral
Plate 12. The Mosque of Amr ibn al-As, Cairo
Plate 13. The Parthenon, Athens
Plate 14. A picturesque Athens alleyway
Figure 2. A map of the Holy Land and region
Figure 3. A plan of Jerusalem through the ages
Plate 15. In the Old City of Jerusalem
Plate 16. The Western Wall and Dome of the Rock
Plate 17. The Jerusalem British War Cemetery
Plate 18. Gethsemane; the olive press
Plate 19. The Library of Celsus, Ephesus
Plate 20. 'Mary's House' on the hill behind Ephesus
Plate 21. The Thutmose III obelisk, Istanbul
Plate 22. The Hagia Sophia basilica, Istanbul
Plate 23. The entry to the Holocaust Memorial, Jerusalem
Plate 24. A plaque to those who perished in the Shoah
Figure 4. The Daniel 'seven times' timeline chart with intersecting historical lines, including the overarching ages of the zodiac

Introduction

Seventh of the Pharaohs

I scrambled up the last few steps, past the seventh floor – then the eighth. The elevator did not quite reach the rooftop restaurant of the tourist hotel.

A large entertaining area, complete with bar and attendant, was baking in the dry desert heat and the old world air of an ancient land seeped in from every side. The translucent roofing was sculptured in a pleasant, if faded style and the languid flowing lines and graceful colours of the drapes and décor blended well. The combination trapped the rising temperatures of the fast approaching summer though, and there was an unfamiliarity as well and no room for doubt, I was far from home.

It was early evening, a couple of minutes past six and I looked around for the small group who were to be my travelling companions. They were tucked away to the left at a long dining table, where our chief guide had already commenced his briefing. I judged him to be in his late thirties, slightly built of medium height, with aquiline, somewhat scholarly features.

Not particularly characteristic I would have thought of this part of the world. But then I was to learn what a cosmopolitan, multi ethnic country this was.

He was at the head of the table with six young people; four girls and two guys in their mid-20s to mid-30s. They were sitting on both sides, eagerly introducing themselves while soaking up the atmosphere.

I, on the other hand was the only oldie and took the remaining available seat at the other end; taking my time to relax for the evening and appearing, I suppose, like the bearded grandfather I was. Five of the group were from Canada, and one lass of Chinese descent, an accountant from Sydney, Australia.

My roommate to-be was of Sri Lankan heritage and an engineer from Toronto. He was travelling with an engineer friend from university days and his lady partner. The younger girls were from Saskatchewan out on a big adventure. One had already travelled widely, including a trip to her ancestral homeland in the Ukraine.

So this was Cairo, Egyptian to the core, Arabic speaking, but with the diverse mix of humanity expected in a country of over 90 million people. A country where English and French are also taught as a matter of course and spoken by a fair proportion of the population. A nation many times at the centre of great empires stretching to the Euphrates and Anatolia, Ethiopia, Nubia, Libya and beyond.

A nation also visited and invaded by people of many ethnicities and cultures for well over five thousand years. Some looking for land and loot, others looking for a market and an opportunity. Some offering expertise and ability, or simply tourists fascinated by monuments and gob smacked by history.

Unlike any other nation; not quite Middle Eastern, African but unique, certainly not European, and yet with the faded but still vibrant heritage from that continent - just overwhelmingly and proudly Egyptian, with its amazing people, almost unbelievable history and timeless heritage.

I had already enjoyed a couple of days in Egypt by myself and had squeezed in some sites around town; the Citadel, Old Cairo, the river, tourist shops and markets with all manner of copper plate, crystal baubles, aromatics and the ubiquitous papyrus with their 'handmade' factory stamped pictures of pharaohs and chariots, archers and lotus plants. Genuine papyrus of course – never that fake banana leaf!

Then a day trip to Alexandria to catch a glimpse of the ocean and clamber over the Qaitbay Citadel, where the famous Pharos Lighthouse once stood. Later the catacombs and Pompey's Pillar (that my lovely guide Hend advised was *not* erected by Pompey at all) and a delightful seafood lunch on the first floor of a restaurant overlooking the harbour.

Of course I had also been there long enough to be ripped off around town by an opportunistic trader – but you soon learn.

After all they are not unique to Egypt, it is simply the price you pay for an overseas education. Your money is often extracted with a smile and besides; you meet many wonderful, friendly people who more than make up for the occasional pest.

In no time at all, we had the basic trip rules re-emphasised by Hany (our supervising guide) and a few extra survival hints thrown in to make our time enjoyable. 'Don't be afraid to barter', and 'have the right change and you will never be disappointed with the meagre amount offered on the street'.

Of course there were a lot more travel details besides, and a few options to be decided later in regard to our tight and exciting schedule; like a dash to Abu Simbel from Aswan and a balloon ride over the west bank at Luxor.

Things were looking up and I started to feel more at home and relaxed. Dusk was falling as we walked from the Pharaohs Hotel to a small cafe-restaurant a few minutes away.

I came to enjoy my Egyptian fare which was always tasty, varied and nourishing. It was only later, after a few weeks in the Near East and Eastern Mediterranean, that I really understood what chefs mean when they say, 'nothing quite enhances the taste of food like ham and bacon flavour'.

But I survived. It was my own personal Ramadan - with a twist.

But why was I here? A holiday, a pilgrimage, a study tour? Perhaps all three, but apart from needing an overdue work break, I had an overriding desire to follow up ideas from several books; some historical, some academic, some speculative - and thread research interests together in a way that would make the break as meaningful as possible - as well as an adventure.

I guess after many years working in geoscience, there was always going to be a need to string exploration, intellectual goals and concepts together. Sometimes it is a pain not feeling able to just travel for the heck of it. But this type of travel can be very stimulating and for me personally, satisfying at the very top shelf level. It is just another journey, but with a theme - and a theme with a purpose.

Mulling over history for years, particularly ancient history and the Bible, I had discussed many topics with my father, a pastor in his early years. He had his hobby horses, I had mine. His viewpoint was fundamentalist for the most part and when it came to the Bible, he had no time for those 'higher critics'. On the other hand, he had a broad interest in science and enjoyed reading magazines like the Scientific American and National Geographic, when he had the time and cash to do so.

From a biblical point of view, he was fascinated by Bible prophecy and steered me toward the 'historicist' approach to the prophets. A view which seemed the most logical of all anyway, after exploring the alternatives over time.

One of his other interests was the mystery of the 'lost tribes of Israel' and their likely identification today.

As I matured, the question of how much of the Bible could be relied on became more important - especially during my high school years where I pretty much ended up studying geology by accident. Looking back on it now, it is more than tempting to see Divine guidance. At the time it was rather hilarious.

Like many young men I was keen on some sort of professional flying career. However I had already failed to be accepted as a Navy midshipman because of poor eyesight (it was 20-20 vision or nothing in those days) and probably a good thing if you are landing an aircraft on a carrier or helideck. So standing in a line-up at the beginning of year 11 and continuing an academic course of two math, physics, chemistry and English, it was just a matter of finding one more subject compatible with the school schedule.

I had decided that was going to be geography, because of its relationship with my interest in the outdoor world, navigation and a vague hope that flying might still have a place in my life.

Now we had a school Principal who we called 'The Boss'. He was a Scotsman, five foot nothing tall (or close to it) who made up for that with a practiced composure and presence, that was guaranteed to make any teenager, however tall, brash or self-confident, know his place.

The Boss: 'So young man, what have you chosen?'
Me: 'Well Sir, I would like to do geography'.
The Boss: 'That's not possible. It clashes with the schedule for your other subjects'
Me: 'Yes Sir'.

The Boss: 'Geology sounds almost the same as geography doesn't it my man?'

Me: 'Yes Sir'.

The Boss: 'Well stand over there in the geology line'.

Me: 'Yes Sir' - and I trotted over to the geology group.

It turned out to be a fortuitous move, firstly because it was an all-boys class and we had such a great time, albeit getting into a good deal of trouble by way of one of our teachers, the Assistant Principal. He went by the nickname of 'Shambles' because of his gait and other mannerisms.

He taught one of the mathematics subjects and physics, and just happened to be absent on Assistant Principal's duties for a fair proportion of the time. This allowed us to get up to considerable mischief and make a hell of a racket.

We were all threatened with expulsion by 'The Boss' on one occasion, after a frozen orange passed cleanly through a window pane. Constant visits by teachers from adjacent classrooms were also a feature, because of the noise we were making.

Some bright spark decided to rearrange our personal lockers early on. They were situated at the back of the classroom and were brought forward enough, so that a couple of seats could be placed in the small enclosure that was created behind.

We took turns to play chess or other games there, or get up to more nefarious activities. Shambles would eventually turn up (although not guaranteed) with kids still behind the lockers and shoes, apple cores and orange peel being tossed to and fro to provoke a response one way or the other.

No one ever got caught, although I am sure our Assistant Principal played along with the chaos to keep the quiet.

In any event, he also managed to teach us a thing or two in a way which was absorbed, despite the distractions of our adolescent minds; something which must rank as one of God's minor miracles.

On the more productive side, by the time I left school I had the benefit of a four year science education, including a geology course which was easily a first year university equivalent; physical geology, mineralogy, petrology and paleontology.

Our high school was one of only three in the mighty state of Queensland, Australia, to conduct the course in those days.

So it was a privilege and I did pretty well, and must have gone close to topping the state, even if that was only three classes at the time. As it happened, it helped shape a different career from that of my choosing, but one that was to be in many ways, so much more meaningful and challenging than I could ever have imagined.

As an adult student I was to go back and redo much of this in the 80s - particularly physical geology, mineralogy and petrology. In 1963 when I started the course at school, the revolutionary data about continental drift, plate tectonics and the resultant dynamics of mountain building and volcanism had yet to arrive in the class room.

Looking back now, how amazing to have studied and worked in exploration geoscience during this time!

So it was in this setting of a traditional Protestant evangelical background, side by side with a 20th century science education, that formed my thinking about the Bible and what it meant in a changing world. Whether its history was reliable or relevant and for that matter, whether it's teaching, principles, laws and relationships were still applicable. A lot to chew on really, and even if the historical milestones were only a connecting chain, they seemed important to understanding. At least to someone searching for truth and making sense of the world, and the rise of the technological society we have today.

While my immediate focus was primarily engineering, as a student of earth science and one who had a strong interest in ancient history, connecting the dots between geological history and that of human prehistory, archaeology and the written word was a natural and obvious challenge. It was just not all consuming; so much to do, places to be, a family to be brought up, a career to cement and life to be lived.

Strangely I never got as far as really questioning the New Testament. Sure, I knew there were minor discrepancies, but other more immediate testimony in my own experience and person, seemed to confirm its veracity. The four viewpoints of the gospels were comforting too, like reading four eyewitness accounts of the same event. Differences obviously, but exactly what you would expect in news coverage today.

Besides, I came to realise those were extraordinary and chaotic times and there was really nothing contradictory enough to diminish the narrative, despite the amazing events recounted. Events almost too incredible and dramatic not to have some basis in fact I decided - despite of course, other times of doubt.

So here I was, the beneficiary of a few weeks break in a hectic work schedule, after shoving together an itinerary at the last moment. I was hoping to get a feel for the truth, or otherwise, of great events recorded in the biblical Old Testament. Events for the most part derided by a sceptical 20th and 21st century academia, while at the same time accepted without due diligence by many fundamentalist believers.

And a look at New Testament places too; cities and towns whose names were very familiar, but otherwise meant little in terms of geography and social setting.

On this trip it was the Exodus saga I was most interested in, although I had also recently read a book about Mary the mother of Jesus. It speculated on her possible travels and final destination during the tumultuous events of the first century.

So I had a couple of threads to follow and I thought that would help to tie my holiday travels together.

However a year or two after returning home - and following a couple of rather remarkable experiences - I began to realise that I needed to write things down. At first I tried the easy way out, by joining a group online and sharing things that way. But the babble that is social media was not going to do the subject justice and in any case, once I started writing and my geoscience interests kicked in, I ended up at the very beginning.

The Creation story, Noah's Flood, Babel and the Exodus to kick things off, and then moving on to Israel's King David, Solomon and the Prophets of the Babylonian exile. This then rolled into the remarkable timeline for the subsequent 2,500 years of civilisation, all built on the backs of the great ancient empires of Middle Eastern and Mediterranean history.

Despite my interest in the events preceding the Davidic Kingdom circa 1000 BC, ultimately I was to see things on my travels which either confirmed my beliefs, or ruminating on them, were to take on a new significance that I did not foresee at the time. This was partly due to new research coming to light; including tree ring dating evidence on one hand and new archaeological and geological evidence on the other.

Furthermore, some of this material was only published well into the second decade of the 21st century.

Other propositions such as language and the potential significance of the Egyptian hieroglyphs also intrigued me, as I eyed off massive inscribed columns and wall after wall of writing.

And then there were the prophecies of the Old and New Testaments. Could you take them seriously as I had as a young man - but then given away as too hard to interpret anyway - or not?

But thoughts were not so deep and meditative, as we finished our first evening meal together and strolled back to the hotel chatting and getting to know one another. The morrow was planned, straight into a tourist highlight; a visit to the Giza Pyramids and a camel ride, followed by the Cairo Museum in the early afternoon!

Fortunately, between my professional duties in geological survey operations, engineering support and broader capability issues in national geoscience, I had time enough to acquaint myself with the various dynasties of Egyptian history. The Old, Middle and New Kingdoms followed by the Late Period - divided respectively by the First to Third Intermediate Periods.

In particular I was interested in the Second Intermediate Period and the New Kingdom, because this is the time generally associated with the biblical Exodus, i.e. by commentators courageous enough to write on the subject.

So the next morning dawned warm and dry and after breakfast we headed for Giza, a few short kilometres from the Nile. Apparently the old course of the river (or at least a canal) had been immediately adjacent to the pyramids when they were constructed and allowed easy access to them and the temples. Easy access these days was by Toyota minibus!

The wonder of the Great Pyramid greeted us, Khafre's pyramid with its shiny remnant facing stone at the top, stood to the right hand side looking as we were from the north. The pits where Khufu's wooden funerary Nile boats were buried for his afterlife were to the left and east. Later in the morning after a picturesque ride via camel, we visited the amazing and enigmatic figure of the Sphinx, marvelled at the void cut out of the sandstone rock around it, and the obvious craftsmanship of the stone in the adjacent temples.

No one else in our group bothered, but I took time out to visit the tall building at the back and on the south side of Khufu's Great Pyramid. This housed the first of the two wooden river boats to be restored from the pits. They were discovered in the 1950s and it certainly makes one wonder what else remains to be unearthed in this place. Buried around 2550 BC, beautifully restored, complete with the ancient rope that was found with all the timber - I was mesmerised.

OK, I was a mariner of sorts, a seagoing technical and science specialist who had no qualms calling a ship, a boat, to get a rise out of professional seamen.

In fact I had special interest in boats. I had recently spent quite some time on a research vessel that was the same overall length. The difference was that the research ship - although a pocket sized one by today's standards - was nevertheless much broader, higher and heavier than the slim river boat before me. But heck, this one was over 4,500 years old. Incredible!

The significance of this in the context of biblical history was not to hit me until 3 or 4 years later, and neither was the weather-warn body of the Sphinx and its surrounds. But as I mentioned before; recent research which I was not privy to at the time, was to trigger my thoughts and present some big questions and I think answers. The answers impact us all whether we know it or not; from those of a religious fundamentalist disposition who have been beguiled at times by slick, but largely baseless presentations in the church pew, through to the entrenched sceptic who assumes that biblical history is total myth - and of course all those others in between.

Then it was back to Cairo; the early afternoon visit to the Museum was awesome and the Tutankhamun exhibition dazzled. And what was that iron dagger (shining as though it was made of modern stainless steel) doing there dated back to 1320 BC?

Eventually I was reconciled to the idea, that as the richest man on the planet, Pharaoh Tutankhamun was the most likely person to have one of the first iron-age objects ever made.

Interestingly, metallurgical analysis has determined it to be made of meteorite iron. It explains a lot about its especially durable composition and mirror-like appearance.

And so it went on; mummies, sculpture, stele, art, a model bird with outstretched wings - or perhaps a model aeroplane? All from over five thousand years of Egyptian history, that subsequently had me scratching my head and on returning home, racing back to the books.

On the day however, there was little time to catch breath.

It was back to the hotel for a shower and freshen up and a group get-together in the mid-afternoon. Sorting baggage and other last minute activities followed, as we prepared for the overnight rail journey on the sleeper train to Aswan.

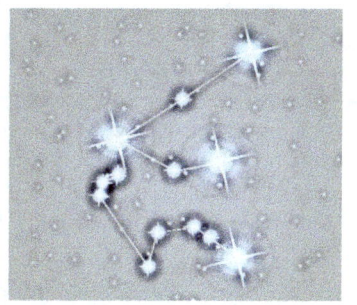

IN THOSE DAYS

CHAPTER 1

Mists of Time

'I travelled to a timeless land
And stood in awe and wonder,
Where once the feet of Israel trod
And our Lord brought yonder,
But for all the lessons I learnt there
The things I saw did not compare
To friendship, love and laughter'

Beginnings

The subject of origins was not part of the research interests of my trip, but having worked in geoscience for most of my life, a short mention of the subject is unavoidable. Besides I suspect one should start at the very beginning (a very good place to start we are told) if only briefly to summarise the issues.

So why are we here? Is life just a hologram as some suggest? Some higher being's virtual reality experiment? A mere mirage that only exists when we explore and observe it, as other quantum physics buffs believe? Or is it, as most of us imagine, real; sometimes too real for comfort?

It could be said that the Bible has a parallel theme; while heaven and an afterlife are not always front of mind, they are nevertheless not insignificant concerns for many, and suggest that life is a proving ground for the future - a test where we learn the craft of living.

And yet life is meant to be lived. We are not meant to be bystanders, but doers; sensitive to others and our surroundings, but true to ourselves, positive and courageous in action. We are meant to be role models filling others and ourselves with hope and happiness. In short, a faith in life and our place in it - and as we mature an awe of being.

Certainly the thrust of Bible teaching and Judeo-Christian thought in general; is Creator God first and the creation (cosmos / universe) second. 'The heavens declare the glory of God' we hear from the Old Testament (the Jewish Tanakh). The Apostle Paul in the New Testament chips in with:

'By faith we understand that the universe was formed at God's command, so that what is seen was not made out of what was visible.'
(Hebrews 11:3 NIV)

These observations apply equally to any creation model, whether 'Big Bang' or any other possible concept of creation which might be proposed now, or in the future. They are very basic statements of philosophy.

But is it instinctive to acknowledge our unseen Creator God as Paul suggests elsewhere? And is it just as valid today as it was to Paul, despite our Big Bang Theory, great strides in cosmology and its quantum mechanics perspective? Or does it even matter? We are not meant to wonder endlessly about the big questions are we?

Then again a little reflection is inevitable, particularly for those of that disposition. Occasional meditation is beneficial and even necessary for a healthy mind and outlook on life.

Now some popular science writers and presenters have turned this cause and effect scenario on its head. They suggest the cosmos first and out of the cosmos has developed the senses and intelligence (us) to observe and understand the universe.

Beguiling, interesting to contemplate, perhaps enlightening in some ways, but does it really tell us anything about the past? Does it do anything to explain existence? And why are we always considering ourselves to be the highest and best there is going around? Seems a little bit self-centred and illogical to say the least, and especially so, measured against contemporary knowledge, where evolution suggests the possibility of other life forms, someplace - somewhere in time.

Other modern hypotheses almost parallel the Bible and its emphasis on the underpinning of the creation by a Divine Force.

For example, the idea that those sub-atomic particles and their inherent quantum forces are self-ordering, and that biological life is the end result. Even the idea of the universe and life as a mere hologram begs the question; if so, who, where and how?

I will touch on philosophical thoughts briefly from time to time, but will primarily focus on the biblical evidence and the statement 'In the beginning God'. While sceptics have other ideas, do any of these ideas make more sense than this statement? The first verse of Genesis 1 says:

"In the beginning God created the heavens and the earth."

It acknowledges the conundrum of existence by introducing the First Cause. The following verses continue:

"Now the earth was formless and empty, darkness was over the deep, and the Spirit of God was hovering over the waters." 3 And God said "Let there be light" and there was light. 4 God saw that the light was good, and he separated the light from the darkness. 5 God called the light "day" and the darkness he called "night". And there was evening and morning – the first day." (NIV)

OK, from verse two onward we might be tempted to see some divergence from a 20th and 21st century script of the formation of the cosmos. The focus is on the 'earth' which is described as formless, empty and dark before light appeared.

But is not 'earth' simply matter? And did not the heavenly bodies form from a coalescing of interstellar gas, dust and dark energy under gravity? If we do not allow ourselves to get stuck on words unnecessarily and appreciate the limited expressions available to the early writers, then we might equally use a word like 'matter', 'waters', 'gas', 'star dust' or 'stuff' instead.

We then have a similar picture to the 'Big Bang' formation of stars and galaxies out of the 'water' of cosmic debris, gas, energy and apparent chaos.

However there is an anomaly later, where in verses 14-19 we have lights being created in the sky; one large one for the day, another for the night. These, together with a host of smaller ones called stars. These verses seem really to expand on verses 3, 4, and 5 - and undoubtedly should follow them, based on our science age knowledge. In fact they should probably precede verse six.

Perhaps two early versions of the creation story have been amalgamated, or a transcription error resulted in verse 14 being separated from verses 4 to 6. Someone may have gotten their parchment or clay tablets mixed up, or something fell out of a saddle bag and broke in an inconvenient place.

Maybe someone would have received a well earned boot up the backside, if the error had been picked up at the time.

So verses 14-19 come after the separation of the dry land and waters, and they come after the creation of vegetation (both small plants and trees) following the emergence of dry land.

This is definitely an issue in comparison to our understanding of modern geology and paleontology, but it can be cleaned up with a little thought, especially considering how easily out of sequence errors like this might have occurred. Hand copying over hundreds of years, quite possibly 1,300 years - and perhaps several thousand depending on when it was written - would easily cause such corruption. In fact, to this writer there is a sense that divine revelation played a part here, because it follows today's best science scenario so closely, or would do so if some such subsequent error or scribal mix up was corrected.

If not a revelation, we can say that compared to most early creation stories of a mythical nature and involving animals of one sort or another, there is an attempt to logically and sequentially understand how life could have been formed. And that, from the simplest life forms to the vastly more complex.

In practical and descriptive terms, it seems closer to a 21st century understanding than other ancient stories, where animals feature and turtles or elephants hold up the heavy bits.

Perhaps our Genesis 1 picture is not quite so colourful or entertaining, but there are a few key ideas and a sequence of formation and development which says some thought has gone into this.

Unfortunately, we also have an issue with the 'yoms' of the Hebrew which are usually translated as literal 'days'. But that can depend on context, and again the small subset of words available to the early writers to describe periods of time should be acknowledged in any attempt at interpretation. This is where our (God given) 21st century scientific knowledge can help do a restoration job on the masterpiece. We can call scientific knowledge 'God given' if it is true; at least true in the setting under discussion. That really is the only criterion.

So in fact with a modicum of rearrangement, the verses very nearly fit the modern picture as we understand it.

We just need to see the 'evening and morning' suffixes as beginning and ending, assume geologic ages for the periods of time defined, and the narrative becomes an up-to-the-minute summary of the initial creation, evolving cosmos, the formation of our solar system and planet, the evolution of its varied life forms - and eventually - us.

Whether the story was carried by Abraham from Mesopotamia circa 1900 BC orally or on clay tablets, whether it came later from Moses or even the Babylonian exile, is not a huge concern. In the end the statement right at the outset: 'In the beginning God created the heavens and the earth' has a magnificence and relevance which makes one sit up and take notice. It makes a science / logic statement - one of cause and effect.

We know that creation stories existed before Abraham, and that one of these came from a place called Ebla in today's northern Syria. Ebla is now an archaeological tell situated south of Aleppo, and west of the site of ancient Mari on the Euphrates River to the east. We are told that the Benjamites tended their flocks in the countryside here and on occasion acted as mercenaries for the kings of Mari. As it happens, it is also about the same distance from Harran to the north where Abraham lived.

The story is simple in the style of Genesis 1 and a fragment in the form of verse:

> 'Lord of heaven and earth:
> the earth was not, you created it,
> the light of day was not, you created it,
> the morning light you had not [yet] made exist.'

This verse clearly parallels the biblical account, where we also find a "Lord of heaven and earth" who created the earth and everything in it out of nothing. However it is important to note that Ebla was primarily a pagan culture, where pagan gods such as Dagon, Baal and Ishtar were central to the people of that region. Regardless of who is being addressed, these tablets show an affinity with historical details found in the scriptures.

They also bring to life the heritage of a people from over 4,000 years ago, and the accounts parallel stories found elsewhere around Mesopotamian and the eastern Mediterranean.

When Abraham moved away from worshipping the Baals; the sun and the moon of Harran and the fertility gods of the northern Levant, and settled on an unseen God who created and underpins everything, was this not a step in the right direction? Indeed a significant step?

Eventually the Hebrews understood that there was a conundrum here, a paradox of existence. In the beginning something or someone was, or is, self-creating.

Not merely existing, but 'Was', 'Is' and 'Will be'.

This 'out of nothing' concept is inconceivable to the human mind. We have not been made in a way that we can fully understand this, and it is in some ways a more significant issue than simply acknowledging God through recognising his creation handiwork - as true as that observation is. The Hebrew elite eventually settled on the Hidden One, the Uncreated One, the Outside Time One. Not a pantheistic god, but one separate from the reality that we know with our physical senses.

There are all sorts of conundrums and paradox at the edge of our understanding – chance, game theory, chaos theory, quantum mechanics and the duality of nature. The Heisenberg uncertainty principle (which prevents us from quantifying both position and motion of sub-atomic particles) and more recently, quantum entanglement or 'spookiness at a distance' both highlight this edge of reality.

Is, for example, the idea of endless evolution of universes going back beyond the Big Bang at all helpful? That is what some propose, but where do we stop? I believe Abraham, Isaac and Jacob - and the prophets that followed - got it right; too much hypothesising in that direction without further data is time wasting. OK for scientific research perhaps, because there is always something more to unravel, but pointless as a basis of a personal philosophy. Humbling ourselves and acknowledging our Creator (the Universal Mind) I think is instinctive and ultimately brings personal peace - at least in my experience.

From Aristotle to the present, each generation believes it is just one step away from the TOE, the Theory Of Everything. And just when we think we've arrived, better measuring methods and equipment are devised and constructed, to show our neat looking equation is not as precise as we thought.

$E=MC_2$ doesn't it? Well apparently not. Recent experiments suggest it also is only an approximation. And that will not surprise the philosophers amongst us.

So yes, it may take hundreds of years or more, but eventually we always find our latest mathematical modelling is approximate; useful for the practical and technical purposes of the day, sufficient to push the boundaries of our current interests, but in fact actually demonstrating that we fall short of a full understanding. Even when we know we are delving deeper than anyone before, lo and behold we find our ancestor philosophers have beaten us to the punch. For if they believed in and worshipped an infinite (in every way) God, they surely ensured that their concept of God would endure. That tends to be what happens when you do your philosophical homework!

I hope to show that with this mindset, we open our mind's eye to possibilities that transcend known science. And that can only be a good thing, because it leaves us open to new revelations regardless of our personal view of the First Cause.

So we might well ask: is the Jewish Bible the first book to tackle these problems?

Well no, the Egyptians had a bundle of gods early on at Heliopolis, with the chief one called Re and another called Atum. These were apparently precursors to the Theban ideas of God; Amun, the Hidden One and Ra (Re) the sun's rays - the light and life giver. Put them together as the priests of Thebes eventually did and we get Amun-Ra, a hidden (invisible), creator God who gives light and heat to sustain us.

In addition, Amun had a partner, Mut, who was a hidden beauty - and one looked on by the Thebans as the mother of us all. In fact where we derive the word 'mother' even today.

It means there are issues here of course, but for now it becomes plain that the priests of Heliopolis and Thebes were pretty deep thinkers, and I think that people like Moses would have learnt a lot from them.

Then years later, for a short time under Pharaoh Akhenaton, one god, the Aten, represented by the sun disk, was promoted as a monotheistic deity. Now this seems on the surface to be a move forward from polytheism, but was it?

At one level, yes, but on another, somewhat conflicted and a step back to Baal worship. The term Ba was used by Akhenaton as an expression for the soul and the similarity with Baal is obvious. Indeed he saw the sun god Aten as eventually being disseminated, as it were, throughout nature in a pantheistic fashion. At a guess, he did this on the basis that the sun's energy is transferred to living things.

So there are mitigating circumstances here; pantheism in regard to sun worship on one hand, but some regard for the value of monotheism on the other. In practice it was the 'queen of the night', the moon, which epitomised sin and darkness, whereas the sun is used in biblical imagery to depict the light of God.

The family of Nefertiti, the Pharaoh's wife, are surmised by some to have originally come from the northern Levant / Syrian region. In fact, probably an association with Indo-European culture, as her famous bust now in the Neues Museum in Berlin suggests. She may have even been of Minoan ancestry.

Whatever her heritage, I suggest she could have been influential in introducing the new religion.

Baal worship in northern Syria and Harran in today's south east Turkey was the worship of the Sun and Moon, and it was from within this society that Abraham, the father of the Hebrew tribes emerged, many years before. So perhaps this latter day Eve may well have been the culprit and the one who set the worship of an Invisible (Spirit) God back decades in Egypt, while confusingly, enhancing the idea of a single Creator.

Despite this philosophical hiccup and the ongoing worship throughout Egyptian history of the deities Isis / Osiris (or similarly defined gods who reflected the male / female aspects of the sun and crescent moon) the thing to consider is that in general the Egyptian priests at Thebes appeared to have moved toward the concept of an invisible, Creator God.

Whether in a logical and ordered sense, or in a moment of inspiration from above - or a bit of both - is something we do not know. Certainly the biblical record states plainly that the Hebrew Israelite people of the Bible had considerable help along the way. It is recorded in Genesis that on several occasions from Abraham through to Joseph, that they were subject to inspirations and visions. Despite this, it would be totally logical that their spiritual experiences would be reinforced during their extended sojourn to Egypt, and the intellectual approach and teaching of the priests of Thebes and their associates.

One of the previous sources and experiences for the Hebrews was meeting a priest-king called Melchizedek, King of Salem (Jerusalem). He seems to have had a huge influence over Abraham and one wonders if he was a priest of Amun, or some other recognisable group of the time.

The book of Jasher suggests that he was actually Shem, the son of Noah, who was supposedly still living at this time. This extra-biblical source has come down to us in fragments only, and is quoted in the Bible and other Jewish literature. In my view, the provenance is likely to be from a relatively late period and seems to be a mixture of fact and some fantasy.

Certainly biblical quotes in regard to Melchizedek, e.g. 'he had no father or mother' hint at a mystery, but do not lead one to believe the Shem association.

Some commentators in fact suggest, that the Shem story is Judaic propaganda designed to disrupt the messianic association in the new Testament between Melchizedek and Jesus of Nazareth. Furthermore, that the ages of the patriarchs and the 'begat' ages of their offspring in Genesis, have been adjusted deliberately to make it seem possible that Shem could still have been alive at the time. There are certainly issues with some of the numbers in Genesis and I will address this topic later, but to suggest this was done deliberately is a big call.

But perhaps Melchizedek was simply some God anointed figure standing alone and ahead of his time. In fact one wonders if Melchizedek was an appearance of Jesus himself, and / or a theophany of some sort. His identity aside, Melchizedek is definitely a candidate as a precursor priest, or priest-king to the Judeo-Christian belief system. Here 3,900 years ago a figure or group were the catalyst for the monotheistic faiths.

I will show later, that he studied the heavens and perhaps through the signs he saw there, was able to persuade Abraham that there was meaning which accorded with the visions and encounters that Abraham himself had experienced.

So Melchizedek is an intriguing character and how he acquired his knowledge of the stars an open question.

Returning now to philosophic considerations; if we have concrete scientific evidence and many strands of it at that, we would be foolish not to examine and use it to reflect light on scripture and Genesis in particular. On the other hand there are many prophetic passages in the Bible, which if true, impact us to this day by showing the hand of the Almighty over the ages. Importantly they point to our future destiny as well.

Furthermore when we look at the history of religion overall, including the ramifications around early monotheism, there is a case that suggests it is *not* that monotheism evolved out of polytheism and polytheistic ideas, but rather that monotheism was the original and natural belief system. In other words, that this knowledge is inherent in humans and / or was specifically given prophetically at a very early time.

Many have stated this previously, although it is obviously not an easy proposition to prove historically. St Paul implies it, suggesting there was a corruption of a pure and natural belief in the Creator, which then resulted in debasing human thought. From the innocence of the hunter gatherer to the pressure of the suburban crush and a complex society, life became more competitive and wrongdoing and selfishness seemed necessary for survival. Eventually, a pantheon of gods was worshipped in order to assign responsibility to various areas of existence.

Unfortunately instead of bringing order, confusion resulted, with a plethora of idols and gods littering the landscape, as the Apostle Paul noted in Athens many years later.

Perhaps Paul is a bit more practical in his explanation than those who have an overly literal interpretation of scripture;

those that suggest a perfect world existed in the immediate past. He understands the negative effects of wrongdoing and antisocial behaviour, and actually gets to the heart of the matter.

He does this without supposing too much of the nature of the world, six thousand years ago.

For those believing in a six day creation (and undeniably it is still a strong movement into the 21st century) this makes them diametrically opposed to the 'old age' geological understanding of the current era, and the evolutionary development of the cosmos - both of the heavenly bodies and biological life.

The 'six day' doctrine has unfortunately produced outlandish and impossible creation and flood hypotheses, and resulted in understandable cynicism throughout the scientific and broader communities. The problem with the six day fundamentalist point of view, is that the geological evidence refutes their unsustainable hydraulic model completely.

It is evidence which cannot be dismissed at all, except in ignorance, and shows that death has been a natural occurrence for millions of years prior to the 6,000 years of known civilisation.

It is also natural, that starting with what appears superficially to be a six day creation in Genesis 1 - followed closely by images of a serpent in a garden tempting the first humans with fruit from a tree in chapter 2 - that a person relying on everyday observation and the body of research available in the 21st century, will most likely dismiss these stories as myth and bedtime fairy tales. Something like Grimm's Fairy Tales, the Lion King or Frozen. Good entertainment for kids perhaps, and teaching some basic life truths, but not to be taken as history.

However with a little more thought, the 'garden' story from Genesis 2 onward becomes an allegory for the human condition and the struggle against temptation and selfish desire.

Now it could be argued that the Bible is a rather large parable or story, which has the primary aim of separating the sheep from the goats. That good people of faith know what is right and true instinctively, and follow the precepts that Jesus taught of love and goodwill. That they will do this regardless of outside influences and biblical understanding, and that conversely, those with evil intent will follow their selfish desires no matter what. Under these circumstances, why even discuss the pros and cons of contemporary geoscience and biological evolution and the contrast with six day creationism?

While I believe there is some truth in the above, I also believe we should not be a stumbling block to others. That there is surely a need to combat untruths and poor interpretation, and leave no stone unturned in bringing the Good News to the whole world. It should also be presented in a way that is completely unfettered by irrational theories and complications.

Throughout this narrative I will absolutely discuss and support basic and 'fundamentalist' Judeo-Christian beliefs. There are other times, when I will show that they are better understood as analogical and need to be viewed in the light of evidenced based, scientific research. Other stories I will show as absolute history, even though there is sometimes a strong case for dramatisation, embellishment for effect, or simplification in order to present events and values clearly and memorably.

They therefore come to us primarily as vehicles for presenting life lessons in morality, relationships and spiritual wellbeing. It is history as the writer knew it, and like most of our better cinema, plays and musicals today, entertainment is often the best method to send a message. It is dramatisation for effect.

Despite that, we cannot summarily dismiss accounts of the biblical heroes as myth and I will endeavour to shed some light on the major events in light of recent scientific knowledge.

Summing these thoughts a general proposition can be put:

The ultimate affirmation of a belief system is: **when personal and historically recorded inspiration and revelation align with evidenced based scientific observation and logic.**

It is also a fact that not all truth comes from logical and material based research and deductive reasoning. The great leaps are often inductive. They are the 'what ifs'. These are the Einstein moments where vision and revelation coincide, and are so advanced in perspective that they take years to prove, such that critics have a field day for decades, or centuries to come.

As rare as these moments may be, it is the way we make progress and increase understanding in great leaps.

The Bible is no novel, and it requires careful study if we are interested in detail and truth. Laziness will not do at an academic level where a deeper analysis is warranted. On the other hand, everyday life lessons embedded in the text are there for everyone, and can often be used out of context and stand alone.

So there is no denying that the antiquity of the text can be problematic. Particularly where a story has been told and retold and handed down through many generations. This *must* apply before the days of writing where there was no text, and although there has been the occasional example of writing as yet undeciphered and unconfirmed prior to 3,500 BC - out of Greece for example - stories in the Bible from these early times need to be treated with a degree of common sense, insight and wisdom.

Just because tradition asserts that Moses wrote the books of the Torah (Pentateuch) it should not be assumed that historical documents pertaining to creation and the Patriarchs did not exist from which he, or the Levite priesthood, could work. On the other hand, the assertion of a few scholars that the stories only originated around the time of the Babylonian exile (ca. 588 BC) need not, for the most part, be taken seriously. Many stories may indeed be compiled with data collected at this time, but this does not negate their value, or indeed historicity per se.

In my view such scholars are largely the victims of 20th century scepticism; a reaction against 19th century church authority and a perceived lack of archaeological evidence at that time.

As outlined by Werner Keller in his concise and readable 'The Bible as History' (first printed in 1956) there was a great deal of enthusiastic archaeology in the 19th and early 20th centuries, directed at confirming biblical history and place names.

It was in fact a key driver, behind the establishment of the modern discipline of archaeology.

This push to find concrete evidence of the biblical world, was no doubt a counter to the sceptics who were arming themselves with the new knowledge of geological history; a history that described an old earth and the fossil remains of long since departed animals. And that is apart from the most startling controversy of all; the proposition of Darwinian evolution.

So as Werner Keller recounts, excavations throughout the Levant, Mesopotamia and associated areas, began to reveal the Bible as history and clay tablets gave local versions of events. These included accounts of huge floods, the rise of civilisation and nations, ambitious constructions and developments, evidence of strife and wars; all events which affected or disrupted in some way, the irrigated land of the Two Rivers.

Then in 1933, the city of Mari in Syria was discovered when a sculpture was unearthed, followed by a whole palace and government complex. It was complete with thousands of records on clay tablets, all of which shed light on a kingdom from Abraham's time - and one that he would likely have had some connection. There is every indication that the Mari kingdom's influence extended north to Abraham's home town of Harran.

The one thing missing at the time was much evidence from Israel itself and the Davidic kingdom.

But this evidence is now being unearthed and understood to a much greater degree than ever before. This, as a certain amount of stability has come to that small state, where Israeli academics together with universities and research establishments around the world, excavate more sites and apply the latest analysis to artefacts, inscriptions and manuscripts.

You would almost think that the veil was being deliberately lifted according to some great master plan. I believe it is so.

Notwithstanding, when it comes to manuscripts, revisions occur, scribal errors creep in and anachronisms are deliberately introduced in order to facilitate understanding amongst the readership of the day.

The layman also relies on accurate and unbiased translations and commentary by the few that spend their lives in detailed study and examination. To top it all off, much of our information today comes from the documentary makers who condense these topics to bite sized episodes.

So the first five books of the Bible need to be treated with due respect, but also with caution. Consider the history involved.

The long-legged Aramean

Abraham and his sons were in the right place geographically and right time historically to have picked up copies of creation and flood accounts in their homeland in the upper Euphrates. Failing this, on the subsequent occasions recorded where they commuted backwards and forwards to their Habiru relatives in Harran. Whether they did so or not, I cannot say, but the creation and flood stories were available in Mesopotamia and Syria well prior to 1900 BC.

Now the existence of the Abraham of Genesis is difficult to prove absolutely and many are simply biased against Hebrew (Habiru) history, dating back as it does, to Syria and Mesopotamia. This bias will be personal in many cases, because of the religious connotations, rather than a matter of historical probability. After all, there is an Abraham recorded in the Ebla tablets who does some of the things recorded in Genesis, but of course sceptics simply say the biblical record is plagiarised from there and not the history of the Israelites. They take the concepts of reasonable doubt and the burden of proof to the extreme.

But the dating is entirely compatible with the movement of Hyksos and Canaanites, who began migrating south and eventually into Egypt in waves beginning in the 19th and 18th centuries BC and who are recorded in Egyptian history.

'Hyksos' is a Greek term (derived from the Egyptian) that Manetho, an Egyptian priest, used to describe the Asiatics who invaded the Egyptian delta. Interestingly, the nomadic people in these northern parts were known as *Hurrians* by the Semitic speakers, but called themselves *'the kings who lived in tents'*.

It is therefore highly likely, that they were part of the invasion of Egypt and it is obvious that the city of Harran was named for these people.

This major exodus out of the Levant started prior to the Egyptian Second Intermediate Period (SIP) and we do not lack extra-biblical 'written in stone' narratives, or Semitic language associations for this.

Canaanites are said to have entered Egypt a few years earlier, but either way, Abraham is in the midst of this migration and in no way out of place of this confirmed history. In fact Hyksos means 'foreign rulers' or perhaps 'shepherd kings' and seem to have been a mixed bunch. Some were probably relatives of Abraham, with others hailing from the same general region. Other groups, best described as gangs and militias, came from further east and the Arabah as we will see later.

If Abraham's story is a stylised one from several parallel experiences, or the story of one man as presented in Genesis, is almost immaterial. For the most part I will assume the latter in my exegesis, because of the very personal nature of the literary material and the prophetic impact of his many experiences. The episodes he goes through, seem absolutely the story of his time and place. Nevertheless, it is distinctly possible that his story encompasses episodes from namesake ancestors; such as an uncle or grandparent from earlier generations.

This depends on the actual dating of the Ebla tablet record, compared to the best estimate for the biblical Abraham's journey south to Canaan ca. 1900 BC.

In any event, his journey seems to have been a typical one for the Habiru, Benjamites and other rural dwellers of the time. They were stockmen and graziers and the SIP is the story of the disruption caused by that migration into Egypt. They were almost certainly escaping drought after the floods of earlier centuries, but also the old Assyrian Kingdom which rose to power around this time. It was not the first time Asiatics appeared in Egypt and the Sinai Peninsula either, if the reliefs at Saqqara mastabas dating around 2,350 BC are anything to go by.

Abraham was not an egotistical sovereign who was automatically going to be noted down for posterity. He was not a ruler, like a Ramses II, or an Alexander the Great driven to monumental construction and personal aggrandisement. Nevertheless when we look at this period ca. 1900 - 1700 BC and how Abraham travelled south from near today's Turkish-Syrian border, we see that in general terms he qualifies as a Hyksos.

Regardless, Abraham was a wealthy pastoralist, although one amongst many and if it was not for his spiritual insights, would have passed into history unnoticed. But indeed he does more. It seems he keeps a diary and collects stories from his ancestors, the patriarchs of Syria and Mesopotamia. Either that, or related descendants went on a 'who do you think you are' experience to document their heritage.

The real answer is probably a bit of both. And this is not so different from many of us today, who are interested in our family origins and dig through the photos and letters of our forebears. Sure, these stories may have been compiled by later scholars, or simply added to by subsequent generations of Hebrew scribes. But as long as they are historically grounded and impart the lessons they were designed to teach, is all that really matters to us today.

In fact, one way of looking at Abraham's life and time with respect to that of the Israelites who lived 1,000 years later, is to view his story, as the British today view the story of William the Conqueror in 1066. The time span is roughly the same. Genetically, most anybody in England who can trace their ancestry back a few generations is likely to be a descendent of William, or the following kings and royal lines. Intermarriage, genetics and the application of statistics to lineage, suggests this.

The city dwellers in Canaan who lived in the cities of Hazor, Ai and Jerusalem may have had somewhat different ancestry. After all, these cities existed in the third millennium BC.

But both the citizens of the cities and the Hebrew pastoralists, would have been connected back to their ancestors of the northern Levant, whether they wanted to be identified that way or not. After all, Hazor in northern Israel was only 800 kilometres away from Harran and much closer again to Ebla and Mari.

So whichever way we look at this, the term 'father' Abraham becomes a perfectly acceptable description of Abraham and his immediate family line; this, in terms of the later Kingdom of Israel and its people.

Now we learn early in the biblical record, that Abraham moves with his father Terah to Harran, from either Ur Kasdim (Sumer) or Urkesh (Syria).

Scholars are still wrestling with the identification of his birthplace, so the question remains an open one. I have a slight preference for the Syrian Urkesh, given the connections implied in the migration of the Habiru and Benjamite tribes' people, and the supporting archaeological material at Ebla.

Circa 1900 BC, Abraham would have spoken a Semitic language very much like that of Mari or Ebla. He then sets out for Canaan to the south with his servants and family, flocks and herds, goods and chattels. On his journey he would have passed through, or adjacent to Ebla, south of Harran. Its large library dates to a couple of hundred years before Abraham, but was probably still flourishing to some extent or other when he travelled through. In fact if the records found at Ebla refer to our biblical Abraham, or indeed a close ancestor, he (and his family) obviously had more political clout than I give them credit.

On the other hand, there is a slim possibility that he could have diverted eastward along the Euphrates and travelled the scenic route via the Mari kingdom. Mari would have been interesting, with its vast written record dating back before 2000 BC. Regardless of the route taken, he takes a liking to Canaan.

But times are tough, drought has set in, and at one point Abraham travels to Egypt as noted in Genesis 19. Life was not always 'milk and honey' in Canaan it seems and Egypt offered some respite. By and by things improve and he is able to return again to that land. This guy gets around, he is a genuine Habiru!

Later, under great-grandson Joseph, the Israelites again live in Egypt proper for over 200 years, before they depart in the Mosaic Exodus. At least that is the consensus and by 'proper', I mean in Africa and the Sinai Peninsula. The Bible actually states that they spent 430 years in Egypt, but a valid interpretation is that this period includes their time in the land of Canaan; a region under Egyptian control for centuries over the years.

We know for instance, that Thutmose III established Beit She'an in northern Israel as an Egyptian provincial centre around 1450 BC, and that this continued to the dawn of the Davidic kingdom around 1100 BC. And this poses the first of many questions: are the 430 years meant to be taken as literal calendar years, or are they meant to be symbolic only?

There are some interesting aspects to the number 430 in the Israelite–Egyptian relationship, and surprisingly it flows into Bible prophecy. In any event, it is likely that the Hebrew people would have maintained a composite of much of their former language experience, with Egyptian now an influence. After the

Exodus and return to Canaan, the written language becomes (what is now known as) proto Sinaitic and later paleo-Hebrew through a synthesis with the Canaanite-Phoenician languages. Those languages will also be derived from Ebla and Mari and the parent language(s) in the northern Levant and Mesopotamia.

Hebrew becomes the square script written word we know today, no earlier than the period of the exile in Babylon. This began in the early 6th century BC for most elite Jews.

However the new script was adopted from Assyrian and Persian sources at this time, and puts something of a wedge between the Jews and their natural cultural and linguistic relatives in Phoenicia and the eastern Mediterranean.

Recapping, it is uncertain where the earliest Genesis stories were sourced in the Hebrew journey, but we should be under no misunderstanding; Abraham could well have accessed the creation and flood accounts in 1900 BC. They were available on clay tablets in cuneiform at that time.

If he did not know about them, I would say he was spending too much time with his sheep and goats, and ignoring the people and territory he was travelling through.

As a participant in surveys of one sort or another, I know that the boss has the job of making sure all the personal contacts are made, diplomacy conducted (over a beer if necessary) and sign offs completed. In my view, all indications are that Abraham's life was a heritage story, handed down through the generations and retranslated as required.

There are many examples in Genesis that could be used to indicate authenticity and the likelihood the narrative was not invented. For example, consider the following episode of the sacrificial offering of Isaac by his father Abraham.

It is central to the whole idea of forgiveness of sin by God and is firmly part of the Aries Age of animal sacrifice and the later agreement at Sinai. The story is also an analogical precursor of the basic tenant of Christianity; that sacrifice ended with the one and only atonement through the death and resurrection of Christ (the Messiah and Lamb of God).

Now sacrifice in general was made to appease God, or your particular or personal gods because of your own wrongdoing. It could also be prompted, because of real issues such as drought, flood or war. If sacrifices were not given regularly, they were certainly made when circumstances became dire.

So there are two ways of looking at motive; devotion and sincerity on one hand and the fear of retribution on the other.

It is obvious the biblical version of this sacrifice story is portrayed as the former. However, I believe this is an occasion where only one interpretation of the story is emphasised and that there is also a darker, untold lesson that comes with the package. It is also instructive and potentially even closer to the truth in terms of Abraham's motives, and perhaps even to his conversion to the worship of a one and only Creator.

This other side is something we can deduce today, because we know more about the background of the Canaanite, Phoenician, Anatolian and early Greek civilizations of the Mediterranean through contemporary archaeology. This darker picture is one where Abraham is wrestling with the evil side of Baal worship; the idea that the greater the sacrifice, the more likely your god would hear you and answer your prayer request. I am talking of human sacrifice and child sacrifice in particular.

It seems that this evil practice of human sacrifice was not uncommon around the world. It may have even spread from the eastern Mediterranean because of the mobility (seamanship) of the Phoenicians. Perhaps even to the Americas. The fact that the Spanish still found human sacrifice occurring in the 1500s, illustrates how pervasive this practice had become worldwide.

So there is a case for the proposition that; when Abraham felt that God was saying 'stop' and not to go ahead with the sacrifice of Isaac, we have the first demonstrable biblical break with Baal and other polytheistic religious practise of the region. After all, Canaanite society and practice was as old as that of the Eblaite, Marian and Ugaritic regional states in Assyria.

How ironic, but logical (perhaps inevitable) that the light which was eventually to be shed by the holistic 'love and goodwill' teaching of Yeshua / Jesus and his actual atonement as the 'Passover Lamb' of the Exodus, had to come from the same precise geographical spot.

In my view, this interpretation militates against the idea that the story was invented during the exiles; the latest iteration of the sceptical view of Israel history by some writers.

Their predecessors of the 19th century derided the existence of the flood cities of Sumer mentioned in the Bible, as well as Babylon, Nineveh, Sodom and Gomorrah, the Troy of Greek legend and countless other towns and locations. Today, they are forced to backpedal to more sophisticated arguments, in a desperate attempt to negate the biblical record.

If we say there is no archaeological evidence for a well off sheep and goat herder; one who travelled with his family from a place called Ur, to Harran and then from Anatolia to Canaan and on to Egypt nearly 4,000 years ago - exactly what evidence are we expecting? Broken clay tablets beside a dusty track with the Abraham seal on them? An animal-skin water bag with his brand burned into it? Signature sheep and goat droppings?

I think not. And yet there is a trail, it is linguistic and cultural in a context revealed by archaeology and text. It is a story driven by war, flood and drought and the need to survive - and dare I say, a desire to explore and experience the Divine.

And yet no matter how sacred or inspired we consider the text of the Pentateuch, it would be ill conceived not to subject it to the closest possible scrutiny. So many geographical, cultural, religious and language changes over thousands of years warrants close inspection, no matter how diligent the scribes. Only then can we be certain that the history recorded and the prophetic messages embedded within the narrative, are indeed as clear and valid for us today as the writer intended.

Writing was at the very forefront of technology in those early days and no doubt a specialty of only a few. As a new technology it would have changed dramatically in the early years, as for example, recording and recording devices changed in the latter part of the twentieth and early twenty first centuries. From magnetic wire, to reel-to-reel tape, to cassettes, compact disk, hard disk, memory sticks and the remote 'cloud' based digital memory technology we experience in the 21st century.

On the other hand, everything that was written from the time of the exiles has not undergone this extensive change in the written Hebrew, and therefore is much more likely to be characterised by stability and accuracy in the text. In fact we are told that the revisions in the 11th century AD with the vocalisation marks are the only major technical change.

That as may be, there is still a need to watch out for late alterations - and of course - to understand that contemporary Hebrew speakers and scholars, will sometimes find difficulty in interpreting ancient idiomatic expressions and figures of speech. Sometimes meaning becomes totally reversed and (tongue in cheek) we can blame the imagination of teenagers for that!

So there is always a challenge, even where writing is unchanged. A reliable historical record of codices, recensions (revisions) and translations over time, are therefore critical to the validation of our current biblical versions, originating as they do from various source materials.

Unfortunately much of the intermediate work is missing; such as most of the Hexapla (six translations) of Origen from the 3rd century AD. In addition, translators and copyists of the past, may have had different priorities in the approach to their craft, than that of modern scholars. They certainly did not have the advantages of modern computer and print technology.

More significantly, if their work was lost or destroyed, we are left only with remnants quoted by later scholars. The resulting manuscripts may well be reliable and great for general usage, but decidedly difficult for modern scholars to trace their development over time.

Comfortingly, we have older manuscripts of the Bible in Greek, Aramaic and Syriac to compare with later Hebrew translations, and the discovery of the Dead Sea scrolls from 1947 onward, has added to the body of work.

The Hebrew Masoretic texts are the acknowledged basis for the Old Testament translations and come to us from the 10 and 11 centuries AD, while the Greek Septuagint dates much earlier to the 3rd century AD. The Syriac Peshitta also derives from this general time. However there is an agreed history of the Greek translation, which says the Septuagint was translated by Jewish scholars (perhaps six from every tribe) into the Koine Greek in the early 3rd century BC. This exercise was undertaken through the sponsorship of Ptolemy II in Alexandria, Egypt.

It is known they completed the Torah (Pentateuch) initially, but how many of the other Old Testament books did they complete before the time of Christ? Is there any doubt why such controversy abounds in the critical study of the Bible?

We can also see why most people throughout Egypt and the Levant (including Israel / Palestine) would have been familiar with the Greek version, rather than Hebrew. This familiarity with Greek, the all-pervasive language of the Empire, is attested to by the Dead Sea Scrolls as well. Importantly, it was the language of choice well into the Roman governance period.

It was the common language used to communicate laws and collect taxes, and the Septuagint was most often referred to and used by New Testament writers of the period. This was despite the fact they were nearly all Jews. Today many more translations, codices and fragments have come to light, and add to our knowledge of the old texts and their origins.

If we open-mindedly take all this into account; the carriage of early scripts from the dawn of writing through to the updates by Levite priests, the many language changes up to the exiles - and include the translation and copying difficulties of the last two and a half millennia - we arrive at one conclusion:

The first eleven chapters of Genesis have been on an extraordinary journey, the historic details cover millennia in their own right, and we therefore need to diligently assess and carefully approach the content.

Scripture and science

On a different note, reconciling science and the Bible amongst some Christians is tantamount to betrayal and more than worthy of the title 'liberal', or other euphemism for heretic. A 'liberal' is then a dirty word deriding the individual as an effective non-believer. These critics often have an extensive list of things they maintain are mandatory doctrine, before one is considered kosher or fair dinkum.

But many minds of a more considered and sensitive nature, have taken well established and observable scientific principles, and recognised that these scientific principles - like the scripture that so many hold sacred - are also God given and indeed God created and moulded. For example, consider a possible reconciliation of the creation story and modern cosmology.

A Jewish astrophysicist, Gerald Schroeder of the College of Jewish Studies, Aish HaTorah's Executive Learning Centre, has looked at the mathematics of widely accepted Big Bang theory and like many others, attempted some calculations on the expansion of the universe based on that model. Those calculations suggest to him, that if we look forward from the initiation of the creation, rather than backwards from our perspective today, that time would slow inversely to the stretching of space-time at a predictable rate. When translated into the rate at which time passes today, he suggests this could correspond to just seven days - a Bible view of time if you like and maybe God's view.

We know that both velocity and gravity cause relativistic effects; effects which cause atomic clocks to run faster or slower depending on their situation. This has been demonstrated in practical experiments on jet aircraft and satellites. We would therefore expect someone to age slower at high velocities, or very much quicker in high gravitational fields. So perhaps there could be something in Schroeder's assertion, if in fact there is any ultimate truth to the expanding fabric of space.

I have no idea if this concept has merit, or how the mathematical model matches timewise with the birth of galaxies and their star systems, the age of the earth and the subsequent geologic ages. At the very least, I see the idea as an attempt to reconcile the biblical account with contemporary observation, and others have attempted similar modelling to their credit.

On the surface, it appears to be generally consistent with Einstein's theories of relativity and current experimental observation. This is because the speed of light is a constant regardless of the motion of the light source, whereas the passage of time is not. This is called 'time dilation'. It means that depending on our point of reference, it is time and not the speed of light which changes. Yes, it may not be intuitive, but it is sure amazing.

It therefore seems likely, that time has to be linked to the dynamics of space itself within 'space-time' theory.

In any event, my own view is in essence: what if the word 'day' in ancient Hebrew had a broader meaning; one meaning a defined period of time only? That the period of 24 hours was just the most common example, and that a literal 24 hour, seven day week concept was never meant in the creation account? What if the account was always meant to be solely about the creation sequence, but that sometime after the account was given, somebody also decided to use it to teach the Sabbath rest?

If this is the case, then no physical reconciliation is necessary regardless of the scientific merit of various proposals. The text may have been altered for what seemed like sensible reasons at the time, but has only caused confusion and grief from our perspective today. This issue will be thoroughly explored and discussed in the next chapter and a solution offered.

Now creation, first causes, something out of nothing, and such esoteric topics are largely the playground of cosmologists, physicists, geoscientists and philosophers. I will therefore leave further discussion for now, in order to focus on historical matters that everyday folk can appreciate.

To this writer, the fundamental beliefs are not so much about a six day creation from a text blurred by time and uncertain provenance, but about the teaching of Christ and the words of the prophets. They are also of eyewitness accounts of events and encounters - no matter how extraordinary – which are nonetheless totally and eminently possible, when addressed from a logical and scientific 21st century standpoint.

Put simply: the biblical stories of great events are believable as history, because they do not contradict modern science and research outcomes. They may not be currently proven by science, but that is a limitation of contemporary science, rather than any ultimate reality. Furthermore, much of this is regardless of whether we say we believe in a personal Creator or not.

The stories which are personal, or family anecdotes should also be taken seriously. They may be stylised, but are ultimately based on eyewitness observations which would make us exclaim, 'how on earth would anyone realistically dream this up?'

The answer is, that they are almost certainly based on real experiences. They are often unique historical events, prophetic blessings, otherworldly encounters, messianic messages, night visions or other personal inspiration that cuts across the centuries. It matters not that we do not understand how these things happen, or the processes of inspiration; it simply behoves us not to discount the record.

There is also the little matter of the Holy Spirit teaching of Jesus and the Apostles. This is the glue that holds Christianity together. In fact it is the real story of the whole Hebrew-Christian experience. Fortunately for us, there are enough anecdotes from Abraham to the first century and right up to the present, to know that there is continuity of spiritual experience throughout this impressive timeline.

An awareness and interaction of the Creator Spirit is reported in the heritage of other tribes throughout the world as well; tribes which have managed to retain a holistic understanding in this regard. How far back it goes into the mists of time is anyone's guess. Certainly in the Bible we have a historic link to Adam, but Adam (adamu) just means 'man' in early Semitic languages like Eblaite. The Bible does not say how much time is involved in this genealogy of Adam, but we popularly think in terms of six thousand years. My preference is to call them prophetic years, because it is in this period that we see an awareness of spiritual things amongst humankind. It is also the timeline of the written word and that is an overriding factor.

In regard to the above, I will also show that in biblical prophecy, the singular is often used for a whole line of individuals. This still applies in cases where there are dual interpretations. A specific person might be highlighted in a particular instance, but a whole sequence of individuals who hold that position before or after, can also be implied. It means that a reference to a particular person, almost certainly applies to a whole hereditary line when prophecy is involved.

Nevertheless, there are likely to be some historical difficulties, if we infer lineage instead of individuals, between Adam and Abraham in the biblical narrative. So messing with that timeline should be a last resort. But, if in fact we needed to go back to the time, of say, Gobekli Tepe in today's Turkey to encounter a first understanding and interaction with the Spirit of

God, the obvious answer is that in Adam and the pre-Adamic line we see a continuous lineage, rather than a single man.

Looked at this way, the prophetic age started when mankind first became aware of their Creator through specific spiritual experiences and recorded them. It can no doubt be looked at from two directions; a matter of genetics, evolutionary change and historical interventions and interactions on one hand, but also as an act of the sovereign will of God on the other.

I will leave this small discussion of the Spirit of our Creator God and our relationship to him at that. However it seems something rather extraordinary marks the change from earlier hominids (and perhaps our cousin Neanderthals, Denisovans etc.) to that of the first Adam of the Bible.

Now few people with a 'liberal' tag, would currently espouse direct experience of first century New Testament Christianity. But almost unbelievably a few of us are so blessed; followers of 'The Way' as they were called in Jesus time. Because of this, we have a responsibility to address the issues, and assure other believers that they are on the right track and have nothing to fear from contemporary knowledge. That is, once it has passed the speculative phase and been truly tested.

But after Adam and Eve, what? All those old Bible stories with their vivid and dramatic portrayal of heroes and events that we learnt about at Sunday / Sabbath school, but were forgotten on weekdays; 'all that your preacher is liable to teach you ain't necessarily so'. What about them?

To give the reader a feel for the background to the great world empires of yesteryear, a smattering of modern earth science will assist; a science with all its many diverse strands which help to illuminate geological history, both recent and immeasurably old.

CHAPTER 2

On Creation

*'The city was crazy, the pyramids soared
The hotel dusty and the river broad,
The train raced on its long way down
To beautiful Aswan, a jewel of a town'*

The changing universe

We will now move on from literary considerations and the early days of Abraham and focus on the nature of our earth, its formation and history, processes and possibilities. This is presented mainly as background for non-geologists, with an emphasis on the variety of dating techniques available to specialists today.

Of course the modern scientific picture is incomplete, the job is not finished and there is always more to learn. But that is what excites the true investigators amongst us. Essentially it is no different to the exploration urge of the wandering Habiru of old. What can possibly be over the hill?

Despite the unknowns, great strides in understanding the earth's crust have occurred in recent decades, and genetics and evolutionary biology have come a long way since Gregor Mendel's experiments and Charles Darwin's 'Origin of Species'.

Whilst investigations continue day by day, with the probe Juno recently moving into planetary orbit to sample the Jovian (Jupiter) atmosphere, the model of the earth and the rest of the planets forming out of a condensing gas and dust cloud seems eminently plausible and satisfying.

The interactions of matter, gravity, radiation, electrical and magnetic fields and the chemistry and nature of space in regard to planetary formation, are ongoing research topics. However in the light of the new discovery of hundreds of planets circling distant stars, we know our solar system is not a one-off.

In addition, the study of various belts and clouds of smaller bodies; such as meteorites, asteroids and comets and their interactions through collisions and near misses, also fascinates us.

As for the structure of the earth, the internal layering is not directly observable, but by studying the shock waves from earthquake movements, meteorite impacts and man-made seismic explosions, the overall stratification is known.

The magnetic and electrical properties are determined by instrument measurements on the surface and above, while satellite data also gives us valuable supplementary information on the oceans and land surface. For example, altitude measurements using satellite and aircraft-borne radars, define topography in remote locations and vegetation covered landscapes. Similar systems make ocean surface measurements to millimeter accuracy and use that information as a proxy to map the seafloor.

Multi-spectral analysis of the near-visible electromagnetic spectrum, allows the classification of soil and rock types in arid areas, or the identification of the diverse crop, forest and wild vegetation of the flora covered landscape.

So from the deep geophysical data we have determined a solid, but very hot, inner earth core under enormous pressure and a liquid outer core. These are mainly composed of iron and nickel. The core is surrounded by a viscous semi-elastic mantle. This in turn is separated from the surface crust we live on, by a boundary known as the Mohorovicic discontinuity.

Again, we get a sense of something formed by a process. Differential layering processes are controlled by density, pressure and heat with cooling occurring toward the outer layers.

The result is that the crust at the surface is composed of lighter material which 'floats' on the denser inner layers. The crust is also largely solid, belying the plastic and liquid nature of the material below. We therefore have a setting where; as the earth cools, we have bits and pieces of crust made of SIAL (silica and aluminium, named for the dominant mineral assemblages) coalescing on the earth's surface. This happens because these minerals are lighter than the iron rich material below.

This picture helps us understand the nature of continental drift (or plate tectonics as it is called in the trade) where we have the crustal slag being driven around at the surface, by convection forces in the fluid layers below.

We find a similar situation for example, in a furnace, which is probably the best analogy for what occurred and continues today. As the 'slag' is moved around the earth by convection currents from below, gravity interactions from above and forces such as the Coriolis Effect (caused by the earth's rotation and spherical nature) acting at the surface, we see a mechanism for the earth's crust to break up and move around the globe.

This sets up a scenario where crustal masses crash into one another causing mountain formation, or indeed continental subduction of one continental plate under another. Moreover, this is bound to happen time and time again over countless years.

Sometimes though, there are near misses and glancing contact, as continent and oceanic plates slide past each other. This is the case in California and causes the earthquakes along the San Andreas (and associated) fault lines.

So apart from astral strikes by interplanetary objects of one sort or another, continental collisions are the real catastrophism of today. Here, almost immeasurable forces are at work, and crustal adjustments sometimes cause immense destruction from earthquakes on land and tsunamis at sea.

When subduction occurs, a very unstable volcanic situation exists, where the subducted material is heated as it descends to depth by the very hot, plastic mantle rocks. This causes molten material to rise in the form of volcanoes.

The creation of island arcs are a feature in this instance, and typical examples are the Indonesian and Japanese island chains. There are other cases where subduction is not so obvious and the interaction between geological plates more complex.

Such is the case with the volcanoes of the Mediterranean, where the African plate interacts with Europe and Anatolia.

The real daddy of them all is the Pacific Rim of Fire. Here, as the floor of the Pacific Ocean expands, mountains are being formed right around the ocean from the South American Andes, through the Rockies, Alaska, Aleutian Archipelago, down to Japan, the Philippines, Indonesia, New Guinea, the Solomon Islands and then further south to Vanuatu and New Zealand.

In addition, the upwelling convection currents of hot molten and semi-molten material which cause this oceanic expansion, sometimes appear along elongated sites in the deep ocean. A prime example is the famous mid-Atlantic Ridge.

This ridge extends from Iceland in the far north, to the distant South Atlantic and the material is produced continually, forcing the continents of North and South America away from Europe and Africa. The broad symmetry of the boundaries of the continents to the west and east of the Atlantic, was the clue which first alerted geographer Alfred Wegener to the idea of continental drift in the 1920s.

He had no plausible mechanism for his theory and therefore was ignored for 40 years – in the wilderness if you like – and such is often the life of an innovative and inductive thinker.

Today the earth is believed to be 4.5 billion years old, forming out of gas, dust and rock like the other planets and driven by the forces described above. The obvious question is: how early in the earth's history did floating crust become common?

Research is now showing that the continents sometimes have very ancient tracts, which although now cemented solidly to their host continent, have originally come from distant global sources. This suggests cycles of crustal coalescence and breakup going back into the very ancient past. So ancient in fact, that there are signs that before the 'slag' of the continents had even appeared, the earth was devoid of dry land and entirely covered by bubbling, steaming ocean.

The most recent super crustal mass, where pretty much all the 'floating' crust available collected in one place and into a single global continent, is called Pangaea. This is estimated to have occurred about 270 million years ago. It then began to break up, starting around 200 million years before the present (BP). The detailed history is complex and the breakup was gradual, but the result was the division into two great continents called Eurasia and Gondwanaland.

These two super continents subsequently split into the continents of today, with the latter breaking up into Antarctica, South America, Africa, Australia, India and various other associated bits and pieces. Of course the geographical shapes have changed dramatically while moving around the globe, as mountain chains were formed by collisions, and volcanoes emerged due to subduction. All the while, the usual background processes of erosion and sedimentation continued unabated.

As this merging and separation of land masses proceeded, life as we know it rose both in the oceans and later on land. Vast amounts of erosion caused new sedimentary basins where oil could be generated and where coal from forests was buried. It transformed the planet from one with a sterile, non-oxidising atmosphere to one rich in oxygen, allowing the environment we have today.

Imposed on this, we have a sequence of geologic eras to describe rocks formed in this process and the timeline of life as it was created. The broader divisions include the Archaean and Proterozoic of the Precambrian, starting 4 billion years ago. The Cambrian followed (540 million years ago), Ordovician, Silurian, Devonian and Carboniferous (360 Mya). Next came the Permian (299-251 Mya) - coinciding with the formation of Pangaea at 270 Mya. Then came the Triassic, Jurassic (200 Mya) followed by the Cretaceous, Tertiary and Quaternary (starting about 2.5 Mya). Of course there have been many subdivisions defined within these eras, as further detail has emerged.

Many non-geological folk will be familiar with some of these ages and know for example, that before the Cambrian, only non-skeletal animals were plentiful – and of course hard to identify in rock. Others will recognize the Devonian and its association with fish and marine life, because that is where it was happening earlier, as life abounded in the oceans.

The Carboniferous is noted for coal, which was the result of the growth and eventual burial of massive forests of mostly extinct trees. A few of these still survive today.

Once there was plenty to eat on dry land, the terrestrial animals emerged and dinosaurs became prolific in the Jurassic-Cretaceous. The first flowering plants appeared in the Cretaceous, followed by the rapid rise of mammals and primates in the Tertiary - and flourishing in the Quaternary – us.

In addition, the study of the fossil magnetism preserved in volcanic rocks (paleomagnetism) helps us understand where and when the continents have moved over time. And the geochemistry of volcanic, igneous, metamorphic and sedimentary rocks aids us in determining how they were formed.

To cap this all off, the strata sequences (sedimentary, volcanic, etc.) logically indicate the overall history. This aspect of geoscience is termed physical geology and gives us the relative timing of events; events which can then be finetuned with fossil evidence and radioactive age determinations.

The analysis of radiometric data such as uranium, thorium and potassium within stable minerals (such as zircons in rocks) yields absolute long range dates over many millions of years, and covers the geologic timescale above. Shorter term radioactive isotopes such as C_{14} and Pb_{210} provide the timing required over tens of years, to tens of thousands of years.

The intermediate age gap is largely covered by non radioactive methods, and both rocks and skeletal remains can be tested in this way. It means that researchers can often choose from more than one method to make sure their dating is valid.

Sometimes we might see igneous rocks with large crystals on the surface and wonder, how those crystals have come to be so large? This, given large crystals must cool slowly to grow.

It means of course, that the molten magma from which they have formed, cooled under an immense pile of other rock (often sedimentary) which has acted as a thermal blanket.

This material has subsequently been eroded by wind and water and carried away, thereby exposing the coarse grained igneous rocks at the surface. However the overlying blanket may still remain in other places, and can therefore be traced across the landscape.

We might find that on other occasions the overburden has been pushed up into giant folds, perhaps hundreds of metres thick and to considerable height by the intruding magma. On the other hand, we might also find that the beds have fractured and faulted as well, under the extreme forces applied.

It all means that by examining the chemistry and grain size of the igneous and volcanic rocks and tracing sedimentary rocks across the landscape, we learn a lot about the history of a particular area. For example, which sedimentary or volcanic formation occurred earliest and the sequence of events that followed. By dating the igneous rocks using radioactive measurements, and comparing them with the sedimentary history of the adjacent strata, a complete history of the area can be derived from all sources. This may include fossil assemblages found in the sediments.

Granites are rather interesting and are categorised mainly by their mineral content. Because granites are made from regurgitated material, which previously had a life as igneous, volcanic, metamorphic or sedimentary rocks, they are quite diverse chemically and are often formed by subduction at continental margins. On the other hand, massive limestone and dolomite deposits (with their fossils and original marine structure intact in many cases) were once laid down horizontally, or sub-horizontally in marine environments. They can now sometimes be found thrust up through ninety degrees or more.

This attests to the mighty forces of continental collisions, immense power of volcanic activity and in some instances, bolide (asteroid and comet) impacts from space.

Overall, the extended growth of coral and marine sediments (sometimes kilometres thick) and the complex, but evolving biological material contained within the strata, is evidence of millions of years of activity. Add subsequent thrusting and sometimes overturning of these beds, and we have a complex and extensive history.

There is much more beside, sufficient to convince an open minded, but diligent and persevering student, that there is an exceedingly long earth history extending right back to the earth's early formation. Back to the point where oxygen in the atmosphere was virtually non-existent and therefore unusual rocks and minerals - and even metals - could form without being oxidized as they would be today.

So this all takes time and is wonderfully complex. It is also quite interesting, once you get into the science properly. The interrelationships between the major rock groups, together with the fossil record - the association of environments, the processes of formation and so on - give us a solid understanding of the history of the earth. Moreover fine stratification, such as that seen in lake (lacustrine) sediments, ice cores and even tree ring dating (dendrochronology) allow a direct annual counting, over a range of more recent years. A record of over 9,000 years has now been recorded using tree ring patterns alone; as living trees are matched to those preserved in oxygen poor environments such as bogs, frozen lakes and tundra.

Tree ring dating and annual varved clay layers in lake sediments which contain carbon, or carbonate material, allow cross referencing and correlation with the Carbon 14 dating method.

These measurements can also be used to fine tune C_{14} calibration curves where that is deemed necessary. And indeed, calibration is required to guard against local C_{14} variations caused by unusual circumstances, e.g. volcanic activity, or other rapid climatic variations over decades and centuries.

The checks and balances that result from using combinations of independent dating methods - and the sheer volume of data which is now available - ensures we can rely on the interpreted results to an impressive degree. It is solid evidence to verify beliefs or challenge them.

We only need to look at the story of the Australian Aborigine, to see how various strands of evidence can be brought together to study recent prehistory.

The extinction of many of the Australian mega fauna (large marsupial animals which disappeared about 40,000 years ago) has been tied to either the aridity associated with the height of the last ice age, or from over-hunting. As it happens, 40 to 45 thousand years ago is closer to the start of the last glacial maximum, rather than its height, and so aridity is not likely to be the only cause of the extinction of the megafauna.

Fire - a prime hunting tool of the Indigenous folk - was used to herd animals together and so facilitate capture for food and clothing. An examination of lake sediments has shown when this use of fire began, because the lake sediments show a dramatic increase in carbon at the time above - immediately before the ice age. An example of this is Lake George, near Canberra in Eastern Australia. It is a shallow and sometimes dry lake, now without an outlet to either the eastern seaboard or the western Murray River drainage system.

Moreover C_{14} dating with modern AMS (atomic mass spectroscopy) technology, can now be used for dating carbon containing materials back to 50 thousand years. Other methods can also be used in this space and in some cases for older remains. Optically Stimulated Luminescence (OSL), Thermoluminescence (TL), Electro-spin Resonance (ESR) and even Uranium series dating are alternatives and sometimes more effective.

The electron spin / luminescence methods, have been successfully used on some of the bones found in grave sites at Lake Mungo, in western New South Wales. This was a lake system providing a home for Indigenous people that has long since disappeared. This, as the climate dried out over the last 8 to 10 thousand years. One grave around 30,000 years old, shows indications of the love and respect of the deceased by those bereaved. Other indications of Aboriginal occupation are far older.

Clever analysis of microscopic amounts of carbon found embedded in ochres and patina, has been used to date Aboriginal artworks in cave paintings. Such dating, for example, can also show how artistic styles have changed over time.

The implements used in depictions of daily life in paintings, can also give us insights into ages and migrations and correlate with climate changes affecting the landscape.

The use of hunting tools varies according to the amount of woodland in an area. So in Northern Australian, cave paintings show individuals there tended to use spears and throwing sticks in heavily forested terrain, where boomerangs were not suitable. As paintings show a move from one form of hunting to another in the same location, we see a proxy for climate variability and the onset, of say, an ice age and low sea level during a glacial maximum. Under these circumstances migrations from New Guinea or elsewhere are more likely, and new painting styles might be expected as newcomers move in.

Now some of this evidence is rather flimsy and not conclusive on its own, but in conjunction with other dating evidence, builds a formidable case for history as promoted by contemporary science. It shows us that the earth has been formed by many 'natural' processes, which we can understand to a greater degree than ever before. But we must be prepared to do the work, based on sound principles and using the latest proven techniques.

We also have to approach new findings cautiously, until the results are verified and supported by others.

To say that the earth was created over a very short period as some misguided folk still insist, is to make God a deceiver. Today He has given us the evidence to show how and when, as well as the wherewithal in mind and resources to logically examine that evidence. However it is only coming together in a massive way in this day and age, because of the immense complexity of the task and the lack of the requisite technology in previous centuries. This is all fundamental to true knowledge.

The important point here, is that it is not in conflict with the basic premises underpinning the Bible, such as the creation and formation of the natural world. We just need to come to terms with earth processes and with the scriptural evidence.

There is also abundant fertile ground for further geological research in association with human prehistory - and a mention will be made later of potential marine investigations in regard to the close of the ice age. The last ice age at the end of what geologists call the Pleistocene, ended officially 12,000 years ago, but with repercussions for several millennia thereafter. This occurred as the world's ocean levels stabilised, but oscillated slightly to close at their current, but slowly rising levels.

There is evidence in places such as Magnetic Island on Australia's Great Barrier Reef (GBR) that the sea level may have been about 2 metres above current levels as the oceans bounced around (as it were) under the effects of the extra weight of melt water; this perhaps only 4,000 years ago.

Worldwide, there was a combination of climatic oscillations and earth rebound. As the weight of the ice in high latitudes (sometimes several kilometres thick) melted and allowed the earth beneath to rise, the weight of rising melt water in equatorial and temperate climates forced low lying coastal areas down under gravity - and flooded them. This readjustment is still proceeding in some regions.

It also turns out that there were several ice ages, over many hundreds of thousands of years, but they were not as severe as the last. We know this, because mapping and analysis of drill cores of the submerged continental slopes, deep ocean sediments and ice cores reveal that: whereas the earth's ocean levels were depressed about 130 metres during the last ice age, in previous ice ages, levels were only lowered around 50 metres below present day levels. In addition, each ice age was separated from the others by a glacial minimum, where temperatures and ocean levels were much the same as they are today.

One of the amazing consequences of this, is that coral reefs around the world (including the GBR) have been in and out of seawater several times over many millennia. Because of this, the reefs have formed distinct hard-ground stratified layers when they were exposed to the atmosphere, and these have been mapped and the ages measured. These tell-tale hard grounds formed by exposure to the weather, have been investigated by drilling and coring, as well as high resolution seismic surveys.

Of course, in some places the continents have moved up and down as well due to tectonic effects of continental collisions, unrelated volcanic activity, or to a lesser extent through the weight of seawater flooding dry land as described above. In short, it can be quite complicated determining the overall history of any particular piece of turf. But in the case of Australia, with its very ancient landscape, minimal recent crustal activity, vulcanicity and glaciation, we find greater tectonic stability than most places, and this helps with research into sea level change.

I am fortunate to have a general grasp of the subject, having spent time on the GBR with various marine research groups over the years.

Changes in biology can also be observed in reef populations and at the micro level, growth rates of coral, for example, can be determined over hundreds, even thousands of years.

The reef is a natural laboratory and historical library all wrapped in one, and the research shows the durability of the reef given sufficient time to recuperate. It is a wonderful place, but one that is now unfortunately under threat by very rapid environmental change, with overheating surface waters and undesirable run off from the land. The speed of change is not something that the reef has had to cope with before.

On another tack, ice cores from both Greenland and Antarctica have yielded detailed historical records of atmospheric conditions and volcanic eruptions, back to 100,000 years. More general data to 500,000 years is available at time of writing, and potentially, there is sufficient depth of ice in Antarctica, to record one million years or more of atmospheric history.

So, recent geoscience research has shed much light on the history of our earth through geological and geophysical mapping, borehole drilling and satellite imagery. Much of this has been driven by oil exploration and the innovation of new geophysical techniques, or the refinement of, and up-scaling of older technology. It has allowed the determination of rock boundaries and rock types from the near surface, to deep within the earth. In concert, a myriad of complementary laboratory analyses have become increasingly more accurate and sophisticated in determining physical and chemical properties.

The overall increase in understanding of the crust of the earth is astounding, with an amazing acceleration of knowledge between 1960 and 1990, but still continuing today.

Now for those interested in history and earth science this is all fascinating material. In fact it is eye opening for anyone who is interested in truth. If we cut ourselves off from this, we are losing our heritage as twenty-first century citizens of planet earth. Unfortunately this is happening widely today in some sectors of our global community.

Appallingly, it is rife in many of our Christian churches.

Six day confusion

Perhaps a more appropriate heading for this section would be: 'Why did our Creator God need a rest day?'

This is actually a profound question, too profound it seems for most to ask. In fact why one literal rotation of our humble planet Earth when he is Creator of all?

Dividing the creation sequence up into six (the number of working days in a seven day week) is obvious enough, if you want to emphasise the lesson of the Sabbath rest from the Ten Commandments. We might imagine therefore, that it was designed to set an example from the outset. After all it is a lesson well worth teaching. The freedom of rest, recreation and worship for everyone every seventh day, recharges the batteries after a hard six day slog. However it is a lesson that has little to do with actual history, or an understanding of how the earth came into existence - and this is really what Genesis 1 is about.

Moreover history teaches, that both a seven day and ten day week were derived from dividing the lunar month into four and three respectively, and that both periods were used by various societies in ancient times. The divisions are only approximate of course, but were adjusted in a strict lunar cycle by varying the length of the last week, before the next new moon. There is also little doubt, that the whole concept harks back to a time well before the giving of the law at Sinai.

Because there has been some editing of the Bible throughout the copying process in its early history – and that has been briefly alluded too – there is the suspicion that not all changes to the script have been made to clarify names and locations. Rather, some changes may have been made to reinforce the power of the elite over the rank and file.

So conflating the creation account with the Sabbath rest commandment can also be seen in a cynical light; one of opportunism by a political and religious class who desire to enhance their power. To compound the matter, for much of Hebrew-Israelite history, there appears to have been little separation between legislative and religious power.

In our contemporary era, contrast the enlightenment and revealed knowledge of confirmed geoscience (and what it ultimately reveals about our earth and its Creator) with the six day model of creation still espoused by some 'fundamentalists' in the 21st century. Essentially the model is based on that popularised by John Whitcomb and Henry Morris in their 1960s book, 'The Genesis Flood'. The basic principles taught today seem not to have changed, despite the vast amount of evidence and evidence based conclusions delivered by the geoscience and associated scientific communities, since then.

The details are not elaborated on in the congregational environment, as presentations are usually confined to all that is wrong with the world. This is then attributed to old earth beliefs, the teaching of biological evolution and perceived inconsistencies with science and scripture.

In addition, six day proponents do not attempt to seriously address the creation of the universe as a whole, but focus almost entirely on the earth. To this end, various publications and books provide some detail of the 'hydraulic' flood model, where the 'hydraulic' model is essentially Noah's Flood.

Mostly though, six day creationist literature aims at discrediting scientific thought, by introducing doubt of scientific methods, the objectivity of the science community and the robustness of peer review. To be blunt, their arguments are 'conspiracy theory' based and have consequently caused enormous harm to western society and believers in particular.

Notably, in the 1950's it would have been prudent to use Dr. Libby's new carbon (C_{14}) dating technique with caution. After all it had just been invented then, the equipment crude by today's standards and there was no vast library of data to rely on and use for calibration. But to continue to use the same cautionary quotes articulated by Libby and his contemporaries, as a weapon against the validity of the method in the 21st century is outrageous. It is so irresponsible, as to be completely contrary to Judeo-Christian and most secular values and notions of truth.

Six day teaching is a hotchpotch of examples, purportedly showing how quickly sedimentation or coal formation can happen. They point out vertical logs in coal seams, or the similarity of sediments from recent volcanic eruptions with old ones, and then suggest that if these things can form rapidly, surely this indicates that all geologic strata are very young. There seems an unwillingness to understand that ancient sediments may have formed rapidly and catastrophically in many cases, but that does not mean they were deposited and buried recently, or overnight.

So just because trees are sometimes found fossilised in a standing position in coal seams (with their heavy roots to the base of the seam) does not mean they were buried a few years ago. It says nothing about age, as any reasonable person would conclude, but more about hydraulics and the floating characteristics of big, water-logged trees – something, one would have thought, that 'flood' geologists should have a handle on.

Sedimentary strata shows that many diverse environments can be found in a single vertical sequence of rocks.

Vast marine silt and sandstone layers interspersed by desert sand deposits, windblown coastal dunes interrupted with freshwater lake and river environments complete with fossils, are common examples. Igneous intrusions of acidic rocks such as granites, or expansive basalt flows, may separate sedimentary and metamorphic sequences, or cover coral reef deposits complete with in-situ reef structure buried and preserved 'as is'.

Even buried remains of asteroids have been located on seismic profiles - and the list of possible sequences is endless.

We end up with a fossilised history of the locale being investigated and viewed objectively, it becomes apparent that none of this could have happened in one flood, no matter how catastrophic. There are simply too many individual elements that need vast amounts of time to develop; and coral reefs and coal seams hundreds of metres thick, and buried where they grew (or were deposited) are prime examples.

There are also chemical budgets; such as carbon in the form of coal and carbonate as reef, that are so great, that it is impossible that all the plants buried in coal seams, or marine creatures found in ocean sediments, could have lived on the earth at the same time. Space and time prohibit that possibility.

It therefore becomes obvious to the serious observer, that this complex sequence of events did not happen in one event, but rather over millennia and indeed millions of years.

Despite all this, rather juvenile arguments, unchanged since the 1950's, are still rolled out as arguments for a young earth:

- The rate of increase of the earth's magnetic field means it should be much higher if the earth is as old as scientists say: it ignores the fact that geomagnetism is cyclic even flipping poles on occasion
- The concentration of salt in the oceans should be vastly higher than it is for an old earth:- a fair enough question in the '50s, but ignores the recycling effect of all produced rocks and minerals (including salt minerals) due to erosion, sedimentation and continental subduction

- Meteoritic dust should be ubiquitous: disposed by everyday erosion and the same plate tectonic recycling processes
- Volcanic rocks should cover the continents, if the rate of volcanism has continued at the current pace for millions of years: again all explained by erosion, sedimentation and continuous crustal recycling

One could continue, but it is interesting how often a pseudo uniformitarian argument has been contrived, or a now outdated one applied, to discredit a late 20th and early 21st century understanding of the nature of the earth's crust.

So much for keeping up to date with the last 60-70 years of thoroughly confirmed scientific observations.

But for most of the captive audience in church pews, in an environment where open discussion is either not appropriate - a sermon for instance - or not permitted, the detail of their model is simply not discussed and therefore any focus on the model is bypassed and deflected. In a situation where the discussion format is a little more open – say at a university or other public forum - a minder may attempt to manhandle you, because you have unwittingly planted your bum too close to those raucous sceptics up the back.

At the very least you are likely to be told to give someone else a chance to respond, because you have already asked a question – code for 'shut up, we don't want to hear from you again'. Incidentally, I do not need to quote any sources, not even a pers. comm. for these observations.

But 200 years ago and stretching back to antiquity, mankind was not so blessed with data that the modern model of earth formation was at all obvious. We can now see, that if someone came upon marine fossils in fossiliferous limestone, or dolomite with visible coral, bryozoans, fish skeletons and the like - say high in the Judean hills, or even more incredibly the European Alps - then it would be assumed that since the earth was solid, but the oceans fluid, surely it must have been a huge watery flood that deposited all those sediments and fossil remains at altitude. The idea that land emerged hundreds, or thousands of metres out of the sea was an unreal proposition without a cause.

There was no known mechanism for the latter. No plate tectonics, no mind boggling collisions of humungous continental land masses - nothing. It seemed after all like a good example of 'Occam's razor', which states the simplest explanation is most likely to be the right one.

The problem is, that you have to have all the facts; if you do not, the rule does not work. And anyway, it was never meant to be applied as a universal law; it is just a useful approach designed to streamline an investigation.

There are a lot of practical questions that can be put to 'Six Dayers'. Yes they may seem trivial, even derisive, but they should have answers. The questions revolve around the Noah's Flood story, because most of our global topography is attributed to the Flood by the six day model. Here are a few of many:

How did the animals get to and from the ark from every continent on earth? How was this massive double ferry operation before and after the flood carried out? And why is it not mentioned in the Bible?

Around 2350 BC is the estimated time of the Flood using biblical history and genealogies. Bishop Ussher and more recent biblical chronologists have confirmed this estimate. This date then begs the question; when did the ice age happen?

Where is the evidence for a recent ice age after that time; or for that matter in the last 10,000 years?

Insisting that the ice age came after the Flood is mandatory doctrine for 'flood geology' advocates, because they believe that the vast kilometre-thick sedimentary basins were laid down in the Flood.

However, Antarctica has many deep sedimentary, fossil filled basins buried by kilometres of ice even today, and the surface evidence for ice cover throughout northern Europe, North America and Siberia is indisputable. With all the evidence pointing to peak ice cover 22,000 years ago, the ice age problem for the six day 'model' is simply insurmountable.

Indeed, where were the four rivers of Genesis before they were buried by multi-kilometre thick, oil bearing sediments in the Middle East? How can they possibly be identified with the Tigris and Euphrates and tributaries today? And what about the ancient flood cities of Sumer unearthed by archaeologists near the present day surface, and recorded in the Bible?

Is all this fascinating archaeology simply to be ignored, because it sits on kilometres of supposed 'flood' sediments?

How can Jericho in the Jordan Valley, estimated to be about 8,000 years old, be sitting on 6,000 metres of rift sediment - presumably deposited by the Flood in that same time frame? What of the Giza pyramids? Weren't they standing in 2350 BC?

And standing on hundreds of metres of marine deposited limestone? Deposited when?

For that matter, what about the evidence for 65,000 years or more of Indigenous occupation in Australia, on lands overlying one of the thickest sedimentary basins in the world (the Great Artesian Basin); how is this to be explained away?

This is not the end of six day 'creation' self-deception. It is recognised by some followers, that a Flood alone (no matter how immense) could never upend sedimentary strata laid down horizontally in water to the crazy angles seen in the European Alps, Himalayas and American Rockies. In fact, in some cases the strata has been overturned by folding and faulting due to the huge forces involved in continental and oceanic plate collisions.

So to explain all this, some six day pundits have reluctantly proposed that continents must have whizzed around the globe like Frisbees; this in the short time available since the Flood episode. And that has to be since 2,350 BC, if the biblical genealogical chronology is to be followed.

Naturally, this is where a few thousand years longer is often claimed, because of the obvious silliness of this scenario.

In the following chapter, more will be analysed and discussed in regard to a Flood Epoch and the scientific evidence for the 2350 BC flood event. It will confirm in most minds that six day flood geology is man-invented nonsense.

Frankly, there are simpler and more God honouring ways to understand the early chapters of the Bible, without proposing false historical and pseudoscientific scenarios. Scenarios based on vain imaginations, physical impossibilities and totally devoid of evidence. If we are looking at a proposed earth history of only six to ten thousand years for most geology, one surely has the right to expect definitive answers to these questions; especially a precise date for the Flood which six day creationism and 'flood' geology hinges on completely. Unfortunately, I know from experience it is easier to extract hen's teeth.

In Mesopotamia their world was bounded by the Two Rivers, The Tigris and Euphrates. A wider view would have developed later; bounded by the Mediterranean to the west, the Black Sea and Caucasus Mountains to the north, the Persian Gulf and Indian Ocean in the east and the Red Sea to the south.

Either way, the focus of the Mesopotamian world was a regional one when it comes to geographical extent. Context is everything and when we read that the whole world did this or that, or that 'all the people of the nations will worship at Jerusalem', it does not necessarily mean that the entire population of the globe, or every living soul will participate at one time. Rather, there will be a desire to go there and be a representative.

So we should be careful about forcing our modern global understanding of the earth, on the historic stories of flood recorded by the ancients. If there is a global aspect to an ancient event, a broader descriptive such as a 'flood covered the whole earth' is most likely to mean there were reports of 'flooding from the most distant places imaginable'.

And remember, these misunderstandings would be compounded by the very basic lexicon of the day.

Despite the above, we will see in the next chapter that the 'Flood Epoch' experienced in Mesopotamia and particularly the biblical flood event dated at 2350 BC, did in fact impact the whole of the Northern Hemisphere. In that sense it was global in extent. There was global flooding and catastrophic weather events for years. But the earth was not completely submerged all at one time. It was simply widespread flooding in the way we would understand it today.

Other 'creationist' advocates suggest there may have been a former 'old age' earth; one that accounts for the formation of the cosmos and its galaxies and solar systems – but also some (or most) of the fossil bearing sediments are suggested to come from this time. However, they insist this earlier age was before the biblical creation. In this scheme, the early verses of Genesis

1 are depicted as the creation of the old earth, and the remainder of the chapter becomes a recreation / renewal. This is referred to as the Gap Theory and attempts to get around the 'death before Adam' problem, by placing the early sedimentary fossils in an older creation that was then discarded.

Many other variations on this and other themes have been proposed, outside a strict six day scenario; but all are fraught because they do not address the evidence.

As described above; the essential and pivotal hypothesis of strict six day theory, is to categorically maintain that the deep sedimentary basins of the world (which are often chock full of fossils of mostly long since extinct animals) are a product of the Genesis Flood. Any attempt to imagine the pre flood world on the continents becomes impossible in terms of today's topography, and this is especially so in the Middle East; the centre of our biblical journey.

It is an attempt to dismiss the fossil record which shows the development of plant and animal kind; a key piece of the puzzle for the vast ages and intermediary life forms that once existed.

The six day argument is obviously unconvincing, but it seems because of the philosophical difficulties, the extreme six day view is the one which is flogged to the churches without explanation and sold as a pig in a poke. It is propaganda with no redeeming features.

There have also been misconceived defences of a literal six day creation. One of these arguments came to prominence after Charles Darwin's 'Origin of Species' was published and the protagonists of the new theory lined up against the Church.

When, for example, the supporters of Charles Darwin such as the brilliant, but agnostic Thomas Huxley began to support the evolutionary hypothesis, some of the argument was inevitably directed into a discussion around 'Creation (Design) versus Chance'. 'Chance' being the seemingly random way in which evolution was imagined to work.

This was unfortunately swallowed by many defenders of the faith and is still espoused to this day. They agree with their atheist opponents that evolution is 'chance', and thereby unwittingly assert that it is outside God's domain and cannot possibly be the agency for creation. They then double down and advocate an impossible scenario as a solution. This in turn makes the situation worse and Christianity a laughing stock.

The more discerning Christians and Jews recognised that our God is the God of everything, including 'chance', and that chance theory is just an aspect of the mathematics and apparent chaos that underpins the existence of the Cosmos. They therefore understood that the discussion had to be based on scientific evidence and logic alone.

There are historical examples of 'chance' which are extremely improbable, even unimaginable, but nevertheless do not contravene mathematical theory. In fact the 'creation versus chance' argument reduces our Omnipotent God of the Universe to a bit player - that is, if we assume he cannot, or does not work through things that look like 'chance' to us.

If we do so, it means we are suggesting he has no control over aspects of his own creation. We simply *must* conclude that our Almighty God is also the God of chance, game theory and any other manifestation of life which appears random and chaotic, or indeed impossibly repetitive. In fact it is far *more* likely that chance theory is one of the more direct manifestations of God's power and glory, albeit a conundrum on the fringe.

The inconvenient truth for six day proponents, is that their model is not only unsupported by the evidence, and their philosophic arguments flawed, there is no suggestion in the Bible for such a radical remodelling of the surface of the earth through the Flood. In fact a 'quiescent' flood model, where little obvious evidence for a worldwide event remains, was espoused quite quickly in the mid-19th century. This occurred as soon as it became obvious to researchers that the evidence for an old earth was irrefutable - and it should be noted - that some of these researchers were well educated clergymen.

It is therefore a great pity that this idea was not wholeheartedly embraced by all believers, because the evidence now at hand manifestly supports this view.

And just because we associate sin and wrongdoing with death and quite legitimately so; i.e. murder, suicide, lying, robbery, sexual impropriety, drug taking, chronic abuse, covetousness, etc. Surely this does not exclude death through natural aging?

All that is actually required, is a common sense realisation, that selfish human desires of the past, have caused unnecessary negative cultural and environmental consequences today. This includes death - and specifically premature death - and most of the heartache around the world. This is the death caused by Adam's sin, which in turn gives rise to the concept of 'original sin'. It is the death for which Jesus, our Saviour, paid the price.

So the six day hypothesis is a concocted scheme, that also diverts attention from the basic teaching in Genesis 1; 'In the beginning God'. The first verse is the only foundational part of the creation story, in terms of the Good News message of the whole Bible and our relationship with God and each other.

One may have the best motives in the world for defending the Bible, but if defending means inventing a non-biblical, fictional hypothesis and one that is contrary to the evidence, then we are indeed supporting a lie. We should therefore not be surprised when others use this as an excuse to ridicule the faith and distance themselves from it.

In reality, we are absolutely privileged in the developed world with the wealth of knowledge available today; and it is spreading further every year. It was foretold by the prophets.

Daniel 12:3-4 (NIV):

*"Those who are wise will shine like the brightness of the heavens, and those who lead many to righteousness, like the stars forever and ever. But you, Daniel, roll up and seal the words of the scroll until the time of the end. **Many will go here and there to increase knowledge**."*

And in 2 Esdras 6:18-21

***18** And it said, "Behold, the days are coming, and it shall be that when I draw near to visit the inhabitants of the earth, **19** and when I require from the doers of iniquity the penalty of their iniquity, and when the humiliation of Zion is complete, **20 and when the seal is placed upon the age which is about to pass away,** then I will show these signs: the books shall be opened before the firmament, and all shall see it together. **21** Infants a year old shall speak with their voices**, and women with child shall give birth to premature children at three or four months, and these shall live and dance.** (RSV)*

Note that Daniel did not say 'many will go here and there to increase disinformation and confuse many' – even though this is now happening. And the fruits bear witness we are rightly told. What are the fruits of six day creationism in the 21st century? How about a disregard for science and scientists, even though the comfort and benefits of science and technology are accepted readily - even greedily. Everyday things such as education, immunisation, medical technology, buildings, cars, television, smart phones, aeroplane flights and space exploration bring us enormous advantages, opportunities and interest of which our forefathers would have been truly envious.

Yet many are questioning even recent history. Such things as the moon landings are being denied and twin tower conspiracy involvement by the US government is promulgated as truth.

Basic concepts like the heliocentric model of our solar system are being rejected, and support for flat earth theory with fancy sun and moon imagery (reminiscent of Baal worship from ancient Mesopotamia) can be found on religious social media sites; this supposedly from people of 'faith' and even 'evangelical' quarters judging by their 'status'.

Then there are the slick, but fatally flawed movies which are promoting these erroneous and wayward ideas; again with 'Christian' tags, suggesting something terrible is going on within the faith - something that needs to be addressed immediately.

Is there any reason to doubt why conspiracy theories abound, global warming is denied and citizens are so easily hoodwinked by untruths, 'fake news' and 'alternative facts'. The most brazen liars gather the greatest following, or so it seems - and I believe much of it stems from the popularity of six day theory in the post world war period.

So what are the things we should focus on, in the creation narrative, in order to understand it properly in the light of the overwhelming evidence for an old earth?

There are a couple of basic questions I believe, we should ask.

Firstly, did Almighty God need a rest day after creation? The answer must be a resounding 'NO', of course not! He's GOD.

The Sabbath was made for us to rest and worship, as Jesus understood and taught. Not that God needed a rest, but that we do! It is basically a workers day off; a Workers Union agreement with the Boss which speaks as much about every employer's responsibility, as it does to employees.

Furthermore, why would God not create in an instant – or for that matter over an immeasurably long time? What is the point of creating over six literal earth days anyway?

We are told that our universe is maintained to this day by the word / power of God. That does not seem like resting to me and a literal day makes no sense when speaking of the Divine. The only possible conclusion is that the timing of six 'days' is *not* meant to be taken literally as 24 hour periods.

An objection that is often raised, is: why the emphasis on 'evening and morning' to describe the beginning and ending of each day, if ordinary 24 hour periods were not meant to be understood?

This is a fair enough point, but a somewhat broader meaning can be implied; that of a definite beginning and ending of the time period. Whether a period of time is a day, a week, a year or an age has to be determined by context, because the ancient word used is not necessarily confined to a literal 24 hours – particularly with prophecy.

Prophetically 'days' and 'evenings and mornings 'are used for defined prophetic periods of time in the Bible, as I will outline in the latter part of this volume. They are usually longer, and the context will give some idea on how to interpret the 'day' in each case. The 'day for a year' rule is the most common and even applies where the term 'evenings and morning' is used alone. In fact in Daniel's prophecy of chapter 8 where he describes a time of 2,300 'evenings and mornings' until the sanctuary (holy place) is re-consecrated, he is referring not to 24 hour days, but 2,300 years.

In this case the 'evening' and morning' only serve to define a complete year, from its beginning to its end - from go to whoa. It has absolutely nothing to do with dusk and dawn.

It is the problem of insisting on a strict literal interpretation, when such an interpretation was never intended.

This is all a clue to the origin of the first chapter of the Bible. It has all the hallmarks of a prophetically given statement on how the universe was created. Most prophecies predict future events and tell us what our Father God has planned to come. Genesis 1, on the other hand, is an example of a prophetically given description of what occurred in the past!

Whether first given to Adam, or Moses, or some other prophet is not certain, but even Adam was not around for the first five days, or most of the sixth according to Genesis. In addition, the fact that the picture painted follows our scientific understanding today so closely, suggests a prophetic link is extremely likely.

So with this in mind, the words and phraseology used in Genesis 1 and Genesis 2:1-2 are the same in intent as that in Exodus 20, where the Ten Commandments are described and the Sabbath day consecrated. Exodus 20:

¹And God spoke all these words:

² "I am the LORD your God, who brought you out of Egypt, out of the land of slavery.

³ "You shall have no other gods before[a] me.

⁴ "You shall not make for yourself an image in the form of anything in heaven above or on the earth beneath or in the waters below. ⁵ You shall not bow down to them or worship them; for I, the LORD your God, am a jealous God, punishing the children for the sin of the parents to the third and fourth generation of those who hate me, ⁶ but showing love to a thousand generations of those who love me and keep my commandments.

⁷ "You shall not misuse the name of the LORD your God, for the LORD will not hold anyone guiltless who misuses his name.

*⁸ "Remember the Sabbath day by keeping it holy. ⁹ Six days you shall labour and do all your work, ¹⁰ but the seventh day is a Sabbath to the LORD your God. On it you shall not do any work, neither you, nor your son or daughter, nor your male or female servant, nor your animals, nor any foreigner residing in your towns. ¹¹ **For in six days the LORD made the heavens and the earth, the sea, and all that is in them, but he rested on the seventh day. Therefore the LORD blessed the Sabbath day and made it holy.***

¹² *"Honour your father and your mother, so that you may live long in the land the L*ORD *your God is giving you.*
¹³ *"You shall not murder.*
¹⁴ *"You shall not commit adultery.*
¹⁵ *"You shall not steal.*
¹⁶ *"You shall not give false testimony against your neighbour.*
¹⁷ *"You shall not covet your neighbour's house. You shall not covet your neighbour's wife, or his male or female servant, his ox or donkey, or anything that belongs to your neighbour."* (NIV)

Note how brief and clipped the Commandments are presented. That is, except for commandment two in verses 4-6 (in regard to idols) and commandment four (the Sabbath day) in verses 8-11. Why is this?

Let's face it; both these commandments could have been elaborated elsewhere, like the other commandments which are indeed expanded on and discussed in greater detail as part of the overall 613 rules or mitzvoth. These are explained at length elsewhere throughout the Torah and particularly in Leviticus.

I understand people will accuse me of tampering / adding to / taking away from scripture and that this is explicitly forbidden. But what if the scripture has already been 'added to' and doctored? Is there no room for a Holy Spirit lead and empowered believer to correct the text, and restore it to its original meaning and intent? Is this not 'rightly dividing' the scripture?

The second issue is the addition of the weekly and Sabbath analogy to Genesis. Note particularly Genesis 2:1-2 below and the similarity with Exodus 20:11. Note also how both passages point to each other and are tied together.

¹*Thus the heavens and the earth were completed in all their vast array.*

² **By the seventh day God had finished the work he had been doing; so on the seventh day he rested from all his work. ³ Then God blessed the seventh day and made it holy, because on it he rested from all the work of creating that he had done.** (NIV)

In my final interpretation to follow, I could have left Genesis 2:2 in, and changed the 'day' to 'age' or something similar. But I feel confident that verse 2 is an addition, and has been placed within the original closing verses of the creation story that we see in chapter 2 verse 1 to 4. Unfortunately the Exodus 20 passage shows all the hallmarks of being added to as well, and I therefore suggest, there is little doubt that adjustment has been made to both books at a relatively late time. It is unlikely to be earlier than the Exodus and Mosaic era and has almost certainly been altered during, or after, the Babylonian exile, when the square script Hebrew was brought back to Jerusalem.

Whether it was done with goodwill or malice is not for me to say, but it is very suspicious. Furthermore the seventh day Sabbath can be described as the 'ornament' of Judaism.

The fourth commandment is paramount in modern Judaic thinking and has always had huge significance. Even today we see an extreme adoration of the fourth commandment amongst Orthodox Jews, verging, one would have thought, on idolatry.

Compare with Jesus' summary of the law. Unequivocally he quoted 'The Shema' together with a summary of the commandments 'Love the Lord your God with all your heart and all your mind and your neighbour (fellow traveller) as yourself'.

We should view the Sabbath in that context. Not as some onerous requirement, but for our benefit, while being considerate of our neighbours in all that we do.

In fact Jesus told it like it is: 'The Sabbath is made for man; not man for the Sabbath', he stated and showed an obvious disdain at the callous way the Pharisees policed the law.

Elsewhere the Commandments are recorded (such as Deuteronomy) there is no reference to the creation account. There is also no New Testament reference to sin and redemption, where mention is made of the length of the creation period.

And yet it is inferred by 'young earth' proponents using questionable exegesis and vain attempts to stretch interpretation. We are therefore left with the proposition that 'yom' and 'yom echad' are broad terms which can mean a literal day, but also any other period (special or otherwise) of defined time from start to finish, beginning to ending, opening to closing, dawning to setting.

So this is flimsy textual evidence for a literal proposition, and too much has been made of simplistic interpretations which then stunt our understanding of the message. Regrettably, it is the same throughout Bible prophecy, as I will show shortly.

Summing up the key philosophical and prophetic arguments against a literal six day interpretation, it can be stated:

- There were no human eyewitnesses to the creation (or six 'days') so the account was given prophetically to Moses or someone else
- Because it is a prophecy given by vision, or a word of knowledge, it will abide by the biblical prophetic rules and these will be described in the section 'Visions of Elohim'.
- God never rested. It is against his nature. We are told elsewhere in scripture that the creation is upheld to this day by the power of Christ
- Furthermore Jesus said 'the Sabbath was made for man, not man for the Sabbath'. This also implies that he never rested on the seventh day and never has

Blending the themes of Creation and the Sabbath seems unfortunate now, but perhaps the Bible indeed is a giant parable separating the blind from the truth, as much as illuminating the truth to those who sincerely seek it.

What I am therefore proposing and believe to be true, is neither the rigid fundamentalist and literal view of specific parts of the early Genesis Bible stories, nor a totally dismissive 'myth' understanding of the same events. Surely we can be mature enough to see, that Genesis 2 and 3 is an allegory of some of the big deal moral issues which have always faced mankind; this, from the very earliest times of discerning right from wrong.

The origin of the Garden story will have been inspired in some way or other. Either as a well thought out, but essentially human envisioned allegory, or as an inspirational episode of some sort or other. Regardless of origin, the teaching that is embedded is what is important.

That is not to say the writer did not have a particular place in mind for the Garden of Eden. Many suggestions have been made over the years, based mostly on the identification of the four rivers mentioned in the story. Today's north-west Iran is one, other commentators suggest the Hindu Cush or Pamir Plateau in the east, and even the Nile has been invoked as one of the rivers in a broader scheme. Lower Mesopotamia itself is another credible alternative.

The most likely identification in my view, is still around the Tigris / Euphrates river system and its ancient tributaries - this because of the geography of the four rivers of the 'garden' and the proximity to Harran and its association with far older places.

One very interesting proposition put forward recently, is that the incredibly early temple of Gobekli Tepe found in 1994 in south east Turkey was a temple of the 'Garden of Eden'. This temple with its many columns decorated with flora and fauna from back in hunter gatherer times, shows all the hall marks of being erected by people as a place of meeting. It is estimated to be in the vicinity of 11,500 years old. Significantly Gobekli Tepe is only 50 kilometres from Harran where Abraham lived.

The proposal is; that as these hunter gatherers came together on a regular basis for meetings and in increasing numbers, they found that they could no longer support themselves in traditional fashion, but started cultivating local plants and domesticating animals in order to feed themselves at the site.

It is an interesting theory and I think parallels the Eden story in a way that has considerable merit. Importantly it is to the west of Asshur (Assyria) as mentioned in the Bible. The Old Testament talks about 'the children of Eden which were in Thelasar' (in nearby northern Syria) and Assyrian texts mention 'Beth Eden' in the same Turkish locality west of their capital.

Ezekiel 31 through 'cedars of Lebanon' imagery, also potentially links Assyria with Eden. The location in the headwaters of two of the four rivers of Eden – the Euphrates and Tigris - is in the same general region (Eastern Turkey) that linguists suggest was the centre from which the dispersion of the modern languages of the world occurred.

Summing this up, it is fair to say there is a pattern here and strong circumstantial evidence for emergent, post hunter-gatherer civilisation in the region. In essence, this is what the biblical garden story tells us; the human moral dilemma, the imagery of wrestling with self and the serpent (Satan) and the consequences of failing to deal with new knowledge, self-realisation and 'progress'.

Although too early to be confident about calling Gobekli Tepe as a Garden of Eden Temple, it would not surprise if we have the origins in this general area. In fact there is another site 35 km to the north east of Harran, called Karahan Tepe, which also has T-shaped pillars that seem to represent human figures. Here they are placed in lines which may be ancient avenues.

Little work has been done at this locality so far, but there is potentially something special about the district as a whole.

For our purposes here, the important message the Garden story and the tree of the 'knowledge of good and evil' shows us, is that there was a time when mankind had no further excuse for bad behaviour.

Basically our consciences were developed sufficiently to know right from wrong in relation to our fellow man, and setting us a little apart from the rest of the animal kingdom. It is even possible, that something happened overnight to distinguish those that went before, from the Adamic line of the Bible.

The Apostle Paul contrasts sin coming into the world through the first man (Adam) and Christ (the second Adam) as redeeming us in Romans chapter 5. This is a commentary on sin as it affects mankind. It is not a world or universal history and does not teach a 6,000 year old earth.

I can only emphasise again, apart from the Genesis 1 narrative finishing at Genesis 2:2-4 and Exodus 20:11, that there are no other verses in the whole Christian Bible that are written specifically to teach a six day creation. This is despite attempts to twist words beyond their intent. I am therefore suggesting in regard to understanding the Bible that:

- o The ancient Genesis stories of the Creation, Eden, the Flood and Babel, while based on historic fact, come from pre-Abrahamic sources of antiquity, and must be treated with due caution as to accuracy and extent. Other historic accounts e.g. Genesis 1 are inspired in some way or other and are very likely to be the result of prophetic vision

○ The events, stories and prophecies focussed on and around the family i.e. Abraham, Isaac, Jacob and their offspring, are personal eyewitness accounts written down by family members and historically based. The prophecies associated with senior family members seem truly inspired and the result of visions and true prophecies of God

We should never forget that over 2,000 years between Adam and Abraham are covered in just ten chapters of Genesis (Chapters 2 to 11) and those ten chapters are only a small fraction of the Torah (Pentateuch). Adding the rest of the books of the Tanakh (the Jewish Bible), the apocryphal and pseudo-graphical works - plus the mountain of writing in both the Christian and Jewish traditions, only emphasises the considerable total volume of ancient Hebrew history and inspiration. This is in contrast to the few sparse words of the early chapters of Genesis.

The idea that the whole Torah was dictated word for word by God to Moses is only tradition. They are called the books of Moses, but there is a strong likelihood that a final compilation of these works occurred quite late, and most likely around the exile and post exilic period. That does not discount Moses' inspiration in regard to the law, or the fact that he may have been the one to receive the Genesis 1 account directly.

Now, attempts have been made to codify the possible early sources of the Torah by several scholars in recent centuries, and out of this, Julius Wellhausen and colleagues described what became known as the Documentary Hypothesis in the late 19th century. This approach was popular through much of the 20th century and terms such as Elohist, Jehovist (Yahwist), Deuteronomist and Priestly are used to identify portions of scripture which seem to come from different sources.

Other terms used as the basis for the study of biblical literary sources, are the Supplementary Hypothesis and Fragmentary Hypothesis. These provide alternate derivation for portions of scripture and endless discussion for the experts.

There is certainly validity in tackling the issues of antiquity, provenance, inconsistencies and anachronistic references in a scholarly fashion, even if all too often such analysis is used to discredit the basic tenants and inspiration of the content.

But that pitfall is what life and faith is all about. We need to test the record for better understanding, if we do not, how can we defend it properly?

In any case there are positives; a truly robust understanding of literary characteristics helps to trace the lineage and history of a language. It helps to support its antiquity through multiple translations, journeys, dramas and societal changes.

There is also another issue which may impact the phraseology and nature of the written word. It is the Gematria embedded in Hebrew writing. Without a separate counting system, numbers were assigned to letters for mathematical purposes. The numbers and combinations of numbers were also assigned meanings of their own, and used as apparent confirmation of the script - or alternatively, to impart an additional message.

Now if this is done in an anointed, God directed way by the priest or scribe, who can argue? However it is obvious that such knowledge is dangerous in the wrong hands and could subvert, or interfere with the true message of scripture, whether intended or not. The question is did this ever happen?

I believe a little caution is advised and it is possible some script may have been rearranged to accommodate this practise.

One thing is for sure, the Lord works with those of bad intentions as well as those with good. However in doing so, I believe he *has never allowed his word of truth to be corrupted so badly that we cannot understand it,* given sufficient guidance from above and intent here below.

So watch what happens when the text of Genesis 1 to 2:1-2 is restored in the way I believe it was presented originally; before the summary at the end was added and some of the script was misassembled from old parchment, or rearranged for other purposes.

It becomes a prophecy of our current understanding of the creation / evolution of the universe, albeit in a somewhat poetic fashion.

Genesis 1 and 2:1 with the verses reordered, but the original numbering intact:

*1 In the beginning God created the heavens and the earth. ² Now the earth **(matter)** was formless and empty, darkness was over the surface of the deep **(void / space)**, and the Spirit of God was hovering over the waters **of the void**. ³ And God said, "Let there be light," and there was light. ⁴ God saw that the light was good, and he separated the light from the darkness. ⁵ God called the light **(energy)** 'day' and the darkness **(absence of energy)** he called 'night'. And there was **the dawning**, and there was **the waning—the first age**.*

*¹⁴ And God said, "Let there be lights in the vault of the sky to separate the day from the night, and let them serve as signs to mark sacred times, and days and years, ¹⁵ and let them be lights in the vault of the sky to give light on the earth." And it was so. ¹⁶ God made two great lights—the greater light to govern the day and the lesser light to govern the night. He also made the stars. ¹⁷ God set them in the vault of the sky to give light on the earth, ¹⁸ to govern the day and the night, and to separate light from darkness. And God saw that it was good. ¹⁹ And there was **the dawning**, and there was **the waning—the second age**.*

*⁶ And God said, "Let there be a vault between the waters to separate water from water." ⁷ So God made the vault and separated the water under the vault **(oceans and lakes)** from the water **(vapour and cloud)** above it. And it was so. ⁸ God called the vault "sky." And there was **the dawning**, and there was **the waning—the third age**.*

⁹ And God said, "Let the water under the sky be gathered to one place, and let dry ground appear." And it was so. ¹⁰ God called the dry ground "land," and the gathered waters he called "seas." And God saw that it was good.

*²⁰ And God said, "Let the water teem with living creatures, and let birds fly above the earth across the vault of the sky." ²¹ So God created the great creatures of the sea and every living thing with which the water teems and that moves about in it, according to their kinds, and every winged bird according to its kind. And God saw that it was good. ²² God blessed them and said, "Be fruitful and increase in number and fill the water in the seas, and let the birds increase on the earth." ²³ And there was **the dawning**, and there was **the waning—the fourth age**.*

¹¹ Then God said, "Let the land produce vegetation: seed-bearing plants and trees on the land that bear fruit with seed in it, according to their various kinds." And it was so. ¹² The land produced vegetation:

plants bearing seed according to their kinds and trees bearing fruit with seed in it according to their kinds. And God saw that it was good. *[13]And there was the **dawning**, and there was **the waning—the fifth age**.*

[24]And God said, "Let the land produce living creatures according to their kinds: the livestock, the creatures that move along the ground, and the wild animals, each according to its kind." And it was so. [25]God made the wild animals according to their kinds, the livestock according to their kinds, and all the creatures that move along the ground according to their kinds. And God saw that it was good.

[26]Then God said, "Let us make mankind in our image, in our likeness, so that they may rule over the fish in the sea and the birds in the sky, over the livestock and all the wild animals,[a] and over all the creatures that move along the ground."

[27]So God created mankind in his own image, in the image of God he created them; male and female he created them.

[28]God blessed them and said to them, "Be fruitful and increase in number; fill the earth and subdue it. Rule over the fish in the sea and the birds in the sky and over every living creature that moves on the ground."

*[29]Then God said, "I give you every seed-bearing plant on the face of the whole earth and every tree that has fruit with seed in it. They will be yours for food. [30]And to all the beasts of the earth and all the birds in the sky and all the creatures that move along the ground— everything that has the breath of life in it—I give every green plant for food." And it was so **(the sixth age).***

2 [1]Thus the heavens and the earth were completed in all their vast array. [2]By the seventh age... (NIV rearrangement)

This rearrangement makes sense in terms of our contemporary scientific understanding of the formation of our world.

We could quibble about minor detail and (say) move verse 14 forward a little, but it is very close to as holistic and as seamless as we would desire. It displays the hallmarks of inspiration to an even greater extent than previously.

And to be clear, I am not placing any limits on how the creation took place. The Big Bang theory seems holistic and logical enough, but space-time may have come into existence in a somewhat different manner to the one currently envisaged.

Nevertheless the prophet Jeremiah refers to 'the Lord who made the earth, the Lord who formed it and established it', hinting at the logical and sequential manner the creation took place and broadly aligning with the current 'out of nothing' theory.

Neither am I suggesting that all biological life necessarily evolved slowly on the earth. Our sovereign God is master of all creation; including any evolution, adaption, introduction or 'impossible' intervention he might choose.

In fact there are hints in scripture that the 'sons of God' (Anunna / Anunnaki of Mesopotamian literature) mixed with the 'daughters of men' to produce the Nephilim. If the 'sons of God' refer to angelic or alien beings, rather than God's human chosen (a 'big if' in my view) then some sort of rare event may have occurred rather late in the history of humankind.

In any event, the Apostle Paul gave the simplest and most straight forward summary of the human condition nearly 2,000 years ago. Romans 1:18-25

[18] *The wrath of God is being revealed from heaven against all the godlessness and wickedness of people, who suppress the truth by their wickedness,* [19] *since what may be known about God is plain to them, because God has made it plain to them.* [20] *For since the creation of the world God's invisible qualities—his eternal power and divine nature—have been clearly seen, being understood from what has been made, so that people are without excuse.* (NIV)

The Apostle was pretty blunt. He said it how he saw it. He was that sort of bloke. His is a simple explanation for the 'fall' or degradation of mankind, without invoking unrealistic theories of the perfection of the original universe; perfection which might be implied or interpreted from Genesis or Exodus.

Besides, once we look at the creation holistically, it becomes apparent that our Creator had to use a very clever method to make us, in order to accomplish his purposes. It suggests he had to develop a subtle, stand-off approach; one which would look very much like 'chance' from our point of view.

Creating and designing using an 'evolutionary' method for all living things, could well have been a key part in doing that.

His primary desire was to give us freewill and independence (or as near to that as possible) in order that our response to him, others and our environment, could be through our own perspective and not forced on us. In other words, *we were created in his image* with all the creativity and independence that implies.

It meant the possibility of sin and selfishness of course, but that 'fall from grace' was a necessary evil. It meant both the fall of the angels and the fall of humankind was a possibility, and a hazard that was deemed acceptable.

Yes, he could have created us as dolls or robots, but it seems he wanted companionship and love to be freely expressed between those of his creation, himself and the angelic host.

It is the same in our own experience. We may get some satisfaction from pretend entities, fluffy toys, or more recently robots and automatons using our imaginations. But it is with other adults, children and pets that we get the most love and satisfaction. Yes, they may drive us nuts at times, but it is the independent love, affection, character and quirkiness they display, that is most endearing. It makes up for those other times of annoyance and aggravation.

Now to those who find themselves unable to accept the above and through many years have happily lived with a six day creation as a literal proposition, I emphasise: the only thing that is foundational about Genesis 1 from a faith point of view is 'In the beginning God'. The first verse of Genesis says it all.

The rest of the creation account becomes a diversion away from living life 'in the Spirit' if it is taught as a literal seven days. In fact a literal interpretation can become such a focus, that it ends up as a replacement for the Gospel of the Kingdom and a relationship with the Lord and each other.

In conclusion, I hope this study of origins has been useful, if brief, and sets a foundation for some readers who have been looking for answers that make sense. Answers, in the context of our knowledge based world.

From here we will focus on the major events from chapters 5 to 10 of Genesis and beyond, as we look at the great historical stories and the personal experiences of the Patriarchs and the Children of Israel.

CHAPTER 3

Ice and Flood

'Remember our exploits to the western bank
Nubia, Philae, Abu Simbel, the Dam?
And then the felucca for the trip back north
On the beautiful Nile, that life giving source'

Giza

We were about to begin our rail journey up the Nile to Upper Egypt, but it seemed prudent to start our discussion of 'beginnings' with some philosophising, a discussion of basic earth science and the scriptural issues that follow. A taste of the adventure has been provided in the verse at the heading of each chapter, but I am going to keep you, the reader, waiting once more for the train journey; this time again in the interests of chronology.

In the biblical scheme of things, the next big Genesis event is the Flood - Noah's flood. As I hinted at earlier, while exploring around Giza and the Great Pyramid, riding a camel and viewing the Sphinx and the temples at the foot of Khafre's causeway (and in general absorbing the atmosphere of this most historic of places) surely there was enough interest, excitement and visual impact for our first morning on tour?

But there was that riverboat housed in its display building near the southern face of the Great Pyramid - and the open pits where it and its companion (currently also being restored) were exhumed. I was also aware of alternate theories in regard to the Sphinx, that suggested it's extremely weathered and waterworn appearance was due to the fact that it was not carved by Khafre's artisans, but by some unknown sculptor millennia before the pyramids were even built.

Robert Bauval, one of the alternate historians and speculators on such matters, suggested that the layout of the three largest Giza pyramids was in the form of the Orion star group and that this particular nebula was especially significant to the ancient pharaohs. Both he and Graham Hancock in 'The Message of the Sphinx' also proposed that as the body of the Sphinx appears to be that of a lion, and a much eroded one at that, it must be much older than the pyramids. This was posited on the basis that it was carved way back, when the zodiacal sign of Leo was the star sign rising before dawn on the eastern horizon.

This 'heliacal' rising of a zodiacal sign seems to have been (and still remains) the traditional way of keeping track of the ages. The observation is constrained in a repeatable science sense, to the equinoxes and is therefore called the precession of the equinoxes. It is observed historically in the northern spring and occurs as a result of the wobble of the earth on its axis.

Today, the star sign we are transitioning to and is about to rise in the east at dawn; is Aquarius. Older folk will remember when the theme song from the musical 'Hair' pronounced, that 'this is the Dawning of the Age of Aquarius'.

That was back in the late 1960s, around the year 1967 and is a date worth remembering for later. The reason for that is its prophetic significance, and I will be going into some detail here - not only to set the scene for a discussion on the Flood - but to suggest something significant later about history, a prophecy of Jesus of Nazareth and a sign of the zodiac. In fact, the current sign of the zodiac, Pisces, links together Jesus' statement to his disciples of 'I will make you fishers of men', with the sign of the fish used by Christians since the first century and the great commission to preach the gospel to all mankind.

Now the exact 'dawning' of a new age based on the heliacal rising of a new sign of the zodiac is rather subjective. It varies with latitude and altitude and was observed only by eye in ancient days. The signs also vary in apparent size. This makes it rather hard to judge when one sign is finished and when a new one has arrived. Despite this, we can calculate approximately when say, Leo would have been the sign on the eastern horizon in the pre-dawn back in ages past, because the signs of the Zodiac transition in order as we go back in time, i.e. they precess.

If we use the 12 signs of Mesopotamian tradition; the ones used in astrological horoscope and familiar to the Hebrews of old - and then standardise the process so that each sign occupies 360/12 or 30 degrees of arc, we can start to use the effect in a practical and useful way.

To go through a complete cycle of 360 degrees averages 25,771.5 years and is the shortest of what are termed the Milankovitch cycles, named after the Serbian mathematician Milutin Milankovitch. He investigated these things in the 1920s.

Milankovitch's research revealed two more earth-sun cycles of 41,000 and 100,000 years in length. The 41,000 year cycle relates to the angle of the axis of rotation of the earth with the ecliptic. It varies over the period from about 22.1 to 24.5 degrees. The longest cycle is dependent on the shape of the earth's orbit around the sun, i.e. a variation of the earth's ecliptic geometry. During the approximate 100,000 years, the earth's orbit changes from a more circular one, to a decidedly elliptical shape, under various solar system gravitational influences.

All three cycles contribute to changes in the earth's climate over the period, particularly as the cycles go in and out of synchronisation. In fact they are thought to influence ice ages and the deposition of sediments over thousands of years and the evidence is seen in tell-tale, cyclic stratification patterns.

As far as I am aware, a coincidence of Milankovitch cycles has no direct bearing on the date of Noah's Flood, although the Mesopotamian floods occurred after the last ice age and therefore a loose association can be implied.

Naturally, if we use 12 signs of the zodiac, it means dividing the complete cycle time by 12, to determine the length of each zodiacal age. We then find that the time between the risings (or dawning) of each new zodiac is close to 2,148 years. If we then focus on the proposition, that the Sphinx once had a lion's head and was carved in that star sign, way back in time (as Bauval and Hancock do) we find that we have to go back beyond 10,000 years. That is, from Aquarius (which we are arguably still transitioning into) back through Pisces, Aries, Taurus, Gemini, Cancer through to Leo.

As it turns out, there is new evidence which precludes this idea of an ancient Sphinx - at least in terms of the body. The body was created by the removal of the surrounding limestone, leaving a rectangular depression (enclosure) around the Sphinx body itself. The flow structure and fossiliferous nature of the Giza nummulite limestone (a marine sedimentary rock composed of foraminifera from the Mokattam Formation) is quite distinctive, and we now know that some of the stone was used in the construction of the Sphinx and Valley Temples.

So it is pretty much a given, that the limestone blocks can only have been carved out from around the Sphinx and hauled the very short distance to the temples when they were constructed in Khafre's time. In my view it leaves little room for the idea of a 10,000 year old Sphinx body. However it does not preclude the idea of some sort of head being carved on the original pinnacle of rock, that existed previously.

This geological evidence left traditional Egyptologists (who would have nothing of this 10,000 year thing anyway) fully vindicated. Most expert opinion suggested that the Giza sandstone layers were overall easily weathered; being eroded by rain, percolation and salt crystallisation. If so, there was never a need to postulate a 10,000 year old Sphinx in the first place.

Other pundits suggested that the body of the Sphinx was originally a jackal, rather than a lion, because it has a straight back as against an arched one. But of course if the body did not exist prior to Khafre, that piece of logic is pretty much irrelevant anyway. However the erosion of the Sphinx would certainly be a strong motive for refurbishment and enhancement and some have pointed to the fact, that the head seems small compared to the body for good proportion. Furthermore, considerable speculation has arisen as to whether the face is that of Khafre, his father Khufu, or indeed some other pharaoh altogether. Some even suggest that it was a pharaoh queen.

There is also the added issue, that some pharaohs had no qualms about purloining existing structures for their own personal promotion!

Put all these balls in the air and it is possible that a re-carving of the head took place before, and / or, after Khafre's time. The head would also be expected to be more durable, and particularly so if the uppermost layers were of harder material. This is in line with the observation, that high sedimentary landforms are elevated above their surrounds because of their extra hardness and durability. They simply resist erosion better than the surrounding softer rock.

In respect of the overall landscape; there are quarries to the south, but the original area may or may not have been a plateau. So whether the area was planed-off for building the pyramids or not, the area ultimately ended up as a fairly level surface, dominated by at least one prominent rock outcrop. And that outcrop may have been carved before the pyramids were built and conceivably more than once.

For me this matter has now been clarified, despite the obvious continuing unknowns. But at the time of my visit to Giza, I only had a vague idea of the various hypotheses surrounding this enigmatic figure.

Now in the early days my interest in the Flood would have been muted, except for two things that I can recall.

Firstly, I read Werner Keller's seminal book 'The Bible as History' (mentioned previously) and the discovery by C. Leonard Woolley of the significant flood deposit at the old city of Ur in Mesopotamia - Ur of course, being one of the flood cities of Sumer listed in the Bible. These diggings were conducted by British and American archaeological teams in the early 20th century, on the back of previous 19th century discoveries in the Fertile Crescent, Egypt and Turkey.

These earlier expeditions included field parties from a variety of countries including Germany and France, and were often privately financed and by today's standards not particularly professional. However this was understandable given the times; the science was new and the skills and tools developed quickly in the 20th century to match the enthusiasm.

The flood deposit discovered by Woolley in the summer of 1929, was a nearly 10 feet (3 metres) deposit of clay, with no artefacts. It was obviously laid down in one event, and estimated by the investigators at the time, to cover hundreds of square miles of the Euphrates-Tigris valley.

The dating techniques were limited to studying pottery in those days, but they were astonished to find that while there were Early Dynastic clay pots which were obviously turned on a wheel above the layer, below the clay deposit, only prehistoric Neolithic implements and hand-made pot fragments were to be found. The clay layer did not cover all the adjacent hills, but despite this, Woolley was in no doubt that what had been revealed was firm evidence for the Genesis flood.

As more archaeology and drilling occurred throughout the Euphrates–Tigris plain, a clearer picture of the flood history became apparent. Firmer dates for these events were derived from additional archaeological evidence, and supported by carbon dating as this science developed after World War II.

Woolley bathed in much of the limelight and published early, but other archaeologists were investigating other sites at the same time. Stephen Langdon and colleagues, for example, found smaller flood deposits at Jemdet Nasr (Kish). It was assumed that the same flood was responsible at first, but subsequent investigations showed that the Ur deposit was from 3500 BC, whereas Kish showed two large floods from the Early Dynastic periods ca. 3000 and 2900 BC.

A couple of years later, another site, this time at Shuruppak (Schmidt 1931) revealed more flood deposits and this one was interesting in that it was the home of Ziusudra, the Sumerian Noah. The name Ziusudra is thought to be equivalent to the Akkadian Utnapishtim 'he found life'.

This flood level dates from 2950 to 2850 BC and separates the late Protoliterate and Early Dynastic I periods. It is also likely to be identical to the one at Kish.

An even larger flood dating to the Early Dynastic III period (ca. 2600 BC) was found and the investigators involved suggested that this flood, or floods, were more likely to be the biblical flood than Woolley's earlier one. Gilgamesh, one of our other flood heroes, was the fourth king of Uruk according to the Sumerian king list and lived about 2600 BC. He was also a contemporary of Aga of Kish, the son of Enmebaragesi, the first independently attested king on the Sumerian list. The Gilgamesh story is famous and in the more realistic passages of the text, parallels the biblical account to a recognisable degree.

Well where does that lead us? Plenty of Mesopotamian floods to choose from for a start.

It is our first inkling of a 'Flood Epoch' in Mesopotamia, rather than one massive event. The timing is also smack in the Middle of the 4000 to 2000 BC period of our biblical patriarchs, and so none of this can be ignored and set aside as fantasy.

Reminiscing

The second event which piqued my interest, was meeting a young man who was a few years younger than myself. This occurred after his family began attending a church where I was a member. This church was situated south of Brisbane, Queensland and was one my parents had been instrumental in establishing.

He was an avid creationist in the six day vein of things, to the extent that it made me take note that there was an issue here. Up to this point I was quite happy with the idea that the biblical account was indicative only, and the garden imagery figurative and focussed on morality and spiritual awareness. I was therefore more than happy to move on to other things.

As previously described, I had already completed a geology course at school. In fact by this time I was also the beneficiary of my first adult employment, with some short term work as a technical assistant and subsequently a cadet engineer. This included work in both public and private positions in the mining, exploration and engineering sectors.

I was more a technologist than geologist, so had not gone on to do further geology studies at that stage. Nevertheless, this fellow's insistence that the geological time chart (with which I was quite familiar) was in fact nonsense, stopped me in my tracks. The result of this was that I decided to keep a more open mind on the subject and read as much as time allowed.

It prompted me to buy a copy of Whitcomb and Morris' tome 'The Genesis Flood', to get a better grip on the six day 'creationist' critique of modern geological theory and the alternate hydraulic model they espoused. However I was not convinced, and it goes without saying that one or two interesting 'discussions' on doorsteps and elsewhere subsequently ensued.

He was completing a Science (Biology) degree and went on to do a Diploma in Education, thereby qualifying as a teacher and following in his father's footsteps. He was also a gifted pianist and a fluent speaker, and in the decades after, has become a persuasive orator and notable figure in the six day creation movement.

I, on the other hand, was pursuing diploma level, tertiary studies at night, while working in surveying and engineering organisations. I was still looking for my place in life, and so it amazed me that he already appeared to know where he was going - and without studying geology formally - was already convinced that modern geoscience was a crock and of no value to man or beast.

He later went on to write a book called 'The Lie: Evolution', head up an organisation in America called 'Answers in Genesis' and raise the millions of dollars from his supporters required to complete a full scale replica of Noah's Ark; this according to the dimensions given in the Bible of 300 cubits long, 50 wide and 30 cubits high.

If entrepreneurship and project size matters, then this is surely proof positive that the Flood occurred!

Suffice to say, I still hold out hope that some good may yet come out of this grandiose project and its associated exhibition of dinosaurs. If the correct understanding and associations of scripture and history are one day narrated at the site, one would hope some honour and respect can be rescued for the Hebreo-Christian story, the Bible, and the interaction it reveals between God and man.

Melting of the ice sheets

From the mid-1970s through to 2010, I spent a good deal of time working at sea on various research ships; some small, others large and one in particular called the RV Rig Seismic. It was leased from 1984 onward by Australia's national geoscience agency, and was primarily equipped for multi-channel seismic reflection operations.

In common with most true research ships, it was also capable of a wide range of other scientific investigations. Most of the equipment was added in Australia, and included a deep sea coring system, various sea floor dredges, grab sampling systems and a magnetic profiler. It also carried a marine gravity meter, deep sea side-scan, heat flow measuring equipment, and after 1989, was fitted with continuous-hydrocarbon gas detection equipment. The latter system, based on vacuum and gas chromatography technology, was used for oil and gas exploration initially, but later adapted for environmental monitoring of offshore facilities and inshore coastal spaces.

Needless to say, it was a very stimulating environment working with ship's crew, local and international technical specialists, and a diverse bunch of research scientists with more Masters and PhDs than you could poke a stick at.

We were all focussed on geoscience investigative work around Australia's margins, but also further afield in areas adjacent to Indonesia, New Guinea, the Western Pacific islands and Indian and Southern Oceans. There were also operations in Asian waters around the Philippines and Korea, as part of international collaborative programs and overseas aid projects.

It was all very much 'full on', and somewhere during this time I began to wonder, if the plethora of flood stories from tribes and peoples throughout the globe, were in part due to the rise in ocean levels after the last ice age. This rise of over 120 metres was caused by the melt waters from huge accumulations of ice (kilometres thick) over North America, Europe and Asia during the glacial maximum. The glaciation peaked 22,000 years ago and thereafter the ocean levels rose as the ice melted.

There was one significant interruption to this overall melting event along the way; a refreeze called the Younger Dryas of 13,000 to 12,000 years ago. This event is still being investigated today and causes such as multiple asteroid impacts, posited as possible triggers. Despite this hiccup, the whole glaciation was pretty much over 10,000 years ago, but with some abrupt, absolutely catastrophic events, such as the flooding of the Black Sea still to come. However the Black Sea flooding did not occur until a couple of millennia later, and may be related to post ice age stabilisation issues, or some other unidentified cause.

As recorded earlier, the ice ages (known geologically as the Pleistocene) lasted well over a million years. Each time the world's ocean levels dropped significantly, with the last one being the most extreme. It meant that a country like Australia was joined to New Guinea to the north and Tasmania to the south. It also meant people could walk between these otherwise isolated places, rather than paddle their way across.

In addition, South East Asia was joined to islands like Borneo and parts of the Indonesian island chain. India was connected to Sri Lanka, and highway shortcuts were created from Africa to Arabia and Iran (ancient Persia).

England was also connected to continental Europe and Asia was joined to America via (what is now) the Bering Strait – and that is to mention just a few of the land bridges exposed.

When the ice began melting in North America and elsewhere, vast ice dams held back huge and elevated fresh water lakes behind them. This ice cold water was eventually released and spewed out of Hudson Bay toward Greenland, or centrally into the Missouri and Mississippi Rivers. Indeed it also escaped through the so-called Badlands of north-western USA, stripping the soil off and emptying into the Columbia River basin and the Pacific Ocean. These dramatic events involved a sufficient volume of water in some cases, to raise the global sea level a metre or more at a time. If you lived downstream when an ice rupture occurred, it would surely have been the end of your world.

It may have taken a Guinness or two, after weeks at sea (working seven days a week) for the idea to completely sink in, but sometime in the later 1990s, I became convinced that the flooding at the end of the ice age must have been the Genesis flood. Whereas wet conditions are likely to prevail as ice melts (lakes form and climate zones shift through the latitudes as the oceans transition to a higher sea level) during the ice age itself, climate tends to be dry. That is simply because there is so much water tied up in the ice sheets themselves.

Some of our later environmental work on 'stranded' coral reefs, 20 to 40 metres or so below current sea levels, showed coral had obviously failed to grow fast enough to keep up with sea level rise and seemed to reinforce this idea. These reefs with living coral (but subdued growth) were found in places like the Gulf of Carpentaria. Here, environmental conditions such as temperature and restricted food supply, also played a part. This compared, to say the Great Barrier Reef, where coral grows in more open, nutrient rich, oceanic conditions.

In fact, the Gulf may have had a fresh water lake, or two and been catastrophically flooded with seawater as well. That would mean corals in the area got off to a bad start after the ice age was over, and an abrupt rise in sea level was a hurdle too far.

Then to cap it all off, a book came to my attention called 'Noah's Flood'. It was written by two American marine geologists, William Ryan and Walter Pittman.

In their book, they described survey work in the Black Sea which they conducted with Russian collaborators. It pointed to the likelihood that the Black Sea was once a fresh water lake; one much like the Caspian, the Sea of Aral and an older lake further east which is now the Taklamakan Desert. Their thesis was that the Black Sea was inundated around 7,600 years ago by salt water. They proposed it occurred due to seawater flooding in via the Aegean, the Sea of Marmara and the Bosporus, as the level in the Mediterranean completed its post ice age rise.

Others put forward alternative scenarios; including the flooding by fresh water due to breaking ice dams, or simply by increasing flow of the European and Russian rivers into the Black Sea from the north and west. This, as ice unloading elevated the sources, allowing the rivers to flow into the sea more readily. There was a particularly strong discussion on the pros and cons of the evidence and the dating of the event - and various models were presented in the journal, Marine Geology.

From later work on deltaic sediments, it has been suggested that the amount of inundation may have been less than the 100 metres suggested by Ryan and Pittman. Regardless of the actual scenario - or magnitude of flooding - the result was apparently the same; a salt water connection to the Mediterranean Sea and the conversion of the Black Sea from a freshwater lake many metres below today's sea level, to a salt water sea.

The concept that at the end of the ice age, the world's oceans reached a height that caused flooding through restricted and shallow openings (sills) into deep valleys, is a fascinating one in its own right. This is particularly true around the Middle Eastern countries. For example, the Persian Gulf may have been a fresh water river and lake system below sea level, with the Hormuz Strait at the tip of Oman being a relatively shallow sill. Today the Strait is about 90 metres deep, but would have been shallower before erosion.

This hypothesis has some support from recent archaeological evidence from the islands of the Gulf. Ancient fishing villages and artefacts excavated there, suggest fishing took place as far back as 7,600 years ago. This timing follows the Black Sea story and suggests that the Persian Gulf flooded with sea water at the same time. The Red Sea could have had a similar history, with either fresh, or salt water bodies cut off from the Indian Ocean and probably existing below that ocean's level.

Indeed, the Gulf of Aqaba rift area with its en-echelon faulting and relatively shallow connection to the main Red Sea water body, could be another example.

It is also known that an extensive, now submerged area off Dwarka on the west coast of India, was once an important part of that country's land mass and most likely prehistory. In short, there are interesting opportunities for further underwater investigations, in order to determine the sequence of events and possible fresh and saltwater flows into basins. Floods and unusually moist wet seasons could have turned these gulfs into verdant environments at times, even if only temporarily. Eventually, the wetter transitional climate at the end of the ice age would have stabilised to a new drier normal, while extra tectonic activity such as earthquake could then have been triggered, as vast amounts of sea water were displaced, as it were, overnight.

The Bible refers to the 'fountains of the deep' being broken up (hinting at a memory preceding the Euphrates -Tigris floods) and volcanic activity is very likely being referred to here as well. Regardless of that possibility, sea water rushing through narrow straits into extensive, below sea level basins is almost certainly part of the picture, and the flooding of the Black Sea adjacent to the centre of the old civilisations would not be an event without impact. However, these events would have occurred two millennia or more, before the big flood deposits subsequently found in the Mesopotamian valley from 3500 BC onward. So the question is; did these catastrophic floods add to the collective memory of Noah's flood as some suggest?

In my view, this is almost a 'given' considering the catastrophic descriptions of the ancient texts, and the extent of the historical evidence. One Native American story recounts a gradual retreat by a local lady up slope, as the waters rose. It shows that in a time when formal writing did not exist, historical events have been reliably recorded by word of mouth.

These early stories were often presented in a repetitive and dramatic fashion so that they would not be easily forgotten. They almost certainly would have employed rhyming, rhythm or music and probably a verse-chorus framework, where the chorus contained the main story that would be carried from generation to generation. The stories may sometimes be considerably embellished, but the best ones are still with us, showing how effective this type of storytelling can be.

But this is only the tail end of the saga. Going back a good deal further than the Pleistocene we find an instance that impacted all the Mediterranean countries. This of course includes Egypt, where much of our interest lies in this narrative. So for the sake of some more geological background, and in order to clarify the historical situation, I will enlarge on the topic.

Back in the Miocene, about 5 million years ago, the Mediterranean dried up as the Strait of Gibraltar was closed off intermittently. This was due to the African continent driving into Europe, under the crustal forces of continental drift. It has opened again now, but for a time this resulted in the sea evaporating to very low levels, far below those of open ocean waters. The evidence remains in tell-tale evaporite formations, characterised by desert saltpan minerals forming in what is now the bed of the Mediterranean. Salt production in the Mediterranean is based on some of these ancient deposits. It also meant that any river flowing into the basin had to drop hundreds of metres until it reached the bottom, and such was the case with the Nile.

The result of this interesting state of affairs, meant that the Nile River cut its way down through the existing bedrock, all the way from far off Aswan in Upper Egypt. It created a gorge from there, north through Giza at Cairo, until it cascaded into the Mediterranean Basin. Drilling results along the Nile to Aswan attests to this, and much of the Nile delta would not have existed at this time - certainly not as we know it today.

It was only when the Gates of Gibraltar were breached again and the sea refilled, that many of the more recent marine sedimentary formations in the Nile valley could be laid down. This gives us certainty about the sequence of events, the ages of these formations and their extent. It also underpins the history of the more recent deltaic sediments as well.

So the future investigative possibilities, in regard to the geologic and climate story of these flooded regions is absolutely fascinating - no matter where they are found around the globe,.

As high resolution multibeam sonars and 3D swath, sub-bottom profilers capable of mapping below the seafloor sand and mud are further refined - and applied in these now salt water flooded gulfs and shallow offshore regions - nothing would surprise. Certainly not the discovery of buried coastal towns and ports, dating as far back as the likes of Jericho and the pre 10,000 year old towns and temples unearthed in Turkey.

Although many discount finding significant ports from earlier civilisations, there is a case for something substantial underwater in some instances. While it would have been noted by the ancient geeks, that for thousands of years and generation after generation, sea levels were rising (and that there was not much point in building long lasting ports) it could be argued that in the areas mentioned above; such as the Black Sea, Persian Gulf and the Red Sea, there may have been a stable water level for millennia, albeit below global ocean levels. Therefore more adventurous port facilities may have been attempted.

Although we would not expect ports of the prehistoric era to be large, the fact that water bodies surrounding the Middle East have this interesting history, might also account in part for the rise of the early civilisations in that region.

So the flood history of one sort or another is fascinating, but seeking out the timing and piecing together the various strands of evidence is largely in the hands of geologists and geophysicists, as well as the usual suspects - archaeologists, historians and dating specialists, rather than theologians and philosophers.

The problem with the end of the ice age as Noah's Flood, or indeed any event associated with it such as the catastrophic back filling of empty basins, is that they would almost certainly have occurred thousands of years before writing was invented.

Nevertheless, despite the lack of a written record, significant human activity along the coastlines in the days when people were making pilgrimages to places like Gobekli Tepe, is an absolute certainty. Or indeed much earlier, when life was slower paced and time was measured by the precession of the stars.

Back to 2350 BC

But then in early 2014 a startling blog appeared on my social media timeline that pointed to an article from 3rd May 2013 in the Irish Examiner. It alerted me to a documentary 'The Secrets of the Irish Landscape'.

*"A new documentary series reveals that it rained relentlessly for 20 years in Ireland back in 2350 BC. The new evidence unearthed in Irish bogs means historians can now say it's very likely that the biblical story of **Noah's great flood really did happen.**"*

The article then went on to say, that Professor Mike Baillie from Queen's University, Belfast, made the discovery through examining tree rings in Irish bogs.

Interestingly, this record now goes back 9,000 years; the extent of the current record of living and preserved (effectively fossilised) trees and provides direct evidence of climate change due to the thickness of individual rings. Furthermore the pattern of the rings, when matched over several years, can be correlated across many trees of different overlapping ages. When presented sequentially, this provides a much longer timeframe than that of any individual living tree.

The tree ring dates are confirmed by Carbon 14 dating, thus providing a reliable historical framework with which to work. Exhausting and meticulous work no doubt, like most scientific endeavour, but here is the kicker:

The 20 year flood that Baillie's work has defined, coincides with the dating of Noah's deluge of 2350 - 2349 BC from a traditional historical analysis of the Bible.

By studying the biblical genealogies provided for the patriarchs after the Flood, the time recorded for the Israelites in Canaan and Egypt, together with the period from the Exodus to the first Temple and the further genealogies to the time of Jesus; a solid timeline can be established. This framework can then be locked into common events of the adjacent nations to the north in Mesopotamia and Assyria, or to Egypt in the south and west, and astronomical observations have been used where available.

The chronology has been accomplished in a fairly precise manner as far back as Davidic times around 956 BC - and with less certainty prior to that period. Given we might expect some considerable discrepancy due to inaccuracies in recorded ages and uncertainties about the genealogies, this outcome is quite extraordinary. Having said that, some suggest that the 'begat' ages in Genesis 11 from Shem to Abraham - and as recorded in the Masoretic text - should be replaced with those of the Septuagint which are about 100 'years' longer per Patriarch.

This would extend the timeline by 650 years and place the Flood back to 3000 BC and the floods described earlier. However there are several difficulties impacting the numbers before and after the Flood, and these are addressed in 'The Numbers Game'. The result is that any discrepancy is likely to be a few years at most and the Masoretic text, the most accurate after all.

Continuing with the new research, Mike Baillie then further pointed out, that the 2350 BC date also coincided with:

"....ancient tales of massive monsoons in China and Central America".

Further information provided included:

"According to the ancient Annals of the Four Masters, the whole of Ireland had to be evacuated at this time".

From the Irish bogs and fresh water lakes, not only have the secrets of the weather been revealed over millennia, but through pollen analyses, the crops that were growing at the time.

This record shows that crops disappeared when the wet weather prevailed, farming stopped and the fields were overrun by forests. Apparently the Irish annals agree, saying that Ireland was almost completely abandoned in the same 20 year period, because the weather was so terrible.

So this was a turn up for the books - and returned my attention to what I had seen in Egypt. It took some time to seep in, but eventually the reality shook me. If the biblical and tree ring dating of the flood at around 2350 BC was correct and Pharaoh Khufu's riverboat (which I had so much admired in its housing behind the Great Pyramid) was older, at around 2550 BC, then the funerary riverboat was around *before* the biblical Flood!

I was stunned! It seemed impossible. So many people looking for the Ark on so many expeditions with high tech gear in remote places; gear that as a geophysical surveyor and remote sensing trained specialist, I was thoroughly familiar. And yet here, as a mere tourist, I had unknowingly stumbled on another vessel dating from *before* the Flood.

Of course it was not the Ark. That is still hiding somewhere if it remains at all. Maybe someone buried that as well. In any event, hundreds of thousands, if not millions have seen Khufu's boat since its discovery in 1954. The work that has gone into its restoration is impressive as the display shows. OK, it was not the Ark, but the further possible implications started to dawn.

The next bombshell was that the weatherworn surface of the Sphinx body and its enclosure walls are tangible evidence of the biblical Flood!

Of course not all the water erosion on the Sphinx and its enclosure is due to a flood a mere 200 years after its excavation, but it is true that:

- The region is largely desert and has been for millennia
- Erosion of sandstone surfaces tends to be rapid when it is freshly cut. If there is a lot of lime in the rock it may seal up relatively quickly and harden, if not the sandstone can weather badly
- The wet weather transition from the last ice age to the present is generally agreed to have finished ca. 2100 BC
- The enclosure has filled with desert sand and remained like that for centuries (and most likely millennia). It will therefore have protected the result we see today as fossil surfaces and evidence of the Flood

To say that the moral telling of the Flood 'myth' is of greater importance than the actual history, is no doubt true. After all, the story resonates to this day on many levels. However to say it is not based on historical events is absolutely incorrect.

In the same way as the 'mythical' events of the gods of Greek legend are likely to be based on ancient accounts of heroic ancestors, so in this instance as well.

An example of this is recognised in Ryan and Pitman's 'Noah's Flood' outlined earlier. They recount the Greek story of Jason and the Argonauts, where Jason and his compatriots describe seeing 'cyanic dancing rocks' at the Black Sea end of the Bosporus. As the geologists noted, this is highly suggestive of bluish icebergs bobbing and moving around in the currents, but unable to go downstream through the channel.

The reason for this, is that there is a deep saltwater counter current that flows north into the Black Sea from the Aegean to this very day. It flows below the freshwater current of the surface waters, which flow the opposite way to the Mediterranean.

Seeming fantasy, but tantalising in that the connection with icebergs, breaking ice dams and the end of the ice age puts a strange 'I wonder if' note to the notion - one that is difficult to discount entirely.

So the evidence is building for a 2350 BC flood event and is quite exciting. But if true, there should be a vast amount more information out there; small pieces of the jigsaw puzzle that when fitted in position, should confirm our suspicions. Once we are confident of the overall picture, we should be able to see if the biblical writer(s) exaggerated a bit, or in fact if our own Sunday or Saturday School understanding and imagination has perhaps embellished the account in our own minds.

Words such as 'covered the whole world', 'every animal', 'higher than the mountains' etc. may really have been meant to say 'as far as the eye could see', 'all the animals that could be rescued', 'there was no dry land to be seen anywhere'.

Language can be very subjective, and when it is composed of a limited selection of words from the dawn of writing, we ought not be surprised at the need to carefully (and prayerfully) consider the content.

Now from an Egyptian perspective, if the date of 2350 BC is correct, we are looking at the reign of Pharaoh Unas, the last pharaoh of the fifth dynasty of the Old Kingdom. He may have reigned up to 30 years from 2375-2345 BC, although there apparently is some doubt whether he reigned for quite this long. In any case, we can safely assume that the early 6th dynasty pharaohs were also around during the event; particularly if it lasted several years. In addition, the 5th and 6th dynasty pharaohs, consorts and senior staff were nearly all interred in pyramids or mastabas at Saqqara south of Cairo, or in the general area.

Interestingly, Unas was one of the first pharaohs to have the inside of his pyramid decorated with texts and paintings.

Some Egyptologists have also questioned why there is a break after his reign and a new dynasty started. This is based on information contained in the ancient king lists, which suggests that the hereditary line appears to have continued unbroken into the 6th dynasty, i.e. Teti, Userkare, Meryre, Pepi I and so on. This time has also been described as a 'black period' by others because of a general lack of information. Fortunately though, the 'pyramid texts and paintings' we do have, give some tantalising pictures of life at the particular time we are interested in.

A recent archaeological report describes the tomb of Khentkaus III, a queen of Pharaoh Neferefre who reigned from 2460-2453 BC in the era of the Old Kingdom. This is of course before the flood event. Discovered in a necropolis in Abusir and not far from Saqqara southwest of Cairo, she lies 650 feet away from the burial place of her husband.

Scenes depicted in her tomb and associated ones, show (among other things) starving Semitic workers, who may have been through a severe drought prior to the flood event.

Pharaoh Unas' pyramid, although smaller than most in the area, is approached by a 720 metre causeway (the longest found in Egypt to date) and skirts around a slight depression which is thought to be the remains of a lake – probably ornamental. The elevation of the lake above the Nile and how it was provided with water, would be an interesting study in terms of flood levels. Determining signs of flood levels through the positioning of mastabas, or other man made features (or indeed damage to older buildings) could potentially yield further clues.

Human nature being what it is; if we assume that there were high flood levels for several years and that water from the Nile spilled over into the adjacent Fayum Depression on a regular basis, it would then make sense that on the return to drier conditions, the Egyptians would attempt to keep a good thing going. They would have been determined to see irrigation continue in the Fayum, and do that by deepening the natural inlet to the Depression by way of a canal.

As it happens, the first known attempt to connect the Fayum Depression permanently to the Nile by digging a canal, occurred in 2300 BC! That just happens to be in the 6th dynasty immediately after our proposed flood event and probably in the reign of Meryre Pepi I 2332-2283 BC.

It also seems extraordinary in an otherwise dry environment, that Unas had the entry causeway to his pyramid covered with massive stones - particularly considering its length of nearly three quarters of a kilometre. This is not a usual feature of such causeways in Egypt - in fact it seems unique.

One of the wall paintings in the complex also depicts prayers to Ra (the god of sunshine) with a scene of tumultuous skies; something you do not associate with this dry and sunny region today. In fact, after studying the associated hieroglyphs, some Egyptologists suggest that Unas was totally obsessed with water and rain, and like king Canute, also with his own egotistical, god-like ability to tame the elements himself.

I suggest it was not an idle fantasy of his own imagination and misspent time, but rather evidence of an unsurpassed deluge over several years. The event affected Egypt dramatically and Unas was forced to deal with the consequences. The upshot is; this is all solid evidence for an unusual rain event.

Today the Fayum Depression is a high yielding agricultural area, with water supplied from the Nile and with a serious irrigation history dating back to the 12th dynasty.

Now other desirable evidence which would confirm a prolonged flood event might include:

- Analysis of lake sediments throughout Europe and Asia, but also in Egypt, e.g. Lake Moeris (Fayum Depression)
- Erosion / sedimentation along the Nile and evidence of other water works to mitigate or utilise higher water levels
- Covering of outdoor venues and unusual decoration in protected areas
- Emphasis on sunshine (Ra) or weather events in paintings and prayers
- Distress or starvation in human figures and buried remains
- An Ark

So I threw in the last for fun, but the evidence for the Flood has been cored and logged in Ireland, and similar lake cores should be available throughout many locations. It may well be, that all that is required is to re-evaluate and date existing stored material. In Egypt, Lake Moeris should yield cores showing the extent of the flood and sedimentation in that region.

Moeris is not very far from Saqqara, or for that matter Cairo and Giza. And so the whole of northern or Lower Egypt would have been subject to unusual and extended rainfall.

My assumption at this stage is that the 2350 BC flood was the last great deluge after the Ice Age, and that after this time rainfall tapered off quickly in the region with only regular seasonal monsoonal events occurring thereafter. The Sahara then lost its remaining savannah and ephemeral lakes, to become the windswept desert it is today.

So I see fertile ground for research, if this picture is to be clarified and the Flood event recognised for what it was. Not the Royal Albert Hall, Sydney Opera House or Hollywood event, but one significant and all-embracing enough, to be the basis for a warning to us all about our mortality and accountability.

Indeed, one which also explains observations like the considerable water erosion of the excavated surfaces around the Sphinx enclosure and the body of the Sphinx itself.

However we have a potential sobering stumbling block in Mesopotamia. Archaeology has provided us with information on big floods there, with one flood depositing nearly 3 metres of clay at Ur from around 3500 BC. We have other later floods which were probably even larger and indeed closer to the time of one of our heroes, Gilgamesh. But we are missing any data on a prolonged 20 year flood period a few years later. If such evidence exists, I have not managed to source it to date.

But before we move on to assess the impact of a possible 2350 BC flood in Akkadian and Sumerian accounts (and ultimately the biblical one) we should check the best estimated dates for when these stories first appeared. Not the events themselves, but the earliest written records. If the stories were compiled before 2350 BC and describe widespread catastrophes, it is obvious that they must relate to earlier flood events - potentially including the immediate post Ice Age inundation.

We soon encounter a problem, because the earliest historical evidence on clay tablets currently date no earlier than 2150 BC. Two other stories are from the 17th century BC and other versions such as the Greek tales, are from the 7th century BC.

Tablet 11 of the Epic of Gilgamesh (The Sumerian Great Flood) is the 2150 BC account. It is the earliest one and is usually related to either the flood around 2600 BC (king Gilgamesh of Uruk's estimated reign), or to the earlier one at Shuruppak (Ziusudra's home) around 2900 BC.

An account called 'Eridu Genesis' written in the 17th century BC, probably comes from the same source material, while the other 17th century account is the Atrahasis (the Exceedingly Wise One) version also from the 17th century. This latter account is dated to Hammurabi's great-grandson Ammi-Saduqa (1646-1626 BC) and is also considered to be from a much older time period and passed down by otherwise undiscovered means.

The stories are epic, with the gods who orchestrated the flood even scaring themselves - which adds a bit of humour to the plot. Like the flood of the biblical Noah, the events seem of universal proportions, because the world of the Mesopotamians was everything to them - although rather localised in a global sense to the Tigris and Euphrates floodplain. Indeed despite the gods getting themselves and everyone else into a deal of trouble, parts of the flood epics found throughout Mesopotamia parallel the biblical description to a considerable extent. Here are some excerpts from the Sumerian King lists:

> "The flood swept over (the earth). After the flood had swept over (the earth) (and) when the kingship was lowered (again) from heaven, kingship was (first) at Kish (Cush)."

From the Eridu flood account at Nippur (biblical Calneh):

> "A flood came over the cities to destroy the seed of all mankind..... all the windstorms, exceedingly powerful attacked as one, at the same time the flood swept over the culture centres. For seven days and seven nights the flood swept over the land. The huge boat had been tossed about by the windstorms on the great waters."

From the cities of Asshur in today's northern Iraq:

> "...build a ship, seek thou life.... aboard the ship take thou the seed of all living things.... all my family and kin I made go aboard the ship. The beasts of the field, the wild creatures of the field"

> "Six days and six nights blows the flood winds, as the south-storm sweeps over the land.... on the seventh day the flood subsided in battle... the flood ceased. I looked at the weather; stillness had set in"

> "When the seventh day arrived I set forth a dove. The dove went back and forth, but came back, since no resting place for it was visible. Then I sent forth a swallow....."

The three accounts mentioned above and other fragments come from tablets which all postdate a possible 2350 BC flood. They therefore - whether originally about much earlier events or not - would certainly be expected to be influenced by a 20 year long wet event occurring only a couple of hundred years before.

I believe that a flood event which lasted years was probably part of the flood story. But I defer to popular opinion to the extent that: if there is solid evidence to suggest the Gilgamesh Epic and the parallel accounts are based on the earlier floods, then that has to be accepted. Certainly two of the excerpts above indicate a flood of only one week in duration; they have references to known kings who lived at specific places, and tie in with archaeological and dating evidence.

There is also mention in the Sumerian king lists of the kings who reigned before and after the flood. The king lists come down to us in a series of tablets, one of the more famous being a four sided prism which is inscribed in a spiral manner and so needs to be turned, or walked around in order to read. The earliest king who is independently attested historically, I mentioned before; Enmebaragesi of Kish who lived ca. 2600 BC.

The reigns of the antediluvian kings as recorded on the Sumerian tablets are preposterously long, but that aside, we are probably safe in assuming that these god-like mythical kings are based on some notion of history. But we need to be aware that although they have been recorded for posterity, it may be with some kingly name-dropping and artistic licence to boot.

I will endeavour to tackle the issues surrounding the Mesopotamian numbering system shortly. Suffice to say, that with king Enmebaragesi living around 2600 BC and being the 22nd king of Kish, we have a written record of the first dynasty after the 2600 BC flood. The earlier floods ca. 2900-3000 BC and Woolley's 3500 BC flood deposit would have added to the overall memory of the flood epics, if not the actual one in mind.

However this is all a problem for my 2350 BC flood hypothesis, and as I intimated earlier, this is also deflating - I am unaware of any record of a one-off, extended flooding event in the Tigris-Euphrates at this time. Nevertheless with the evidence in Ireland, tell-tale signs in Egypt and unusual rain events in Asia (which allowed, for example, the rise of the Indus Valley Civilisation around 2700-2600 BC) we can see parallels with the experience of Noah and his fellow Asiatics.

Overall Mesopotamia was fertile, but when drought sets in, irrigation makes the difference. As is usual in these situations, salt builds up in the soil and occasional floods would help to wash this away and restore fertility. So perhaps a long, steady 20 year event, would mean that irrigation was not necessary on the plains of the Twin Rivers, over that period. If so, providing waterlogging was not a problem, instead of devastation there may have been a renewal of the land, a cleansing if you will, and a return to prior fertility. There may also have been some shallow deposition of soil aiding this recovery, in the same way that the Nile floods renewed that floodplain annually.

So perhaps something *did* occur in Mesopotamia as well, although significantly, the biblical description more likely comes from the upper Euphrates region and today's Turkey.

This region could have been affected in a more extreme way in 2350 BC and perhaps for '40 days and 40 nights'. There may have been a great inundation in the Anatolian–Caucuses region, with terrible floods affecting local valleys. That this land is most likely that of the biblical patriarchs also adds weight to this argument. So the tentative suggestion is; that on the Mesopotamia plain, the 20 year flood centred on 2350 BC was an unusually extended wet period which rejuvenated the land. It was not the source of the catastrophic flood epics, but may have been the last episode which contributed to the Flood Epoch.

The biblical account and dating of the story which remains with us today, could have been sourced in part from the earlier catastrophic accounts from the Twin Rivers. But the centre of the saga may well have been in northern Syria and Anatolia.

The effect of this unusually long event was also global (at least in the northern hemisphere) and destructive in places, with monsoons in India, typhoons in Asia and intolerable wet conditions in Ireland. It coincides generally with the beginning of the Akkadian and Assyrian empires around 2350 BC, and also some early migration of Semitic people into Egypt. Prior to this, Akkadian names are found within Sumerian written records, suggesting these people were around before. After 2350 BC Semitic speaking society ruled, and this is potentially an indication of some sort of disruption, or rebirth in Sumer itself.

I believe when we investigate the right places, we will eventually find that the major deposition will have occurred downstream of the old flood city of Ur. It will be geologic evidence which was never part of ancient Sumer, because it was then open water and part of the Persian Gulf. Since the days of the 3500 BC flood at Ur, there has been at least a 200 kilometre progradation of the twin river system into the Persian Gulf, and the distance could be conceivably greater. Now that we know an extended event occurred, this would be the place to look.

By Abraham's time however, things had changed dramatically and he and many others exited the northern Levant and Mesopotamian region for other pastures. This was at *exactly* the same time that the Indus Valley society collapsed around 1900 BC. Their plight would have been driven by war and unrest, brought on by very unfavourable dry conditions and extreme drought following the Flood Epoch.

So the available evidence suggests:

- The 'world' of the Mesopotamians and Patriarchs was local and for the most part confined to the Two Rivers region and immediate adjacent lands
- The 3500 BC, 2900 BC and 2600 BC floods revealed by the early 20th century archaeologists were catastrophic events in a 'Flood Epoch' in the Mesopotamia valley. The floods were immortalised in the heroic adventures of Gilgamesh / Atrahasis / Ziusudra / Uta-napishti / Noah
- The 2350 BC flood event was so widespread and extensive, that it impacted Mesopotamia as elsewhere in the Northern Hemisphere and follows the biblical dating. However it did not leave an easily interpreted flood signature on the Twin River plain and may have only been catastrophic in the extreme north-western Levant
- The region was very wet compared to the norm for several years, causing widespread removal of salt from soils, but with most sediment deposited downstream from Ur
- The flood events together with earlier Ice Age flooding conflated into one story contributed to our historical legacy; this by combining both the nature and severity of the earlier disasters with the date of the later, broader flood interpreted from biblical genealogies and recent research
- The deltaic sediments of the Nile River and other northern hemisphere rivers will also contain a sediment signature in some of the lobes of their distributaries from that time
- The Flood story is based on the great floods after the last Ice Age, when climate changed radically and became flood prone, until reverting to a new, dryer normal

The account, summed and stripped of hyperbole and placed in a historical context, is still an effective warning to us of our own weakness in the face of adversity. It reminds us of our ultimate dependence on God.

CHAPTER 4

Hieroglyphs and Pictograms

*'Back on a small bus for Edfu and Karnak
Luxor balloon ride, west bank at daybreak!
The Valley of Kings and temples of Pharaohs
The desert and farmland and eating with locals'*

Language and writing

Now at last, we return to Egypt and follow our Nile railway journey, the desert landscape and the awe inspiring temples. And there was also the ubiquitous written word in stone; the hieroglyphs. There was something about those glyphs which I started to wonder about. I began to ask if the hieroglyphs had anything to say about that universal language the Bible tells us existed before the Tower of Babel - and for that matter - the pre-flood world. After all, much of the writing in Egypt predates most biblical events and emanates from the dawn of historical times. It is just so incredibly ancient!

But I am getting ahead of myself. It took a few days to ask some questions, and in reality weeks and months to ponder on the significance or otherwise of the hieroglyphs themselves.

This chapter is focussed on my musings on language, and on what I suggest are rather interesting characteristics of Egyptian hieroglyphs and Semitic languages in general. The discussion will also be necessarily bound up with the relationship of the Semitic languages to Greek and Latin, and the transliteration of the Semitic words into those languages - and eventually of course into English. It is not my intention to suggest these latter languages are anything but Indo-European in terms of their general categorisation, but rather to show examples of influences coming from the Semitic languages through the ages.

There is also the possibility, that some common language roots date to the initial breakup and dispersion of these languages in the distant past.

Those readers whose first language is not English, will have further questions and comparisons to be made. So this is about possibility thinking and what others might like to pursue, rather than an attempt at definitive linguistics. To that extent the biblical Tower of Babel story is only a secondary consideration, but one which I will discuss early to set the scene.

In the Bible, the Babel story consists of just nine verses tacked on immediately after the Flood epic. It emphasises moral and spiritual aspects of Hebrew teaching, while implying a language breakup. The nine verses probably come from a different source to the Table of Nations in Genesis 10, but have been set where they are because of the fairly obvious historical association with the Table, and the Flood description outlined immediately prior.

Before modern travel and the exploration and colonisation from Europe, there were countless dialects and languages existing in often close proximity, right around the world.

The original diversification of language was a natural outcome of tribal and geographical separation. From Britain and Europe to East Asia, from the Americas to Africa, Australia and the Pacific Islands, there were thousands of distinct forms of communication - and the hundreds of tribal languages in the highlands of New Guinea are legendary.

Many tribal languages still exist today, but in keeping with our contemporary shrinking world, communications between these disparate groups are now often facilitated by the use of an over-arching 'pidgin' language. Such a language is often derived from colonial times and therefore has a European base, with copious words and imagery of native origin integrated into a cohesive whole. This allows people to converse more widely across local tribal barriers and expedites communication through mass media; the printed word, radio and television.

A recent linguistic study at Cornell University, has analysed the sounds and meanings for common objects of nearly two thirds of the world's languages. Their findings suggest, that all languages were once linked back to a common denominator, going back beyond the 'out of Africa' date of 70,000 years.

They say the similarities are statistically improbable, without a linkage to an ancient mother tongue.

But of course this cannot be the language breakup and diversification spoken of in the Babel story. However it can teach us something about how language and writing is formed and how it diversifies, while still retaining hints of the original tongue.

Around 70,000 years ago there was another significant event; a mega-volcanic eruption, the like of which we have not seen or heard since. It occurred in South East Asia and was eventually tracked down to Lake Toba in Sumatra. It left a lake over 100 kilometres long and about 40 kilometres wide. The amount of ejecta (rock, pumice, ash, gas etc.) was so great, and the dimensions of the lake left behind so enormous, that no one realised the water body was once a super volcano, let alone one which affected the climate of the whole world.

It was known that a super volcano must have erupted somewhere, because distinctive ash had been noted over thousands of square kilometres in the Asian region and beyond. But no one knew where the ash had originated. There were weather-worn crater walls in parts, but the sheer size of the lake seemed beyond the dimensions that could reasonably be expected of a volcano. The development of a tell-tale (but huge) island in the centre was also a hint, because a central island is a characteristic of volcanos and calderas, reloading as they do for the next eruption. The scale was simply beyond anyone's imagination.

Volcanic eruptions and other natural disasters such as asteroid impact, have affected climate globally and are likely to be linked between the 'out of Africa' migration to all points of the compass. Therefore the Toba eruption may have accelerated migration and subsequent language splitting and diversity.

The next major possibility for a language breakup, is the disruption caused by the 130 metre rise in sea level after the last ice age. This ice melt was described in the previous chapter, including the potential catastrophic flooding of enclosed gulfs and basins. This event cut off some land routes and caused trade and communications to shift to maritime transport in places. One land route disrupted was a more direct one between the

Indus River Valley and Africa, and options to Sumer were reduced as well. The Arabia - Persia land bridge at the Hormuz Strait and the land bridge from Arabia to the horn of Africa were submerged as sea levels rose.

However in my view, this ice age event occurred way too early to be strongly associated with the Flood and Babel stories. Nevertheless, some native stories suggest we have oral accounts from this time; accounts that have added to the overall picture.

I mentioned an American story earlier and Australian Aboriginal song-lines also hint at anecdotes from this time.

The next candidate can be roughly equated with the spread of the Indo-European languages and diversification across Europe and Asia. When did that language family start, and how and why did the languages spread anyway? Was it tied to the rise of the Egyptian, Mesopotamian and Indus Valley cultures, where writing, urbanisation and Bronze Age tool making and production rose to prominence?

The beginning of this period is the beginning of civilisation, and apart from writing is also characterised by the development of counting, astronomy, the wheel, metal working, baked bricks and mortar, irrigation and so on.

If not associated with the spread of the Indo-European language and culture, then these three civilisations and the subsequent dramas of flood and war are our fourth possibility.

A fifth possibility is that the Bible is pointing to the diversification of the Semitic languages only. This follows because the descendants of Noah and their languages were part of the disruption of the Flood and Babel epics, commencing 2350 BC. The timing of the Genesis 10-11 sagas leave no other option.

There is no real drama though, Noah was not Adam, he was simply a later Semitic speaker in the midst of change (and a flood). The evolution would be from some early Anatolian, Mesopotamian, or Arabian Semitic source, to Akkadian and the subsequent language diversification into Assyrian, east and west Semitic and the Marian and Eblaite cultures and dialects.

Later languages such as Ugaritic, Aramaic and Phoenician (Canaanite) then appeared, and eventually modern Arabic and Hebrew. However the very ancient Egyptian hieroglyphs also appear to be Semitic inspired and the later Ge'ez tongue of Ethiopia is considered to be part of the language family as well. So there are a whole bunch of daughter languages, with some originating very early in the history of the written word.

The problem is that the Egyptian hieroglyphs are so ancient, that they rival their Mesopotamian written counterparts. Therefore trying to determine the sequential history is intriguing to say the least. To the layman this all tends to be something of a mystery and that would have been a problem for the Genesis writer(s) as well. In any case, the history of the Flood and Tower of Babel are easily conflated, in order to deliver the moral lesson which is at the heart of this briefest of narratives. So one might well ask, why bother to dissect the Babel story at all?

I guess the answer is a matter of history and trying to understand it as best we can. I have therefore introduced the topic with a broad summary of events starting 70,000 or more years ago, but the Bible is pretty specific; tying in the Flood and Babel events directly to Mesopotamian and Syrian history.

Towers and tables

Bringing archaeology and the written word together, we see that the Flood - Babel episode must have ended within a century or two of 2350 BC, with the flooding of 3500-2650 BC and the flood cities of Mesopotamia (including the old city of Babylon) prime participants in the period.

As previously intimated, the people of the Twin Rivers may have later considered the whole of the Middle East their extended playground. But even if we include Egypt and the part of Africa immediately adjacent to the Red Sea, it is still a limited world compared to our global view today. Importantly, these boundaries effectively limit the extent of the languages we can include in our discussion; and these are the Semitic languages and a few language isolates.

The Genesis description is short and to the point. Mankind did wrong and God judged. How this was accomplished is not laid out in detail. This is a matter of morality and faith to the writer. Intermediate causes and effects seem not important, even if the author was aware of them. It paints a picture that would be unusual in our experience and emphasises that the author is not focussed on detailed history.

However knowing the rest of the story may help our 21st century minds come to grip with what we have here, used as we are to linkages and cause and effect.

Genesis 11:1-9 tells us little more, than that there was a single language before Babel. However in the previous chapter we have a precursor account and are given the 'Table of Nations'.

Here we learn about the descendants of Noah, their names and how they came to have their own separate languages. The Genesis 10 'Table of Nations' therefore outlines the three separate genealogies of the sons of Noah; Japheth, Ham and Shem through several generations.

We now know, that this can only be a small portion of the human population at the time, albeit a part of the heavily populated centre of the ancient world. The key to the saga, is that by following the Semitic names, we can see the dispersion of the people and the evolution of their languages. This, as they move away from Syria-Mesopotamia to places as distant as Africa and western Europe.

Genesis 11 only talks about 'the tower and the city', but other verses, extra biblical links and commentators (including Flavius Josephus and Herodotus) associate it directly with the city of Babylon and also, to the rather mysterious Nimrod the Hunter. Mysterious, because searching through the Mesopotamian king lists, we find no name which corresponds to Nimrod.

One potential association is through the city of Nimrud. It was located in the old Assyrian Kingdom, centred north-west of Sumer in today's northern Iraq. It was probably built shortly after the city of Asshur (named for their god) dating to around 2000-1900 BC and in Abraham's time.

Now Jewish tradition suggests that Nimrod was buddies with Terah, the father of Abraham, and was a king of ancient Babylon. This would line up closely with the ca. 2150 BC Assyrian period and later establishment of the city of Nimrud, rather than the earlier Akkadian era.

Whether a Nimrod of this period also became a king of the southern kingdom of Babylon by default, is the question.

An earlier candidate for Nimrod is Sargon, who founded the Akkadian Empire in northern Sumer ca. 2270 BC. He was formerly a cup bearer to King Ur-Zababa of Kish, and so could have been the legendary figure. His claim is bolstered by the timing; directly after the 2350 BC Flood, established from the biblical sources and the Mike Baillie research in Ireland.

On the negative side, it would mean that the city of Nimrud was built a century or so later to honour him, or his royal line. Given the old Assyrian Kingdom emerged from the earlier Akkadian, it is arguable whether subsequent Assyrians would have had reason to celebrate him after so much time. He is also known as a legendary founder of Babylon, although that city was apparently built as early as 3000 BC.

Another association with Nimrod is with the constellation Orion, the Hunter. This shows that the name Nimrod was a nickname (hypocoristic name) of the monarch and in this case was a description of prowess - something many leaders and kings have aspired to throughout history. It is no doubt the prime reason he is not immediately identifiable in the king lists.

It means the identity of the legendary Nimrod, remains somewhat of a mystery and could even have been adopted by a whole line, or house of royal figures.

All this confirms our suspicions; to take the view that a king fairly late in the Semitic speaking Assyrian Empire was Nimrod, is to demolish any pretence that the language breakup had anything to do with Indo-European and African people. Even a king in the earlier 3000 BC period of Babylon has that problem and this really locks in the answer; that Japheth, Ham and Shem and their languages are not associated with an original dispersion of the Indo-European, African and Semitic languages.

It implies that any assumption that Genesis 10 is recounting the dispersion of these major language groupings, or the population throughout the globe is incorrect. The dispersion and origin of this plethora of languages must be attributed to an earlier time and consequently, the assumption that only the sons of Shem spoke the Semitic languages is also incorrect.

The 'Table of Nations' in Genesis 10 tell us that Japheth's sons included Gomer, Magog, Madai, Javan, Tubal, Meshech and Tiras. Grandsons included familiar biblical names such as Ashkenaz, Togarmah, Tarshish, the Kittim and Rodanim.

Now these are obviously only a subset of the Japheth line, and rightly or wrongly do conjure up an Indo-European association. The initial lands they occupied however are mostly within the old world, particularly northern parts of Mesopotamia, Anatolia and areas close to the Caucuses - with one or two later identified with Eastern Europe adjacent to the Bosporus. Those further afield like the merchants of Tarshish (Spain) were almost certainly included because of known family linkages.

So the Japheth line probably integrated with Indo-European speaking people after the flood era. But immediately before, they were almost certainly speaking a version of early Semitic, or other language isolates associated with Mesopotamia.

It means languages such as Hurrian cannot be excluded, given the geographic proximity, Semitic associations and writing using the cuneiform script.

If we then go to the Hamites we find his sons were Cush, Mizraim (Egypt), Put and Canaan. Grandsons included Seba, Havilah, Sabtah, Raamah and so on, and finally Sheba and Dedan. We then learn Cush was the father of Nimrod the Hunter, whose kingdom included Babylon, Erech, Akkad and Calneh.

Here we have the associations of Akkad and Babylon again, along with other regional states. They are also all Semitic speaking kingdoms, and particularly so later on.

In regard to Nimrod, it tends to point to the later Assyrian Empire, but again, hardly definitively. What *is* becoming certain, is that all this is a significant problem for language dispersion beyond that of the Semitic family of languages. We also know that those sons and grandsons who are associated with Arabian and African names also spoke a Semitic language.

This is particularly the case for Egypt, Cush (Sudan and Ethiopia), Arabia and Yemen.

I accept that the association of the Japheth line of people with the Semitic language line is a little tenuous, but the Hamite association appears particularly strong - to the extent in fact, that there is no real difference between the Hamite and Semite people in terms of their language, early territorial associations or biblical origin. It means the biblical Noah was simply an early Semitic speaker, whose offspring dispersed after Babel into the various east and west Semitic languages in the region.

It shows the language breakup we are looking at in the Bible is a Semitic one, despite preconceptions we might have of a broader picture in terms of the Indo-European and other languages. A proto Semitic language may have preceded the Indo-European languages out of Anatolia or not. But that would have occurred much earlier, if in fact it happened at all.

We can also be fairly confident that the biblical 'Table of Nations' picture and the Tower of Babel story is simplistic and broadly applicable only. The Bible tells us that the one-world language was broken up, such that people could not understand each other. The result was a splintering of the Mesopotamian – Assyrian society of the day with the result that people were forced apart into several new language groups. It also caused the cessation of construction on the city and tower at the centre of the story. The story is short and simple. Genesis Ch 11:

"Now the whole world had one language and a common speech. ² As people moved eastward,[a] they found a plain in Shinar[b] and settled there.

³ They said to each other, "Come, let's make bricks and bake them thoroughly." They used brick instead of stone, and tar for mortar.
⁴ Then they said, "Come, let us build ourselves a city, with a tower that reaches to the heavens, so that we may make a name for ourselves; otherwise we will be scattered over the face of the whole earth."
⁵ But the LORD came down to see the city and the tower the people were building. ⁶ The LORD said, "If as one people speaking the same language they have begun to do this, then nothing they plan to do will be impossible for them. ⁷ Come, let us go down and confuse their language so they will not understand each other."

⁸ So the LORD scattered them from there over all the earth, and they stopped building the city. ⁹ That is why it was called Babel[c]—because there the LORD confused the language of the whole world. From there the LORD scattered them over the face of the whole earth." (NIV)

There is such limited detail here, it is certainly not worth getting our knickers in a knot over. However confirming the timing of the event and why it is tacked on to the end of the Noah's flood narrative, may give us a little more background.

One clue from the story itself is the use of baked bricks and tar for mortar. Paul H. Seely in a scholarly on-line article, points out that there is no archaeological evidence for baked bricks and bituminous material in general construction prior to 3100 BC. And so even for a major innovative structure, unlikely before 3500 BC. The evidence suggests, that the use of a combination of baked brick and tar for mortar was constrained to the historical period, and immediately we see the chronological relationship with the flood stories of Sumer and Akkad.

Undoubtedly, the erosional effects of rain and floods on unbaked bricks would have prompted the new technology.

The term 'plain in Shinar' is also interesting. There is a reference in Egypt from the time of pharaoh Amenhotep II to the land of Shinar. He reigned from 1425-1400 BC. This establishes the usage of the word to at least that period and the term seems to be Semitic in character, meaning 'a land of pools of water' using the logic I will explain shortly.

It was almost certainly a very old term even then, given the age and origin of the Semitic languages. Furthermore, those 'pools' of water still exist today in this very flat, low gradient landscape. They include Hawr al Hammar, 32 km wide and 80 km long on the Euphrates floodplain and Hawr as Sa'diya and Hawr as Saniyalt near the Tigris River.

Several other regions have been proposed for the 'city and the tower' as Genesis simply puts it, but the consensus is that the old city of Babylon ticks the boxes. Babel may mean 'gate of God' from the word 'babilum', but from the Hebrew language perspective it is associated with the word 'balal' meaning to jumble or confuse. One could be forgiven for suspecting a little subversive tongue-in-cheek jousting here.

Most of what we know about the tower at Babylon, comes from the neo-Babylonian period, when Nebuchadnezzar II and his father Nabopolassar (610 BC) renovated a ziggurat that is said to have been founded in those earliest times mentioned above; 3500 - 3000 BC. In regard to the actual ziggurat or tower, there are two proposed in the vicinity of Babylon.

The first is the famous Etemenanki, the 'temple of the foundation of heaven and earth', which was 91 metres high and located within the city walls on the banks of the river. A 'temple of the foundation of heaven and earth' could well be a 'tower that reaches to the heavens' and since it was *in* the city, and next to a building dedicated to Marduk, seems like a pretty good prospect. It was also where the members of the Jewish exiles who were selected to serve the Babylonian kings, in fact lived.

Another possible location is outside the city at Birs, near Al-Hillah (ancient Borsippa) 19 kilometres to the south. All in all, there were many ziggurats built in the larger cities of Sumer from around 3000 BC. Other sites proposed for the tower are as far away as the Black Sea region, based on the association of languages. Given this is where our biblical patriarchs are likely to have originated, it could mean we have an amalgamation of their history and that of Babylon after the Jewish exile.

In the end, Etemenanki seems the most likely ziggurat for Babel, and the association with the exile may explain why the story seems tacked onto the biblical flood account. Its provenance may come from the exiles and the Judah captivity.

Having said that, this writer does not believe that this is true of most of the early events recorded in Genesis and Exodus, and the following chapters will continue to address this assertion by some. In the chapter on the Exodus, I will show how this event is presented in unbelievable detail - to the extent in fact, that the account could only come from an eyewitness on the road.

However with the Babel account, we have an example of a story where we need to dig a little to see what mechanisms were in play, and in fact, the pivotal role of the Flood.

Once we understand that Babel was predicated on, and intimately connected to the events of the Flood Epoch - and that ziggurats were not only astronomical and astrological observatories, but refuges on the vast floodplain of Shinar - the drama becomes very much more understandable.

So writing began around 3300 BC according to most contemporary archaeological evidence, but as they say 'the absence of evidence is not the evidence of absence'. Rather, this approximate estimate could be overturned by further investigation.

So very recently for example, writing from Greece – undeciphered at present - is thought to be potentially older.

One thing that is obvious about writing and numbering, is that it developed out of a need for permanent records. The requirement for numbering systems almost certainly initiated the whole technology, but then mushroomed into business dealings, tracking transactions and prices, historical records, international agreements and eventually to storytelling.

The evolution of Mesopotamian counting systems is also understood today, and so too the likelihood of transcription errors. This, as the numerals became more abstract and indeed overly simplified and ambiguous. It means that changes in the Mesopotamian counting system itself are one potential issue in regard to textual reliability. To me, there always seemed a likelihood that biblical scribes and translators may have been in error over numbering, simply because of the diversity of the Hebrew experience through travel, with the consequent cultural distancing and language difficulties.

There is also little doubt, that when introduced, so revered was writing by the masses, that it was considered to be the work of the hands of the gods. Only a few knew the art, a few more how to read, while most could only listen to public and private readings. Such a new technology, as with all new inventions, had great power for good, but also misuse by elites.

On the road

Of course, philosophical ruminations were the last things on my mind as our overnight train went clickety-clack, mile after mile and the sun burst into the carriages the following morning. The locals along the Nile are not fazed, that barely a foot of track has been added to the line, since it was built by the British starting from Alexandria in 1854.

They are certainly proud, that after the UK, their country was the second anywhere in the world to have a major railway line - at least that is what the locals say, and I am a fan of Egypt, so I will go along with their version of events.

In any case, it was Egyptian muscle that made it happen, and a credit to that nineteenth century society's determination.

We were soon through Luxor, the once-upon-a-time priestly city of Thebes, with its modern hotels, bazaars and amazing temples. A few hot air balloons were hanging in the sky - as if to frame the view. Out of town, the extraordinary landscape continued, with the luxuriant vegetation and the seemingly endless farmland along the Nile. It stops abruptly, where the desert begins; often along the railway line itself, where sand, rock and arid mountains of the eastern desert, continue into the distance.

This is truly remarkable, even for someone used to the contrasts of the vast arid landscapes and tropical rainforest of my own country. There also, the same abrupt changes due to irrigation in often unexpected places.

Later in the morning we pulled into Aswan. Beautiful clear water dotted by feluccas and other craft, crisp but quickly warming desert air, a mix of old world beauty and modern hotels; a jewel of a town. This is a town awash with history - the first Nile cataract, the magnificent Aswan granite exposed by the river, royal temples and kingly engravings, Nubian villages and ancient Jewish texts. All in all a tourist highlight; a beautiful remote stop on the Nile.

We booked into a delightful hotel; perfectly presented and situated right on the banks of the river. After lunch at a waterside restaurant, we headed off for our first boat ride. With such lovely views of town and the islands and new friends, it was incredibly relaxing as we made our way around the waterways.

The Cataract Hotel and Elephantine Island - once a British outpost and door to the Sudan - remains as a reminder of the colonial past. It is also the hotel where Agatha Christie's 'Death on the Nile' was filmed a few short years ago. As our motor boat glided along, the hotel slipped by on one side, while on the other, archaeological excavations and the Aga Khan's residence in exile, peeped over the west bank, desert hillside.

A heady mix of ranging desert and Nile vista, timeless temples, a Nileometer here, a cartouche of a long gone Pharaoh there, submerged us in history. Elsewhere, a granite column that never made it to Thebes or a Pharaoh's temple, remained partially cut from the surrounding rock.

We stopped for a while on the west bank to meet some of the Nubian population and enjoy their hospitality. A chance to hold a juvenile crocodile, take some refreshment, try a water pipe and then race up the steep west bank sand hills for the splendid view back toward town. Well the younger members of our group raced, I took a little longer.

There was also a 3:00 am start for an escorted bus convoy south through the desert, past Lake Nasser to Abu Simbel. This is the magnificent Ramses II temple with its story-telling drawings and hieroglyphs of the Pharaoh's conquests – all the way from the Sudan in the south to the Euphrates River in Assyria.

It was then back to the High Dam, built by Russian engineers in the 1960s and later, the Philae Temple rebuilt on an island that was created by the old dam, built by the British. Philae was my third introduction to hieroglyphs after the Cairo Museum, but interrupted unexpectedly, when I was motioned aside and shown some British graffiti carved on a side wall. It was the work of some of General Kitchener's men in 1884.

They were on their way to the Sudan, to wreak revenge for Governor General Charles Gordon's murder, and I wondered if a young, Lieutenant Winston Churchill looked on askance..

And of course there was plenty of time to cruise around the bazaars, bargain with shopkeepers and take time out to walk along the river. There was a lovely botanical garden, where the landscape rose to the north and (a little unexpectedly) a view through the trees, of the beautiful Coptic Orthodox Cathedral of the Archangel Michael.

All too soon though, we left Aswan and headed downstream in our felucca. Our first over-night was at a Nubian Chief's motel-like home, with beds on the floors covered by insect nets, old time stone ovens, a courtyard for entertaining and some football in the late afternoon.

All rather simple and traditional and very laid back, but with our intrepid guide Hany still hard at work. This time, he was giving the Chief's daughter a hand to configure a newly acquired laptop, direct from the Education Department.

You have to smile. We live in a wonderful and diverse world today - or at least we do when we live in peace.

The following morning we were back on the water and sailing slowly downstream. There was time for some water sport and a rope-tow behind the felucca and a little later on, the crazy Aussie was offered a paddle by local teens in their rubber ducky. Next, a stop to buy local goodies from children on the riverbank, followed by another dip in the middle of the river. This lead to an evening camped on the boat - and a beautiful sunset somewhere near Kom Ombo, the crocodile temple.

Eventually our blissful cruise was over and it was time to embark. We then continued on our next minibus ride north to Edfu Temple, before pushing on to Luxor and it's plethora of ancient Egyptian temples, palaces, monuments and burial sites.

Two days in Luxor and surrounds were crammed with exploration. A balloon ride, commencing before sunrise over the western bank, kicked things off. There were views to the Valley of the Kings, Queen Hatshepsut's remarkably modern looking temple at Deir el-Bahari, the Colossi of Memnon and much, much more. Then time on foot, by donkey and bus to see things up close and personal; ancient architecture, obelisks, sculpture, wall paintings and inscriptions.

And yet the hieroglyphs were still so much meaningless, if fascinating writing; with repetitive figures mixed with everyday pictures of boats, chariots and war scenes. The signature cartouches of long gone pharaohs were spotted here and there. It was becoming apparent that it was time for our Egyptian hieroglyph education to begin in earnest, with a local guide explaining in some detail, the history of the various places we visited.

But he also gave us a basic understanding of the rules of writing the glyphs and what they meant.

OK, it was only pitched at a tourist level, and yes, we had a lovely lunch at his private residence where we all forked out some cash to have a pendant made.

The idea was to have our given names inscribed in the hieroglyphs. A bit of fun only, but later it got me thinking. You see, using the basic hieroglyph alphabet, it was suggested that our names might actually mean something.

If your name was Cora for instance, then because 'C' was a bowl or basket in the glyphs and 'R' was the lips or mouth, from this a meaning for that name could be derived. This might suggest for instance that Cora was someone who did not mind drinking from large bowls, or slightly more abstractly, did not mind a drink in general - or some other equally amusing picture.

But the mouth could also represent the word or an agreement, so that Cora could be interpreted as a voice with the bowl representing that she was eloquent, or even talked a lot!

Again, it could have a more meaningful and spiritual tone to it, being equivalent to the Greek logos (word) which Jesus of Nazareth was likened to. Here we are looking at the symbol of the lips as the ultimate expression of the mind, where all creativity is sourced. Jesus was then being associated by the writer with the Creator and Abba, Father. In a similar fashion, Cora could be said to exemplify imagination, creativity and action.

So there are a variety of meanings that could be assigned and these would depend to a great extent on context. Context is everything they say and that is particularly true of the early writing business. You could add in the vowels of course, but then in ancient writing vowels were not used as they are today and seem to have been a later writing enhancement. Perhaps it was originally written as a kind of shorthand.

The vowels therefore could be suspect in terms of meaning as I am about to propose here. But then again, could any meaning be ascribed at all, even to the consonants - and was I just simply being carried away with an amusing game?

One of the keys to this exercise is that the word has to be very old and part of the ancestral language we are trying to unravel. So a very old personal or place name could have meaning in this context. If the name is only of relatively recent origin, without an ancient counterpart, then it will be meaningless.

Here we are also focussed on Semitic names, or those likely derived from early Semitic sources. That is the key.

I realise that the following could be considered cherry picking and / or to have meanings derived by chance or coincidence. It may also be considered self-evident or common knowledge by expert linguists, or indeed, the very opposite and controversial. But it is a starting point at least, for some historical analysis that I would like to pursue and something that hopefully will contribute to our biblical discussion.

Back in 1799, a granodiorite stele was found by the French at Rosetta, close to the Mediterranean in the Nile delta. It is known as 'The Rosetta Stone' and is inscribed with three versions of a decree from Memphis. The upper script is in the hieroglyphs, followed by the same message in Egyptian Demotic script. It finishes with a third inscription in Ancient Greek.

This was a breakthrough for linguists and The Rosetta Stone proved to be the key to understanding the ancient Egyptian hieroglyphs for the first time.

The glyphs are stylised pictures of everyday objects, animals and the like and equivalent to alphabetic letters. It means they therefore had sounds attached - but they also had meanings; which when bunched together (or read in a sequence) conveyed the message that the writer wanted to impart. This latter property is something we have largely lost in our alphabetic languages today, because we only recognise the message of complete words, phrases and sentences - not the individual letters.

This is not a new concept or particularly surprising, but perhaps can be explored further. It stands to reason that at the beginning of writing, the actual glyphs did mean something in their own right and collectively and in context had an embedded meaning. In other words, the ancestor to alphabetic language was more like written Chinese or Japanese, in that the seemingly abstract letters were once words or phrases full of meaning.

I understand 'logographic' is a term for such languages, and can arguably be applied to Egyptian hieroglyphs. In fact we seem to have returned to a form of this communication with the emojis employed on social media. Given that other ancient words come from different language groups, their pictograms and letters probably had meaning too. However they would almost certainly be different, unless somehow related to the Semitic family of languages through a common source.

Because we are focussed on Semitic languages in our study here and looking for a mother tongue, the aim is to show that words in various daughter languages have something in common, with their meaning embedded in the individual letters.

If we can relate them to the Egyptian hieroglyphs as well, we may gain a significant insight into their derivation.

I will give more examples of words and names that I think are at least tantalising, but at the same time being cognisant, that most of the Semitic words we will recognise have usually arrived in the English language through the Greek or Latin.

If they did not, we would not see any correspondence with the hieroglyphs, or early Middle Eastern languages in English.

In fact Greek is our basic European source, simply because the Greeks occupied the Middle East and Egypt after Alexander in 334 BC. It remained the universal language throughout the four Greek speaking kingdoms that followed, and well into the Western and Eastern Roman Empire era. The Greeks not only absorbed so many Semitic and Egyptian names into their language, they often did so by transliterating the words letter by letter into Greek. If not exactly, then close enough that we can still see the similarities to the Egyptian consonant equivalents.

Ancient words in Semitic daughter languages derived from the pre flood 'one world' mother language of the Mesopotamian - Egyptian civilisations have **their original meanings embedded in the individual pictograms / hieroglyphs / letters.**

The interrelationships of these languages are nothing new and have been the domain of linguists for centuries. However you never know until you try; there might be something to be gained from a fresh investigative effort.

So returning to our Egyptian adventure, the next day I received my pendant as we toured around Karnak Temple; a temple dedicated to the Theban Triad headed by Amun.

Because my name is David, I got to thinking. I knew that the generally accepted meaning was 'beloved' from the Hebrew, but here was something potentially older and decidedly different. As our guide explained, the first and last letter 'D' in hieroglyphs are the 'helping hands'. In the case of David, we can suppose the right and left hand. As the letter dalet / delta (triangle) they come down to us through the early Phoenician, paleo Hebrew and Greek in a stylised form.

Now the 'V' is a relatively recent addition to the early alphabets, because in the old days there were fewer basic glyphs / letters, and the first alphabets had around 22 letters only. So the 'V' was represented by 'F' which is the viper - the same one that was part of the pharaoh's headdress.

Now they say the viper was not meant as a menacing or controlling symbol, but one of responsibility on the part of the pharaoh to protect his people. To this day David is spelt with an 'F' in some languages – Welsh for example, and perhaps hinting at the antiquity of that language.

Certainly it has been pointed out that some Welsh words and phrases parallel Hebrew ones quite closely.

Other languages use different letters in the middle, perhaps an 'O', 'U' or 'W'' for instance, which to the English speaking eye and ear are not hard consonants, but either soft consonants or vowels. So ignoring the vowels for the moment, we can see that David means helper and defender.

Wow, I thought, what a legacy to live up too! And although I understood the ancient significance of the name through biblical Hebrew-Israelite history, perhaps for the first time in my life, I really appreciated the name my parents had given me.

If we also consider what we call vowels today, the 'A' is the Horus bird (Egyptian eagle or vulture) also associated with the watchful eye of the pharaoh. The 'I' in this context, is usually associated with a glyph meaning romantic love and progeny and so we have in total an association with kingship, that seems to sum up what we would expect of leaders and powerful figures.

King David comes to mind, and potentially describes all royal males, leaders, chieftains and pharaohs.

There is one other, rather strange association I came across in my later travels. On a beautiful day travelling around Lake Kinneret (Galilee) in northern Israel, we stopped by Capharnaum (Capernaum) on the north-west shore. There we saw the remains of the limestone synagogue, built around AD 350 by returning Greek speaking, Jews or Samaritans. Interestingly, the building rests on the basalt foundations of the original synagogue from the time of Jesus and his fishermen disciples. In fact this is where Mary and her family lived.

On the approach road in, there is a limestone wall with some interesting and ornate patterns - one a very familiar six pointed star. The star is interlocked top and bottom in the fashion of two equilateral triangles. These 'deltas' in Greek may have been a clever symbol (perhaps only from Greek times) to denote 'David' and the union of the northern kingdom of Israel with the southern kingdom of Judah.

The symbolism is central to the Israeli flag and Ezekiel's 'two sticks becoming one' prophecy also comes to mind.

In fact a little closer to the shoreline than the old synagogue, a Memorial building now stands. It is suspended in the air on columns like a large spaceship. Protected underneath is an ancient home built of basalt on a central hexagonal plan. It has other rooms tacked on around the centre one, and is thought to have once been the home of the disciple Peter's mother-in-law. It is also directly linked historically to the original synagogue.

The six sided plan with its six faces and six corners may have already been a symbol of the nation in those days. Beyond that, if you want to walk where Jesus walked, there is no better place to go in the whole, wide world.

CHAPTER 5

Cuneiform and Clay

'Once more to the City to see the Old Town
The markets and history with Hany and Hend,
Again to our transport, Moses footsteps or bust!
To scale Sinai Mountain where the Prophet once trod'

Mari where are you?

Because we are looking for a Semitic source language, it is instructive to reflect on northern Syria's ancient kingdoms of Mari on the Euphrates, and Ebla closer to the Mediterranean coast at Tell Mardikh. It was noted earlier, that at Ebla (55 kilometres south of Aleppo) many clay tablets have been found with biblical sounding names. They include personal names and other descriptions which invite comparisons with the Hebrew of later years. In fact Habiru is said to be a name from this part of the world and this general time.

The Mari kingdom existed hundreds of years before the great cities of Assyria, such as Nineveh were built and the Mari Dynasty is thought to have commenced as early as 2500 BC. The capital had a royal palace complex with a throne room, processional ways and religious chapels in one section. Elsewhere there were passageways leading to courts, inner courtyards and an auditorium which would have held hundreds of people. The walls were covered in brilliantly coloured frescoes, some of which have been preserved today.

Administrators, secretaries, scribes, couriers, ambassadors and other officials would have hurried from room to room on state business, or concentrated on their work in offices and vestibules, with the equivalent of a Board of Trade and Foreign Office being just two of their Departments. Other departments dealt with construction projects, canal development and irrigation infrastructure and Mari was part of a night time fire signalling service that could send messages from Babylon to the Harran area (today's Turkey) via Mari in a matter of hours.

Not bad going considering the distance is a thousand kilometres, and in good conditions, a message could be sent and a reply given in one evening!

It all seems quite sophisticated in operation and the scribal output on their standard clay tablets was prodigious. Some of the commentary translated by French Assyriologists in the 20th century and which is of interest to our study here, are passages referring to a particular group called the Benjamites. Check these passages out as recounted by Werner Keller:

"The year in which Iahdulim went to Hen and laid hands upon the territory of the Benjamites"

"The year that Zimri-Lim killed the Dawidum of the Benjamites"

"The year after Zimri-Lim killed the Dawidum of the Benjamites..."

Now Zimri-Lim and his predecessor were kings of Mari, and it appears that the Benjamites who lived on land outside the city were neighbours, who the Mari rulers found quite troublesome. They were pastoralists and artisans at best - or raiders, mercenaries and opportunists when it suited. They also lived on territory that the Marians considered under their jurisdiction.

Elsewhere there is correspondence from Bannum of the Mari Desert Police (MDP) who spent the night in Benjamite territory and reported:

"Say to my lord: This from Bannum, thy servant. Yesterday I left Mari and spent the night at Zuruban. All the Benjamites were sending fire signals. From Samanum to Ilum-Muluk, from Ilum-Muluk to Mishlan, all the Benjamite villages in the Terq district replied with fire-signals. I am not yet certain what these fire-signals mean. I am trying to find out. I shall write to my lord whether or not I succeed. The city guards should be strengthened and my lord should not leave the gate".

This is a fascinating exchange and shows that fire-signalling was a common occurrence to fill in the long evening hours, and not restricted to big city-state communications. Those pesky Benjamites obviously had their own code which the police chief had yet to crack. In all this we find names such as 'Benjamites' and 'Dawidum' - and in the tablets more recently recovered from the Tell at Ebla - we have other Hebrew sounding names as well. To top it all off, what would a woman of Mari (or a man in some cultures) be called; Marion, Marian, Miriam, Mary perhaps?

Here, over 4,000 years ago, we have early forms of later Semitic languages such as Aramaic, Hebrew and Arabic and commonality with early Egyptian to boot. In addition; Aram, the place and Arameans, the people, come from the same general location centred on northern Syria today. Aram is a near anagram of Mari, retaining the core consonants of 'M' and 'R' which are the essence in this study. The Dawidum of the Benjamites is referred to as 'the Dawidum' suggesting it was not a personal or first name, but a title - most likely meaning 'Chief'.

In line with Eastern European tradition, the 'W' may have been pronounced 'V' as well. It makes one wonder if 'King David' is not a case of tautology, or whether by that time, 1,200 years later, the name had already become a first or given name.

In any event, I present it as further evidence for my proposition that the meaning of ancient names is embedded in the letters of which they are composed; at least for Semitic names.

Besides, I rather like being called Chief. Where is a smiley face when you need one?

But the language also provides evidence of real ancestry for the Hebrews from northern Syria and Turkey. The very place which the Bible assures us was the home of Abraham.

There is also the Dravid – Druid ruling / priestly class and elite priesthood in the ancient Indo-European world, broad as it is from India to Ireland. Were these names originally Semitic, or simply loan words, coincidence or locally derived?

But let us not stop there. We also have the familiar 'Dad' for father. I wonder if this is also related. In some languages the 'F' or 'V' are replaced with softer sounds as in Dawidum, or of course shortened. 'Dad' is essentially a nickname (a pet name) and a term of endearment. However could it also be a name as ancient as language itself and predate the usage of Abba (father) in Hebrew?

I emphasise again that English is not immune to linkage back to the Middle East, whether by original association, loan words directly, or through other languages. For example; Akkadian > Marian / Eblaite > Greek > English, but with copious amounts of Phoenician / Hebrew directly cross-linked, through later Judeo-Christian faith association. I am not in a position to say if, or how the early Semitic is related to the Indo-European languages, or exactly what the association is. But as hinted at earlier, it seems possible to me that a primal Semitic language could underpin more than the Semitic language group.

Now, the other centre of writing at ancient Ebla (Tell Mardikh) became an archaeological site in the 1960s and in the 1970s a huge quantity of clay tablets were discovered among the ruins there. These have become known as 'The Ebla Tablets' and they were originally discovered under the direction of two professors from the University of Rome – Dr. Paolo Matthiae and Dr. Giovanni Pettinato.

At this point, about 17,000 tablets written in two languages from the ancient Eblaite Kingdom have been recovered, and they appear to have been written during the last generations of ancient Ebla from a time around 2300-2250 B.C.

What is remarkable about the Ebla tablets, is not just how old they are, but rather the amazing parallels to the Bible that they appear to contain. For example, scholars were very surprised to see how close much of the language on the tablets was to ancient Hebrew. The vocabularies at Ebla are distinctively Semitic: the word 'to write' is k-t-b (as in Hebrew), while that for 'king' is 'malikum', and that for 'man' is 'adamu'. The closeness to Hebrew is surprising considering the 1,200 years to the Davidic kingdom and the way languages tend to evolve.

A considerable array of Biblical names, that have not been found in any other ancient Near Eastern language, have been reported in similar forms in Eblaite (one of the two languages found at the site). For instance, the names of Adam, Eve, Abram / Abraham, Bilhah, Ishmael, Esau, Mika-el, Saul and of course David, have apparently been recognised on the tablets.

Now, it is important to note that the tablets are not necessarily referring to specific biblical identities. Rather it demonstrates, that those names were commonly used well over a full millennium before the Davidic Kingdom.

In addition, quite a few ancient biblical cities throughout the Levant are also reportedly referred to in the tablets. Giovanni Pettinato says that he also found references to the ancient cities of Sodom and Gomorrah; a description which parallels the Genesis 14 account of a military campaign in the Arabah.

Of course, this all depends on whether these early Syrian records have been translated correctly, and there is now considerable controversy surrounding the translation of the Ebla tablets. It is difficult to know the truth of the various opinions.

On one hand, there seems to be some genuine professional argument against some of Pettinato's translations, but there also appears to have been pressure exerted by Syrian officials on the researchers. The latter apparently to prevent them making comparisons with Hebrew, the Bible - and any links to the Israelis or Jews in general. My understanding is that this resulted in a forced declaration to that effect. In other words, the project has been politicised to the extent it has become a tool in the anti-Zionist and anti-Semitic discourse and disrupted the research.

Notwithstanding, it appears that *an* Abraham is attested to by the Ebla library, although if the current dating of 2250 BC is not amended in the future, he is unlikely to be the man we know from Genesis. However he could well be related and a close namesake ancestor. If the translations of Giovanni Pettinato and others are correct (or reasonably close) why should we question the campaign against the cities of the Arabah?

The distances are not continental in magnitude, as some critics appear to imagine. They are only regional and distances which a disciplined militia on a forced march could cover in days - not weeks or months.

So Mari and Ebla, their languages and people (including the Habiru - Benjamite pastoralists) lived ca. 4,500 to 3,900 years ago. This is before and during Abraham's time, but including the postulated and extended rain event of 2350 BC. It is also after the big Sumerian floods of a few years earlier that gave rise to the flood epics. Indeed, it covers the time when Mari was defeated, and their kingship taken temporarily to Kish, around the turn of the 25th to 24th centuries BC.

If a flood was the main physical reason for the disruption of the nations and language, then it has to be the 3500 to 2350 BC floods as we concluded in the previous chapter. It also confirms the hypothesis; that we can rule out both the Mari and Ebla languages as the mother tongue of the pre-Babel world and arrive at Akkadian, or a proto-Semitic precursor.

This all points to a time during the early Mesopotamian flood episodes, for a major exodus from the region and a dispersion of people and language as far as the Nile valley. This was followed by more migrations, which could well have included a significant one in 2350 BC, based on Egyptian paintings from this period. The next wave of well documented migration began as early as 1900 BC by the so-called Hyksos and / or the 'kings who lived in tents' of Hurrian fame.

The first statement in Collier and Manley's bestselling 'How to read Egyptian Hieroglyphs' emphasises that the 'picture-signs are used to convey the sound (and meaning) of the ancient Egyptian language...'

Now I do not pretend to be a specialist of the subject, but given the clues and evidence presented above, I would like to continue to explore this idea of meaning, rather than sound. The focus will be on the very basic hieroglyphs only, rather than delving into the development and meaning of the later, more sophisticated glyphs. The latter, would be expected to lose this underlying meaning over time.

To summarise, we have a potentially much older meaning embedded in the scripts by virtue of the spelling. Older, because the hieroglyphs go back beyond 5,000 years and are much earlier by far than the Phoenician and Hebrew alphabets. Since we are only focussed on writing, the names as used orally could be much older again, and have been common across the ancient tribes and peoples of prehistoric times. Given this historical backdrop, what else can we say about the hieroglyphs?

Apart from my name, a name of one of the more notable Egyptian gods from antiquity comes to mind; Amun of Thebes, - also spelled in variants such as Amon or Amen.

Amun is normally translated as the "Hidden One". 'M' being the owl and 'N' the horizontal wavy line denoting a flowing stream of water. Thus the 'N' is probably meant to imply power or movement. This is in essence a synonym of 'spirit' and part of the Amun ceremonial practice (and temple layout) was a conduit where water was poured for the god.

Thus the 'Hidden One' was invisible, but active and working behind the scenes. As mentioned previously, the Egyptian priesthood later decided that an amalgamation was in order. After all, there were a lot of gods in the overall Egyptian sphere of influence and they decided to join the god Ra with Amun – a step therefore toward monotheism. The 'R' in Ra being the mouth denoting a source, breath (or as in the Greek, the logos, or creative power); imagery not lost on too many Christians in relation to Jesus. And so Amun-Ra became a combined deity; a little incomprehensible, until the combination of attributes that flow from this intellectual exercise are appreciated.

The invisible, intangible spirit (the hidden one) combined with life giver and sustainer (light and warmth) are still key attributes of the one and only Creator God we understand today.

The interesting philosophic concept here, is that neither Amun the hidden spirit, nor Ra the energy source are particularly, physically tangible to us. We can see and feel the result of sunlight and know how vital it is to life, and we can imagine that beyond the invisible / hidden world of subatomic physics and quantum mechanics there is creative power - but we cannot handle it like we handle an object, or see it like something of substance. It is more intangible than that and closer to what the Bible describes as 'spirit'.

Today we clothe these scientific concepts of light, heat and energy in modern terms, but they are not easily quantifiable. That is, until we measure them with comparatively recent, sophisticated scientific equipment. But we can understand that both Ra and Amun are conceptually compatible, and complementary, and 'flesh out' our understanding of the Divine.

So here we have a prototype of later Hebrew and Christian theology, where many names and metaphors are used to fully describe the attributes of our Creator.

But let us move on to another word which is central to the whole of the Egyptian world and has been since time immemorial – the Nile. Can you imagine over 90 million people living along and provided for, by one single river in a mostly desert country?

It just happens that it is also recognised as the longest river in the world by many people; despite claims to the contrary by fans of the Amazon River. If we start with the most remote tributary feeding into Lake Victoria from its source - in your choice of Rwanda or Burundi - it must be a worthy contender, and certainly flows through an amazingly diverse landscape. In fact it is absolutely stunning in every way.

Here again we have 'N' the wavy horizontal line denoting flowing water and 'L' is Leo the lion in the hieroglyphs - the animal most commonly recognised as the mightiest, greatest or top of the pecking order.

So Nile simply means the 'river mighty' or great river. Even the 'I' and 'E' are, or can be represented by a single or double reed. And there are plenty of those in the deltaic waterways to provide the much used and loved papyrus.

'N' is also associated with our favourite water guy; our river boatman who built himself an Ark to save his family and flocks. It can be pretty safely assumed that 'Noah' is again a description of who he was, and what he did, rather than a given name bestowed by his parents.

One point of difference with the northern Semitic language sources, is that in those languages they used the 'L', the crescent moon, rather than the lion to represent the 'highest' or mightiest. This is the same cursive 'L' found by early archaeologists, carved into rocks around the turquoise and copper mines of the Sinai. The workers of course, were moon worshipping Canaanite or Hebrew miners and therefore Baal devotees.

The other Baal is the sun. In fact Baal is another word where the letters describe the original meaning. 'B' represents the legs denoting movement in the hieroglyphs and a cursive 'L' is the crescent moon. So essentially, in the spelling we have a description of something which moves across the sky in the shape of a crescent. If we add in the 'AA' it could also be said to represent the eagle (the Horus bird) and apparent flight, with the double letters representing both heavenly bodies. 'Baal', as a title, was also eventually used colloquially, for master or boss.

In addition, I would imagine that groupings into words as we know them only came later, as did anything suggesting nouns, verbs or adjectives and that an unbroken succession of glyphs / pictograms was the way writing began.

Although a little off topic, this discussion illuminates something else as well; the veneration of cattle, such as Hathor and the Apis bull in Egypt, the sacred cows of India and the head-wear of the Anglo-Saxon Celts. It also flows to other animals such as sheep and goats.

It was common for societies to want to have a representation of heavenly bodies, signs of the zodiac, constellations, nebula and star clusters, in some earthly tangible form. It could be cynically described as idolatry of idolatry, and I believe the worship of cattle is such an example. Simply put, in the head of the animal we have the face (the sun) and the horns (the crescent moon). So on Sinai, when Moses was having a spiritual experience on the mountain like no other (and the people below grew impatient for some action) they succeeded in having Aaron organise the melting down of jewellery to make a golden calf.

This was not about making a gold calf for the sake of it. This was Baal worship; the worship of the sun and moon. It was a return to the idolatry that Abraham had turned his back on centuries before. Unfortunately it remained commonplace throughout Canaan and much of the Middle East deep into Christian times. The symbols still adorn the flags of Islamic countries and the pinnacles of many mosques today.

Many of the ancient gods, like the Babylonian Marduk (aka Bel) are also representative of sun and moon worship, as are a variety of Canaanite, Greek and other deities. Sin, the moon god which gives its name to the Sinai, probably derives from a non-Semitic source (Sumerian perhaps) but was certainly a god in Mesopotamia and a rival for Marduk.

Not the least example is my football team; the Raiders. A common name for football teams as it turns out, and a name associated with the Vikings. I have a cap with the head of a

'Viking' warrior on it (as you do) horns and all. It is exactly the same as the head of the ox or bull when it comes to its ancient meaning.

The only problem with this as an example, is that the Vikings wore helmets without horns and in fact only one or two have ever been found. It was actually the Celtic Anglo-Saxons who wore headgear with the 'angles', aka horns. So this is another example of sun and moon adoration and most likely derived from a group of 'Sea People' of pre Davidic Kingdom times (1200 BC). They were a seagoing tribe who apparently wore the horns in their travels around the Mediterranean.

Indians, with their many sacred cows, probably have little idea that their animals originally represented the Baals, and that those that are polled (without horns) are really imposters. But that is the way history works. Our understanding of things mutates and evolves and we often lose the original intent.

Another representation of the sun and moon is likely the figure of a woman and in particular the breasts. So here is the idea of combining the sun-moon imagery with fertility, and for example at Mari they had a fertility goddess, not far distant from Harran and the centre of Baal worship.

This again ultimately leads back to Gobekli Tepe, a few kilometres to the north of Harran, but millennia timewise, where basic flora and fauna imagery suggests a more grounded focus on nature and the landscape around. Certainly the association of Mari, Harran and Gobekli Tepe (currently dated around 11,000 years BP) is significant. Particularly as the figures carved on those ancient columns at the Gobekli site are almost impossibly old, dating to the Mesolithic as they do.

Harran is of Hurrian origin one would imagine, but the name would seem to have Semitic roots. It therefore may not have been the indigenous name, but one used by Semitic speakers for the shepherds and herdsmen of the area and their town.

It suggests that the Hurrian culture had a central point for trade and contact. At any rate 'H' is the place of, 'R' the mouth or source (perhaps several sources) and 'N' water, which could well describe the setting at the headwaters of the Balikh River, a tributary of the Euphrates - or indeed the Euphrates itself.

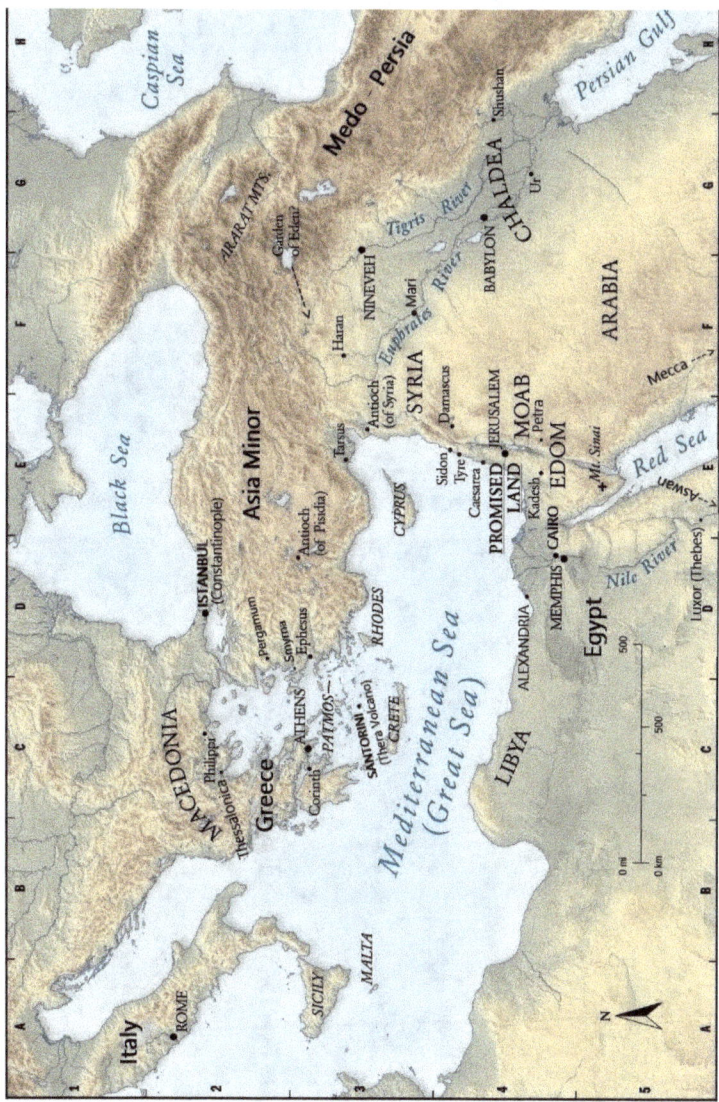

Figure 1: A map of the Eastern Mediterranean, Middle East and Egypt including Israel; the Holy Land.

Plates 1 and 2: The Giza Pyramids of Egypt above and the Sphinx below, with Khufu's Great Pyramid in the background. Some of the erosion to the Sphinx and enclosure is likely due to the 2350 BC Flood.

Plates 3: Khufu's funerary boat dating to 2550 BC; 200 years before the 2350 BC flood that impacted the Northern Hemisphere.

Plate 4: The 121,878 tonne Celebrity Equinox anchored in the vast harbour of the Santorini super volcano. A mega-eruption here in the Aegean Sea triggered the plagues of Egypt and the Exodus.

Plate 5: Jebel Katrine (Mount Catherine) Sinai Peninsula (the highest mountain in Egypt) named after the Christian Saint of Alexandria. The photo is taken from Egypt's Mount Sinai.

Plate 6: Hiking past the famous Saint Catherine's Monastery where ancient literary treasures are still held for safekeeping.

Plate 7: Through the keyhole; the Siq at Petra looking toward The Treasury. Is this area the Kadesh Barnea of Moses?

Plate 8: The mortuary temple of Queen Hatshepsut at Deir el Bahari at Luxor (ancient Thebes) on the west bank of the Nile.

Plate 9: The monumental exterior of the Ramses II temple at Abu Simbel on Lake Nasser, located well south in Upper Egypt.

Plate 10: Amazing Aswan at the First Cataract of the Nile River, just downstream of the old and high dams.

Plate 11: The Aswan Coptic Orthodox Cathedral of the Archangel Michael as seen from the Nile River.

Plate 12: Inside the Mosque of Amr ibn al-As in Cairo. The original structure was constructed in AD 642; the first mosque built in Africa.

Plate 13: The Parthenon in Athens with tourists and political activists mingling on the day.

Plate 14: It was in Athens that the Apostle Paul came upon a statue to the 'unknown god' and used that to introduce the 'unseen God' and the arrival of the Jewish Messiah.

In the Hebrew worldview, I suspect that Elah became Elohim in order to distance the Hebrews from Baal. We are told that Elohim is just the plural of Elah in Hebrew, and that is apparently true. But what I am endeavouring to unearth here, is far more ancient meanings and associations.

Given the ancestor Semitic languages from Mesopotamia date back to 3000 BC in written form, and to earlier times as language, a tentative hypothesis is that the 'M' was added to Elah later in Hebrew history, to change the 'moon god' association of the word to the Hidden One, or Spirit Creator God. The one in fact who created both the sun and moon.

A more personal name was given to the Hebrew God at an even later stage, in order to further distance the original association with the moon. YHWH (Yahuwah / Jehovah), the 'I AM', is the sacred name linked to the Hebrew story from Moses' time (around 1550 BC) onward.

Muslims of course still use the Arabic form of the old name with direct moon connection and are therefore conflicted – at least from a literary point of view. And remember Muslims see Abraham as their ancestor as well and would argue strongly they worship his one true God. Opposed to this is the ritual at the Kaaba in Mecca. It is steeped in the tradition of the moon, with an asteroid fragment embedded in the corner of the structure. This is an obvious indicator that the religion, while espousing monotheism, has not entirely broken away from its association with ancient Babylon and 'moon' worship.

Their revolving lunar calendar, which has no association with the solar year whatsoever, is also conflicted in that respect.

More pointedly, devout Arabic speaking Christians most definitely know who they worship and will not be swayed by language ignorant Westerners, who criticise the use of Allah as the Arabic equivalent of God. It therefore goes without saying that we should not confuse literary development, with the reality of the faith of the individual. However associations can be clues to the origins of nations, ideologies and religious movements and so it is a little unfortunate that Allah (the God) is the equivalent of saying 'the Elah' which harks back to the moon imagery and Baal worship of the ancients.

There is another animal representation that may have some association with the sun and moon. It was also popular in ancient times and is the eagle / falcon / vulture, known as the Horus bird to the Egyptians. In fact in the Book of Ezekiel chapter 17, we see a description of two adversaries and their manoeuvrings for control which is sometimes called the 'prophecy of the two eagles'. Ezekiel was writing at the time of the exiles and the two eagles represent two empires: Babylon and Egypt.

Israel, of course, were the unfortunates in the middle being torn apart as prey. So when looked at directly from the front, the eagle symbol is a body (the head) with a wing on either side. Again we have the head representing the sun and the raised wings representing the crescent moon.

For another example from the Semitic languages we have the mighty river 'Euphrates'. The key consonants are 'P', 'H', 'R' and 'T'. The 'H' denotes house, place, or being, but 'PRT' is essentially covenant or agreement.

How many times do we see in ancient history, including scripture, where the Euphrates River is the agreed boundary between the 'king of the south' (usually Egypt, but for a while Israel) with the 'king of the north'? The king of the north was at any particular time the Akkadian, Hittite, Mitanni, Assyrian, Persian or Greek Empires.

In the Hebrew version; 'B' is substituted for 'P', where the 'B' are the legs in the hieroglyphs denoting movement. This variation is particularly apt in respect of the Habiru, whose own name denotes people on the move.

Other names of places and people such as Arabia and Bedouin are likely to have similar connotations.

So the heavenly bodies, natural landscape features, and names of Semitic peoples are good targets in regard to this linkage of letters and glyphs to their meaning. Man-made features such as ancient cities may also be candidates, but are more likely to have had their names changed, as invaders rename them in order to stamp their authority in the region.

Kinneret, the Hebrew name for Lake Galilee is obviously derived from a Semitic name. It is probably ancient Aramaic or Canaanite, but possibly Egyptian given that Pharaoh Thutmose

III and his successors used the adjacent town of Beit She'an as a provincial capital for centuries.

I have heard that Israelis suggest the name comes from the wind that blows over the lake, or perhaps from the similarity of the lake's shape with a kinnor, a musical instrument. However the ancient origin of the name is much more enlightening and interesting, and for example most likely found its way as far afield as Ireland in the word Connor.

So we have 'K' meaning a bowl, 'NN' for water, with the double 'N' suggesting a broad expanse or lake, 'R' the mouth, of uncertain meaning, but probably denoting a source of water and 'T' (the loaf of bread) something choice, or special, which it certainly is; fresh water in a dry land, fishing, recreation and climate moderating.

Anyone who has been to Galilee will instantly appreciate the description.

It is tucked away like a bowl between the hills and high country to the west and the Golan Heights to the east. The word is also instructive in that it may have originally been read from right to left - T, R, NN, K – and reading something like: 'the choice (beautiful) source of lake water in the bowl of the earth'.

The term 'Canaan', for eretz Israel may come from a similar idea, meaning the land between two waters, or seas 'N'; the western sea (Mediterranean) and the eastern (Jordan, Galilee and the Dead Sea). The bowl or basket 'C' does not quite equate, except that at a pinch 'basket' could be seen as an analogy for a pleasant land flowing with 'milk and honey'.

Olives, bread, cheese and wine - or anything else you would pick for a picnic lunch.

As mentioned earlier, the land of Shinar is a term used in the Bible and Egypt to describe the plain of Sumer; the flat downstream part of the Euphrates-Tigris valley. 'CH' (sometimes 'SH') denotes a pool, or sometimes a well of water and the 'N', like the Nile, flowing water. 'R' in this context is a source or place where water abounds. I think the associations show that the name describes the place. A place where irrigation was relatively easily accomplished, creating a productive paradise and similar in fact to the Nile delta; the major food basket for Egyptians to this today.

Sometimes there are hints underlying the meanings of personal names; Abraham and Sarah, Isaac and Rebecca, Jacob, and Rachel are of interest here. One that fascinates is the latter name of Jacob's favourite wife, Rachel.

Many of us know the story where Jacob goes back to Harran from Canaan to seek a wife from the extended family there. He meets Rachel, seemingly by accident, drawing water at a well. He works for his father-in-law-to-be, Laban, for 7 years for her hand, but is subsequently deceived by him and ends up with Leah, the older sister. He is then forced to work another seven years for the hand of Rachel. Such is true love!

Now 'R' is word or agreement, 'CH' is pool or well and 'EL' is our designator, the crescent moon, for the most high, or God.

Again the name spells out the derivation, and in fact tells us the story of their meeting and the famous marriage agreement before God. It also seems to indicate that some names were nicknames. They might not have been the names that the individuals knew themselves by at the time.

In respect of the use of the glyph / letter 'R', we can see it features in many of the ancient names for peoples and nations who lived in the northern Fertile Crescent region.

HuRRian, URartu, ARmenian, ARamean, SumeRian, and MaRian are examples. These names are pretty obviously Semitic in origin and the proposal here is; that we are looking at a distinction of national groups by language rather than ethnicity. The obvious association of the mouth / lips 'R' with language suggests this and HebRew would appear to follow the pattern.

A classic example in terms of the Bible and the Pentateuch is the word 'Torah', or the Law. 'T' again, is the loaf of bread on a plate which represents something special and necessary for life. 'R' are the lips which could mean the source of the word, or commands, but also that the Law should be consumed and taken to heart.

There is no question that 'Torah' is early Semitic in origin and that its meaning is obvious and self explanatory.

While I am brainstorming, a New Testament example may also have special meaning. Peter the fisherman, disciple and eventually Apostle had the common Hebrew name of Simon (Shimon). In the form 'Simon' particularly we have a sense of locked (the 'S') and hidden water (the 'N') or alternately a pool of hidden water, where water is one of those metaphors for the spirit. Either way, it seems a slightly negative name in that context, more about potential than obvious demonstrated ability. And so could this be why Jesus calls him Cephas instead?

Cephas is a form of Peter, and not particularly enlightening in that form, but Peter and his brother Andrew, together with Phillip are said to have lived in Bethsaida (John 1:44) a fishing village north and east of Capharnaum on Galilee. It is an area known from archaeology to have a strong Greek presence.

So Peter would have been familiar with Greek like most Jews, Galileans and Samaritans, and would surely have appreciated his new nickname.

It was positive in the sense of its association with rock and firm foundation, but also in regard to its roots, 'PTR', the same as 'PRT' and 'BRT', keeping in mind groupings rather than order (or reading from right to left, or left to right) mattered. We therefore have someone who is in covenant with the Almighty – 'on this rock I will build my church'. It is a conversion from hidden potential to a powerful expression of commitment.

Now I have not researched non-Semitic language groups to any meaningful extent. It is simply too big a discussion in the context here. However a possible Semitic example in the relatively distant eastern world is the Himalayan mountain chain.

The Himalayas are the highest mountain range in the world, situated as they are between China and Tibet to the north and India and Pakistan to the south. Smaller states like Nepal and Bhutan are squeezed in between. Interestingly the Hindi and Sanskrit meaning of the name is 'house (or abode) of snow'. Dissecting the name we have 'H', place of, 'M', hidden, 'L', high or most high, and 'Y', reed, or indeed perhaps from 'YA' the more recent, shortened version of Yahuah. The simplest meaning would be 'mighty hidden mountains' based on their cloud and snow covered nature. Or they could have been seen to represent God as creator, because of their 'most high' meaning.

One wonders where this name actually came from and how old, it in fact is. It just so happens that some Afghani-Kashmiri peoples have a heritage which they say is from the Lost Tribes of Israel. Is this Hebrew connection a potential source for the name? Or is it way older than that?

There are other possible associations from the east which I have identified, but require further research. In any case I have speculated way too much as it is.

Suffice to say, there are even greater possibilities from Europe, where words and names have linguistic roots from Celtic, Saxon, French, Scandinavian, Latin and Greek sources, which although predominantly Indo-European, could in turn well have been influenced by ancient, precursor Semitic languages.

I also find it interesting that the ancients must have travelled and traded far more extensively than we normally give them credit. And that war and persecution, weather, natural disaster and fortune seeking, actually prompted them to move from the ancient empires to the four corners.

Now some linguists may disagree with what I am saying, but I believe there is potentially other support for the idea that the Greeks transliterated many Semitic words directly.

It is based on the great migration out of both old Assyria and especially the neo-Assyrian Empire of the 8th century BC. I will discuss this later, but it is in the 8th century that we see Greece really beginning to prosper as an artistic culture. It is also when we see the Greeks become mathematically and scientifically empowered and when their philosophers appear on the scene.

The proposition here is that the emigres (which included Israel tribes) made a huge impact on arriving in the Greek lands and that the ancient Mycenean Greeks gained a lot from their arrival, including a connection to the knowledge of the empires.

To sum up; this excursion into Semitic language origins, influences and ancient meanings, is presented by way of background for the continuing historical study going forward – and also the prophetic revelations to come. Given that the latest findings from Cornell University link so many languages back

70,000 years, it would not surprise if further research throws light on the subset of languages from the Semitic world and the 'how, why and when' of its influence elsewhere.

For English speakers it is of obvious interest. But that goes for a great many languages with similar roots and associations, right around the world.

The numbers game

The Sumerians have always been famous for their sexagesimal counting system. Sexagesimal is a 60 based method of counting, compared to our decimal or 10s system. It means that in the first column instead of counting from 1 to 9, they made it all the way to 59. By adding the second column, instead of counting to 99 as we do in decimal, they counted to an astonishing 3,599!

Having said that, they did use a 600 designator / column in some iterations of the system, but that was still a dramatically huge step from one column of significance to the next compared to decimal. All in all it is a truly remarkable numbers story.

In current computing technology, digital systems use a method where each memory 'bit' is either a 0 or a 1; i.e. there are only two possible states and these are referred to as binary. But to make computing practical, the octal or 8 system was introduced and then a 16 bit (hexadecimal) system, 32 bit, 64 bit and so on. These are all built on binary, and two state memory devices are still used as the basis for these derived systems.

However the decimal system does not reduce down by factors of 2, like a system built on binary. If you count your fingers or your toes (providing you have not lost any, or are especially endowed with more) it soon becomes apparent why the decimal system prevails. There are many other, rather odd systems out there amongst tribal peoples; 20 base systems, 27 base and so on and most of these too are based on counting body parts.

In ancient times many cultures such as the Greek and Hebrew did not have numbers, so they designated letters of their alphabet for the dual purpose of spelling and counting. Others, like the Egyptians and Sumerians adopted glyphs and tokens of one sort or another, specifically for the purpose.

The 60 based system seems unusual to us, but also broke down to many factors, 2, 3, 4, 5 and multiples thereof, making it very versatile. Importantly for us, it is where we get our time keeping from; 60 seconds in a minute, 60 minutes in the hour, and one of the factors, 12, is also incorporated in the hour count where 12 by 2 gives us the hours in the day.

Moreover the system is also built into the 360 degrees of the compass, that is still used in navigation worldwide.

Despite the differences compared to decimal and the plethora of factors above, it is also apparent that 6 x 10 are also factors. In addition 10 is a basic unit of counting within each column. This suggests that sexagesimal may have risen from a decimal system somewhere in the distant past.

Another attraction for the Sumerians was that 360 was very approximately the number of days of a solar year (and I stress approximately) and equal to 12 lunar months of a nominal 30 days each. However the lunar month was their basic time keeping period rather than a solar year, and is actually closer to 29.5 days. This equates to 354 days for a 12 lunar month calendar.

The problem for their timekeepers, was how to use lunar months, while keeping the solar year aligned for crop planting. They knew that the solar year was longer than 360 days and so decided to add an extra lunar month in every now and again, to keep crop planting and harvesting synchronised to the seasons. In short, even though the solar year was not their primary time keeping method, 360 days was a convenient number to use for a 'standard' year and fitted their 60 base counting just nicely!

With its great array of factors, it is easy to see why the Mesopotamian system persisted in the Imperial System of money, weight, volumes and length in the British world, so many years later. In Australia it lasted until the late 1960s and the introduction of the decimal monetary system, with other decimalisation phased in thereafter. I will not name names, but a visit to a hardware supplier (particularly automotive) is enough for the average handyman or woman to realise, that the old Imperial (and Babylonian) measurements still persist with nuts, bolts and other bits and pieces, examples to this day.

Abraham lived around the time the sexagesimal counting tokens (symbols) were simplified and changed in character - and as it turned out - were probably simplified too much, given the relative inexperience of users. This of course was roughly at the turn of the 3rd and 2nd millennia BC, and it caused an increase in ambiguity and error in practice.

Before this streamlining of the script occurred, we know from the Sumerian clay tablet record, that the sexagesimal counting system used symbols and extra tokens which were more cumbersome to write, but not as easy to confuse.

In addition, those who actually emigrated from Mesopotamia would have understood the history of these changes. However it is highly likely that later scribes and translators living all their lives, say, in Egypt, would not necessarily know how to correctly render the old Mesopotamian counting script.

Consequently, when the symbols were simplified, there remained just two counting symbols, a tack like figure 'T' which represented '1' and a left pointing arrow '<' representing '10'.

To complicate things further, like many other ancient counting methods, the sexagesimal system was also without a '0' or a 'place' token to show where a zero should be placed, or a gap indicated. This meant a lot was placed on the shoulders of the scribe, in terms of accurate positioning of the characters by hand. Unfortunately, known examples on clay tablets sometimes show a hurried and untidy style by some scribes. It also means that even more was required of the reader or translator.

Given the absence of place tokens and zeros to clarify numbers, understanding which column was which, was difficult and the proverbial headache. It meant that when using a sixty based system, if you misread the column of significance, you were guaranteed to be wildly out in your interpretation of the number. There was simply too much potential for error and mistakes by later copyists were assured.

Suffice to say, that when we come to unusually large numbers, or unusual combinations of numbers recorded in the book of Genesis or elsewhere, then this is the first place to look. It does not take away from the inspiration of prophecies and the overall attention to detail by the scribes, but simply recognises that God has used the frailty of human beings to record it all. It also happened at a very early time in the history of writing, and well before the invention of printing and automated systems that would later minimise mistakes.

With this background history of the written word and the ancient counting systems under our belt, it becomes apparent we would be foolish to take anything at face value. So I hope the reader can forgive this merry dance through the numbering systems, because a cautious and scholarly approach to the results we see in the Bible is absolutely necessary.

This subject could take up volumes, but one example of numbers that I will now address, is the age of the ten patriarchs listed in Genesis before the Flood. A perusal of the scripture shows, that except for Enoch who lived for only 365 years, the other nine pre-flood patriarchs lived around 900 years, with Methuselah living to the ripe old age of 969. The question is; 'how is this so'?

In the discussion above, I mentioned that the Mesopotamians used a 360 day year as standard and that this is somewhat more than the actual 12 lunar month year of 354 days. But I also stated that the Mesopotamian system of counting time was not a solar one, but based on counting lunar months.

This happens to be the key issue in so far as the ages of the patriarchs before the Flood are concerned. It is also something which has caused confusion in biblical circles generally and also limited our understanding of Bible prophecy.

As discussed earlier, the Mesopotamians used a rigorous scheme to keep the seasons synchronised by adding extra lunar months over a 19 year cycle. This system was introduced early in the first millennium BC around the 8th century.

Before that time, the king and his advisors added the months as they saw fit; that is, when they felt the seasons were too far out of kilter with the lunar months.

In essence though, the lunar month was everything, and this means that if the ages of the patriarchs were actually recorded in lunar months (as they surely were) we are then compelled to do the calculation to translate their ages to solar years. If we divide the age of Methuselah by 12, we should get an approximate value for his true age in solar (tropical) calendar years.

If we want to be a little more precise we use:

---

969 (his biblical age) x 29.5 / 365.2 = 78.3 (actual age in solar calendar years)

Is it that simple? Or is there more to it? In fact there are a whole host of problems that follow. It has been pointed out, that the 'begat' ages, i.e. when the antediluvian patriarchs had their first offspring, would make them impossibly young in some instances. A combination of factors probably contributes to this mystery, but the most significant appears to be the text.

The begat ages found in the Masoretic text average around 100 'years' less than that in the Greek Septuagint within the first few generations, e.g. for Adam 130 years against 230 to Seth; 105 against 205 for Enosh; 90 against 190 for Kenan (Kainan); and 70 against 170 for Mahalalel. Given the Septuagint is generally acknowledged to come from the 3rd century BC and well over one thousand years before the earliest copies of the Masoretic known to date, it would be foolhardy to ignore it.

Using the calculation above (or the equivalent 19/235) we find that 100 lunar months are approximately 8 solar years. This makes a significant difference and results in much more believable numbers. Some fathers of the time were still teenagers, but hey, the education of the day was largely 'on the job training' and other distractions were limited. In any event, the oldest parent by far was Noah at 500 'years' and this only equates to 40 in solar years. This is not unusual by any standards.

Jokes aside, it still leaves the 'begat' ages in the very young category and so the lunar month assumption is unlikely to be the complete solution to this problem.

Now I have never laid awake at night wondering whether Shem, Ham and Japheth were triplets, or stressing over the many unknowns in regard to Noah's life. The account after all is an abbreviated one and given the dire circumstances, not an every day occurrence in the northern region (Anatolia-Caucuses-Assyria) where Noah likely survived the flood.

However the issues causing confusion with chronologies before and after the Flood, will include some of the following:

o   The impossibly long life spans of the pre-flood patriarchs
o   Records lost in the Flood / Flood Epoch
o   Two generations before the Flood (at least) and their children would have drowned except for Noah's family
o   The difficulties reconciling the much shorter life spans of those that followed the Flood, with those that lived before
o   Age differences in the Masoretic, Septuagint and potentially other versions of scripture
o   The simplified sexagesimal counting system of 2000 BC
o   Making sense of multiple records of questionable provenance

Later writers and compilers are likely to have had a reasonable idea of one specific family line, but not of more distant cousins and family connections. There would also be an almost guaranteed attempt by later scribes to correct obvious inconsistencies, despite having insufficient data at their fingertips.

Attempting to construct meaningful genealogies out of this mishmash would have been difficult, particularly once scribes convinced themselves that the patriarchs before the Flood lived ten or more times longer, than later generations. For example there would have been the confusing situation where someone born say, around 3500 BC and the Ur flood would have still been alive for the 2600 BC flood. Or if born a little later in say 3200 BC, could have perished in the 2350 BC flood.

According to the Masoretic text, Methuselah lived for 782 years after he had his son Lamech and that Lamech lived a total of 777 years. Based on the numbers, both Methuselah and his son Lamech may have died in the same circumstances i.e. the same extended flood episode 5 years apart, with Methuselah passing away last. On the other hand, if their ages are rendered as lunar months, nothing changes except that they die 5 months apart instead. Five months difference in an extended flood event (if that was the case) arguably makes more sense than 5 years.

The Septuagint on the other hand, says Methuselah lived 802 years after having Lamech and that Lamech lived a total of

753 years, i.e. Methuselah outlived his son by 49 years, which would translate as 49 lunar months (or 4 real calendar years).

This scenario is far less likely to have them dying in the same flood episode. It also illustrates that we cannot rely on the ages given in Genesis chapter 5 to the 'nth' degree and that adjustment has been made to one, or both texts. This is pretty obvious, given the total number of years lived by both men is the same in both versions, despite differences in the 'begat' ages.

To complete the picture, the ages of the patriarchs after the Flood are recorded thus: There were four who lived between 500 and 403 years and then six more from Peleg to Abraham who lived between the ages of 209 and 119 years old.

I do not have any complex solution to present for the many possibilities here, except to say the result of the original problem of lunar months being confused for solar years, together with the ambiguities of the sexagesimal system will have had copyists perplexed. It will potentially affect not only the ages of the patriarchs, but also birthdays, overall genealogies and other numbers such as the dimensions of structures mentioned in the Bible. It means later historians and writers, including the Greeks, are likely to have been confused and assumed solar years, when lunar months were meant.

Interestingly, after the Flood the 'begat' years to the birth of the first offspring seem realistic in the Masoretic. This suggests better sources of data and / or informed retrofitting of numbers, e.g. 30 years for a mature Hebrew male and working several years for a bride. This should not surprise, because if the deluge was as destructive as the record suggests, very little parchment would have survived. After all, we are told the Ark was a floating zoo, rather than an Amazon book depository.

By contrast, if we assume the begat ages in the Septuagint are lunar months as before, they become way too young (under 10 years old on average) if converted to solar years. Again, there has to be further issues with the recorded numbers.

Now one of the proposals suggested in the past for these huge life spans, is that the earth and its atmosphere suffered some sort of collapse and degradation after the Flood, i.e. everything was near perfect before the Flood, but thereafter human beings were subject to significant negative effects, caused by greater cosmic radiation and the like. This presumably as the protective canopy of the 'firmament above' disintegrated.

In this regard it is obvious that the climate was very wet through the Flood Epoch and there would have been unusual amounts of cloud cover, i.e. 'firmament above'. The current scientific evidence of sediment deposition (and certainly tree ring thicknesses) supports this. The question is; would this be sufficient to prolong the life of humans to any real extent?

There are a sufficient number of people working and living in protected environments, who we would expect to live to great ages if this was true. Laboratories, military installations and hospital complexes for example, that have radio frequency and radiation shields of one sort or another. Some of these might be expected to provide some protection from gamma and cosmic rays, but also various types of sub-atomic particles.

If there is any evidence in this respect then it is being kept very quiet and seems unlikely. In addition we might expect to see a difference in life spans between people living at high latitudes compared to those near the equator, or those in wet zones compared to those in deserts.

Neither has anyone, apparently, ever discovered the bones of an individual showing evidence of such enormous ages, and bones and teeth of early Sapiens, Homo-sapiens, Neanderthals and more ancient hominids give us nothing in terms of extended life spans, as far as I am aware. My limited biological understanding in regard to this matter, is that the human body has a built in DNA clock which also regulates our life spans and overrides other environmental considerations.

So the answers may not be entirely set in concrete, but the obvious and straightforward likelihood is that the text is faulty due to human error, and we have a case where Occam's razor applies. Once we understand how the ancient Mesopotamians counted and reckoned the years and the scribal difficulties that abound, we do not have to invoke fantastical numbers, or theories and make a laughing stock of our sacred text.

In the Sumerian king lists, the length of reigns of the antediluvian kings translated into calendar years by modern scholars are truly ridiculous and extend from 18,600 years to 36,000 years, i.e. 5 sars 1 ner to 10 sars where a 'sar' is 3,600 and a 'ner' is 600. It does not take much imagination to see that in a later copy after the tokens were simplified, a '**T**' (or 1) which is interpreted to be in the third significant column where it represents 3,600 in decimal, may have been originally meant for the 60s or digits column only.

Remember, there are no zeros, or place tokens to minimise this sort of error and ages are most likely to be in lunar months rather than calendar years, anyway.

In the Indian scheme of things, I understand their deities and super characters live even longer and extend back millions of years according to their texts. This suggests even more elaboration and distance from reality. It would not surprise that ultimately the confusion and miscounting started in Sumer originally, and migrated to the sub-continent in much the same manner as the sacred cow. Indeed, did this writing and counting ability originate in Sumer, or in fact further west in northern Syria, Anatolia or the Black Sea region? We read:

*"Now the whole world had one language and a common speech. As people **moved eastward**, they found a plain in Shinar and settled there."* (Genesis 11:1 NIV)

Rightly understood the Bible really is a 'go-to' book for much historical understanding, as well as its unsurpassed moral and spiritual guidance. It is also a testament to the Hebrew scripture, that a relatively simple and obvious transformation is all that is required to arrive at the likely ages of the antediluvian patriarchs and kings of yore.

It means that the provenance of the text they worked from is truly ancient, but still reliable for most practical purposes. Certainly more accurate in terms of ages, than the cuneiform tablets and official king lists the Mesopotamians kept.

Now it could be argued that God could have revealed much of this history prophetically at a later time and thus avoided the shortcomings of scribal record keeping. This, in much the same way I argued Genesis 1 was given prophetically.

The fact is; he obviously did not do that, and this seems to be the case for most of the biblical historic record. It seems to be a case of 'you can work it out for yourselves, I did not make dummies when I made you'. The significant positive that comes out of this, is that the prophecies of the future are by comparison, considerably enhanced in our eyes.

In summary, the Tower of Babel serves as a useful historical hook to hang the dispersion story on. There appears to have been a major migration out of the area, starting no later than 3500 BC and continuing through the Flood Epoch thereafter.

This dispersion to all parts, including Egypt and India, is very likely directly attributable to this wetter than average, transitional period at the tail end of the last glacial maximum.

It was followed by an equally dramatic climatic reversal to drought, which also drove the population from Mesopotamia and the northern Fertile Crescent. It leads to the disruption of Mesopotamian society and the demise of the Indus Valley civilisation. It also fits perfectly with Abraham's journey to Canaan and Egypt around 1900 BC.

This chapter on language is presented as possibility thinking rather than dogma, and as biblical apologetics in terms of the numbers and the ages of the patriarchs. I know that I have probably overindulged my imagination, but hope that this discussion serves the purpose of establishing a certain confidence in the ancient text and its associations, as well as an understanding of its historical brevity and limitations.

I stress once again; up until this point we have been investigating a period of 2,000 years or more, with the corresponding Genesis narrative ending over 4,000 years ago. Given the topics addressed are crammed into a mere 11 chapters of a book of 1,189 chapters - and over 1,200 if we include the Apocrypha I refer to later - we can say that it is truly a remarkable document. One that abounds with moral teaching, spiritual enlightenment and prophetic inspiration - but also very useful historically.

Yes, it has been heavy going thus far, and we have seen that our childhood stories were idealised. But we have barely spun the wheels so to speak, and are about to accelerate into more rapidly paced adventures of our ancient Hebraic ancestors. They are also better documented and now scientifically verified.

CHAPTER 6

Exodus Cataclysm

'We slid down the mountain at evening by torchlight
A hand kindly guiding, as I stumbled at night,
Next day St Catherine's and then travelling on
To our Red Sea beach camp to relax in the sun'

See you in Santorini

It was time to depart Cairo again in another trusty minibus. We made an early start for the eight hour dash into the heart of the Sinai Peninsula and headed due east for Suez. After a morning tea break, we dived into the tunnel under the canal and then south-east along the eastern shore of the Gulf.

My photo album for this segment of the trip I entitled 'Out of Africa; Mount Sinai and the Red Sea'. A tongue in cheek reliving of the 70,000 year old migration, our genetics experts suggest occurred with our mitochondrial 'Eve' way back when. You have to have a sense of history!

We had four 'Eves' on this occasion, who were all eyes passing the seaside resorts, but eventually snoozing while the guys (including this student of geoscience) remained wide awake looking at the unique landscape. It was a lazy, hazy day with big ships out to sea, hotels baking in the sun, and if I recall correctly, at least one petroleum drill ship exploring in the gulf. Crazily laden trucks slowed us down, but eventually we arrived at a point that turned away from the sea and pushed into the mountainous terrain of the southern Peninsula.

I was mindful we were racing past historic places, including the old copper and turquoise mines up in the hills. Artefacts and rock carvings have been found there and analysed by early archaeologists such as Sir Flinders Petrie. He was looking for signs of Egyptian and Canaanite / Hebraic occupation – and found them.

As we turned away from the coast, we were swallowed by the raw beauty of the desert ranges and the aridity of the valley. Huge dykes of volcanic rock cut across the hills, leaving scars as if from giant scimitars, and in the wadi we occasionally spied an odd scruffy tree, or a camel or two. Sometimes we passed a well, or watering hole, or an oasis of desert palms. There were occasional signs of life and every 20 or 30 kilometres, a security checkpoint reminded us of the terrorist issues on the Peninsula.

This was mid-2010 and 9 months before the 'Arab Spring' uprising, but with all sorts of trafficking going on to Gaza. The various Egyptian security forces were under constant pressure, and you felt for the travellers in the handful of private cars held up for hours at the checkpoints. Fortunately our tour managers and chief guide seemed to have our paper work up to date.

And so, except for displaying our passports, we sailed through security and arrived at the town of Saint Catherine at the base of Egypt's Mt Sinai in the early afternoon.

That gave us just enough time to acquaint ourselves with our accommodation, a bite to eat and a short siesta, before heading out for the climb to the summit in the mid-afternoon. Ours was to be a sunset view on top of the mountain.

What a climb! Most of the younger members of the group took the shortcut, virtually straight up. I chose the easier walking trail and a couple went for the Rolls Royce camel ride.

Actually, I confess to being offered a few minutes on a camel toward the end of the trail; an offer I gladly accepted.

It gave me a much needed break before the last stage; the '749 steps' to arrive at the top. No camel ride for that - definitely mountain goat territory!

The mountain is a few metres higher than Mt Kosciusko, the highest mountain on mainland Australia. It is also much steeper - and so there was a sense of relief for this grey nomad when the summit came into view. The desert mountain scenery stretching toward the Red Sea was amazing, as the sun set over Mount St. Catherine; the highest mountain in Egypt towering above us.

Rather ironically, Jebel Katrine and the monastery, are named for the Christian Saint who was killed and burned by the 4th century pagan Roman, Emperor Maxentius. Bound on a large wheel and set alight, she is unwittingly celebrated every year on fireworks nights.

Then an unexpected interruption by a film crew in a rather large helicopter, abruptly reminded us that this was the 21st century and not the distant past. Much later, as the sun set over the rugged mountains and evening closed in, a descent by torchlight under a starry sky.

The following day we visited St Catherine's Monastery at the foot of the mountain. It sits snugly in the valley and is famous for its ancient texts housed there for safety sake; texts preserved in the desert for centuries, away from the dangers frequently evident in Alexandria and its library. Amongst those texts is the iconic Sinaiticus Codex; written in Greek; a copy of the Septuagint Bible.

All too soon we were back on the bus and headed north, eventually catching a glimpse of the sea and the Gulf of Aqaba (Eilat) as we headed down the incline to the coast. As we did so, we passed a lone column on the left, with an interesting modern history care of a rather eccentric, 'Indiana Jones' character, with his controversial theories on the Exodus. No time to stop though, the port town of Nuweiba appeared directly ahead, and passing by, we headed further up the coast to our beach destination at Sawa. It was time for a relaxing, end-of-trip couple of days; dining, cocktails, sunbaking and walking on the beach were the order of the day. There was also a choice of diving and snorkelling in the crystal clear waters of the reef.

There was no chance of getting sunburnt though; the 2010 volcanic eruption in Iceland, which shut down much of the air traffic in Europe, also provided just enough haze to mute the effects of the sun's rays. Slightly surreal in fact; and subtly mimicking the 'cloud' of 3,500 years ago, that hovered over the Sinai and was faithfully recorded in the book of Exodus.

It was not only the stone column mentioned before, but a mountain called Jabal al Lawz (just 15 kilometres across the Gulf in Saudi Arabia) that reminded me there were many other contenders for the Mount Sinai of the Bible. So it was out with the 300 mm lens, in order to shoot as many pictures under different lighting conditions as possible. Jabal al Lawz (mountain of the law) has been considered by the locals and some Europeans as the real Mt. Sinai for many, many years, but was popularised by the colourful character of Ron Wyatt.

Back in 1978 (the story goes) he found an old stone column lying on the ground near Nuweiba, and convinced local officials that it was the remains of an Israelite commemoration of the Exodus. He suggested it marked the spot where Moses brought the people across the Red Sea, and that there was a submarine land bridge extending out from Nuweiba at a depth of 275 metres (900 feet). The idea was that a relatively moderate event – earth movement, tsunami, east wind, sheer miracle or whatever - was enough to make this an escape route for the Israelites.

In fact, the Gulf of Aqaba is a series of en-echelon faults on the Syro-African rift, with very deep troughs of well over 1,000 metres separated by shallower sections. This unusual geometry is caused by the splitting apart of the continental masses by geological forces along the rift system.

But the truth is, the shallower sections are not *that* shallow and the depth at Nuweiba is in fact still 800 metres deep - as I once showed Ron Wyatt's chronicler Jonathan Grey, at his then home in the Adelaide Hills of South Australia. I had previously copied an article from a journal I found in the Geoscience Australia (then AGSO) Library in Canberra. It included a description of a marine survey in the Gulf, and copies of seismic and hydrographic sections obtained by an Israeli survey team.

However this evidence did not provoke any reaction or response; either in person or in their following newsletters. This suggested to me at the time, that mere facts were never going to get in the way of a good story. And this was the problem; the newsletters were accepted by non-critical folk like my family as gospel. They had no reason to doubt the claims, or indeed the science background to check them out. I, on the other hand, was part of the establishment and therefore could be safely ignored.

Of course, an earth movement of 800 metres is not totally out of the question in extreme circumstances. A violent earthquake for instance at the edge of a steep section of continental slope, might cause the shelf to collapse. More modest elevation caused by continental plate collision, or general earthquake activity, could under some circumstances result in tens of metres of uplift in a relatively short time.

A bolide impact would be even more dramatic, potentially rearranging an extensive area of the earth's surface instantly.

But as it happens, there is a much better explanation for all the complex historical detail recorded in the book of Exodus. Unbelievably, despite the extensive passage of time, the detailed biblical sequence of the plagues, the drowning of pharaoh's army, the column of smoke by day and the column of fire by night (not to mention the persistent cloud described earlier) can all be attributed to ONE single event!

And it did not occur in the Gulf of Aqaba, or anywhere in the Red Sea. In fact a careful analysis of the logistics and time available to Moses gleaned from the biblical accounts, precludes this possibility to a considerable degree of certainty.

What I am about to describe, personally rocked me more than anything I had ever learnt about such things as Big Bang Theory and evolution; whether of galaxies, solar systems and an old earth, or the millions of years of biological life. It is hard to explain, but I think there were many things that I had adjusted to as you do when you mature; where you gradually learn more about the natural world, and realise that childhood imaginations are not always the reality you once thought.

But for some reason, Moses and the Exodus had escaped the sort of scrutiny required until quite late in my life. And of course with my general science and geology background, I was more 'liberal' than some and open to the idea of naturalistic explanations of biblical events; albeit on a case by case basis.

Certainly more so than most biblical fundamentalists - of whom I knew more than my fair share.

But I was rudely awakened when reading a book entitled 'Act of God', written by the British investigative journalist, Graham Phillips. This book came to my attention sometime around the turn of the century. I was not particularly taken with Phillips' theories on the timing of the Exodus, or the pharaohs he fingered as players in the drama. But the cause of the event was a major eye-opener.

I was fortunate subsequently, to also come across an extremely well-illustrated and detailed publication which had been written earlier by Ian Wilson, another British journalist who published in 1985. His tome is called 'The Exodus Enigma'.

I mention these two publications particularly, because they are very readable, and succinctly describe the events and the geological background in considerable detail.

Now as a geoscience survey professional and technical specialist working in a research environment, you accumulate a fairly good feel for the veracity or otherwise of new proposals over time. Because of this, I knew instantly that the detailed sequence of otherwise seemingly disconnected events, was too precise and related to be anything but the truth.

Yes, I was shocked, but at the same time here was proof positive, of the historicity of the Exodus account. One of a natural disaster; or as our insurance companies quite correctly describe - an act of God!

Here on the Sinai, I was really looking at the aftermath of the disaster, in what was the tail end of my fortnight in Egypt. So after heading back westward to Cairo across the northern peninsula, and negotiating even more checkpoint security on the road that roughly follows the old Sinai Caravan Way, we arrived safely back in Cairo (and Africa).

Time to catch up with the rest of the world at an internet cafe, a last evening meal together at a restaurant on the Nile, then a few evening drinks and a sing-along in the hotel bar.

The following Sunday morning started with a quiet breakfast and a teary farewell. An hour or so later I was headed to the airport for a flight to Greece. That was my destination, in order to further follow up on something that is described by those in the know, as the most significant, single volcanic disaster to affect mankind, since the dawn of civilisation.

After the hectic time in Egypt, it was nice to catch breath and relax in Athens and admire the city, the classic architecture and visit the museums. On the other hand, there was a deep void after parting with my Egyptian crew and that was despite the beautiful night time view from my hotel rooftop restaurant, extending as it did, across the city to the floodlit Parthenon.

I managed a one day trip to Corinth, the canal and the old town high on the slopes, where the Apostle Paul preached to the Jewish émigrés and local converts.

The little museum too was fascinating, with remarkably modern looking surgical and dental equipment from 2,000 years ago. Instruments that I am sure the Apostle himself would have faced with the trepidation we all do, but minus the benefit of modern anaesthetics and pain relievers.

Back in Athens there was more site seeing, before turning my attention once more to my planned visit; the focal point and origin of the Exodus story. The effects are so vividly described in the Bible, and yet do not give the faintest hint that the cause is a very distant one. Nope, the location was not in Egypt, the Sinai or Saudi Arabia - although the effects were seen and felt there - but way off in the Aegean Sea and the Greek Islands.

In ancient days the Minoan seafaring culture was based in Crete, but extended to Santorini and other islands in the Eastern Mediterranean. Santorini, or Thera as it was once known, has also been fingered as the site for Plato's Atlantis.

There are strong arguments against that proposition however, and a reference to 'beyond the Pillars of Hercules' in Plato's account, does not seem to be consistent geographically, and at the very least is ambiguously described.

Plato also stated that the event occurred 9,000 years before his time. I am pretty sure that like many others, he was guilty of being misled by rumours; and that somewhere along the line, a Mesopotamian number meant as lunar months, was rendered as solar years. Given Plato was born in 427 BC and apparently received this story through his father - who in turn was quoting a historian-writer guy called Solon of an earlier era - who received the information from some Egyptian priests.... it was probably a classic case of Chinese whispers.

If I happen to be right, it means that he should have rendered the 9,000 as approximately 727 solar years before his time. It is therefore reasonable to conclude, that the Atlantis events describe something which happened in the early to mid-12th century BC and 300-400 years after the Thera eruption. Of course, if the Atlantis description had been very basic, it might have been an oral tradition dating back 9,000 years to the end of the ice age. But the story is quite detailed and therefore cannot date back to a time over 5,000 years before the invention of writing.

There is also at least one recent documentary going around, that describes the Thera eruption episode with only Atlantis in mind. Not a mention of the biblical Exodus whatsoever, despite the concept being discussed for decades at the time of writing. Sometimes you just have to shake your head.

Fortunately there are other documentaries which *do* link Thera with the biblical account and do the proposition justice.

Now I know from personal experience, that flying into Santorini provides a great bird's eye view of what a super volcano surrounded by ocean looks like - and that after most material has been blasted to kingdom come. What remains are concentric islands, some with very steep inner slopes surrounding a deep, circular water body. In the centre, more recent islands continue to grow, as pressure is applied from the magma body far below. Santorini is truly spectacular and will almost certainly explode again and again as it has in the past. It is the nature of the beast.

Once my aircraft had landed and I had gained a sense of scale, I realised at the terminal that a hire car was a necessity for a short visit. Fortunately one light blue Nissan Micra was still available, and so I took the opportunity to drive up the very steep and winding road to the ancient Greek archaeological site of Thera. There I admired the ruins perched on top of the ridge, the ocean vista and the magnificent views of much of the island.

On the road a little later, the sight of several huge cruise liners anchored inside the natural harbour, testified to the depth of water. They looked mere bath toys in contrast to the watery expanse and soaring village-topped cliffs. Then, after passing the Thera quarries on the left which expose the volcanic strata in detail - and a couple of stop-offs to take in this extraordinary sight - I arrived in the vicinity of the main town of Phira with its jam-packed lanes, multitudes of tourists, shops and restaurants.

And although we no longer lived together, I had also arranged to meet my wife and her girlfriend on the island. Their cruise ship had a short stopover scheduled to explore this spectacular piece of real estate and the tourist delights on offer.

Now with the car in town - and after spotting my cruise liner visitors at the top of the cable way - we were ready for a tour around the island to see the sights and sounds of this fascinating place. On return, there was time to enjoy some banana boat ice cream and coffee, while overlooking one of the most spectacular sights you will ever see. It was a beautiful day.

Interestingly, before the Mosaic eruption it has been suggested there was only one access by ship from the central caldera lake to the sea. So in its heyday, it must have seemed the perfect natural harbour for the seafaring Minoans.

After the eruption, all this changed and a few years ago the bough of an olive tree was found buried in one of the pumice and ash layers comprising the steep inner cliffs. The layer had previously been identified as part of the big eruption of Moses' day and an analysis of this and other remains, produced radiocarbon dates back to the 17th century BC.

Together with tree ring patterns, that can be matched to other tree ring records in Europe, this allowed one scientific group to date the strata to 1628-1627 BC; a truly startling result!

The date for the Thera eruption has been traditionally placed by archaeologists at no earlier than 1500 BC, but with most writers suggesting much later dates, covering just about every possibility within the Egyptian New Kingdom 18th and 19th Dynasty period (1550-1190 BC). These estimates are based on recorded history in Egypt, with some intuitive interpretation and archaeological evidence unearthed in modern times. But the evidence for an earlier eruption is relatively new information, and was not something to which previous researchers were privy. So naturally, they accepted the wisdom of former years and indeed the Bible reference to the 'City of Ramses'.

This ultimately suggested that Pharaoh Ramses II of the 19th Dynasty (1279-1213 BC) was the pharaoh of the Exodus.

Hollywood has popularised this view in the public mind; archaeologists assumed it (where they did not consign the whole episode to mythology) and so Ramses II was embedded as the villain for much of the 20th century. Other historians have not been so sure, and have suggested the most likely time to be in the earlier 18th Dynasty of the Egyptian New Kingdom; this with a fair degree of circumstantial evidence. In this era, the pharaohs include Amenhotep III and Akhenaton around 1390-1334 BC, while the evidence available includes the use of Hyksos / Apiru / Amu slaves for excavating and cutting stone for construction work, south of Giza. They stand out pictorially in wall paintings because of their lighter coloured skins.

Amenhotep reigned at the peak of Egyptian glory by most measures and although the god Amun-Ra was the official number-one deity at the time, Amenhotep had hundreds of effigies of the goddess Sekhmet erected during his reign. He even re-dedicated a temple he was building to Mut (Amun's partner) reassigned to Sekhmet. Because Sekhmet was the goddess of war and strife (and ironically love) it has been suggested by some, that the 'strife' could have been the event which caused the Exodus. On the other hand, he could have been inspired by an earlier event which grabbed his imagination.

Later of course, Akhenaton worshipped the Aten sun god as supreme. So putting all these ideas and influences together, it has been proposed that the dramatic change of attitude of both Amenhotep and Akhenaton was caused by the Thera eruption; an eruption which caused terrible plagues, weather disruption and extreme visual displays by day and night.

However there were other eruptions that could have caused darkness and climate chaos other than Thera, and there is evidence from Greenland of an acidic layer in ice cores (known to be caused by volcanic ash and gas) from a volcanic eruption which was dated to 1390 BC.

But where was this volcano located? One would have thought the most likely source to be Iceland.

Certainly I and other air travellers, experienced the effects of such an eruption throughout Europe in May 2010. So earlier events would surely have left their mark as far afield as Egypt and the Sinai also. Indeed, another Greenland ice core segment showed signs of an earlier volcanic eruption in 1642 BC; a few years before the period of interest here. The composition of the chemical signature has been analysed and is suggestive of an Alaskan eruption. This timing is also in the general Egyptian Second Intermediate Period, when the Hyksos reigned in the eastern delta. It is also a 'dark' period historically, due to the sparse documentation available for the period.

Thutmose III, an earlier 18th Dynasty pharaoh is another prime candidate for the Exodus pharaoh. Dubbed the Napoleon of Egypt, Thutmose III reigned from 1479 to 1425 BC and fought against northern states in the Levant. These included a battle at Kadesh and the occupation of Beit She'an in northern Israel; a town that subsequently became a permanent provincial outpost of Egypt for 350 years.

His candidacy is helped by the biblical dating to be outlined shortly, and he certainly was a tough player in Egypt and the Near East - and an aggressive one at that.

But then we cannot forget the earliest New Kingdom pharaohs Ahmose I and his brother Kamose, the last of the 17th Dynasty Theban pharaohs. Nor should we ignore their father, Tao II (Seqenenre). Seqenenre's mummy shows he died at the hands of the Hyksos and this occurred in the central delta region where his involvement in such battles is recorded. And why are the 17th and 18th Dynasties separated in what appears an artificial manner between two brothers, Kamose and Ahmose?

It is reminiscent of the previous division I alluded to, at the time of the 2350 BC flood. Another disaster here perhaps?

The period of the reign of these three pharaohs is from about 1570 to 1525 BC, with the possibility they could date a decade or so earlier. This is directly at the end of the Hyksos period, and well positioned for an earlier determination for the Thera volcanic eruption and Exodus drama.

Weeding out this historic 'noise' to identify evidence of the Thera eruption is painstaking and a second tree ring investigation has provided further evidence; this time from the ancient sequoia trees in the White Mountains of North America. This dataset points to the slightly later year of 1613 BC for the eruption, but otherwise backs up the evidence from Ireland and Europe. Taken all together (including tales of catastrophic weather and crop failures) there is an impressive case building for an earlier disaster, beginning in the late 17th century BC.

More data is always welcome of course, and archaeology gives us many further clues. It moves the event from the Late

Bronze Age back to the Intermediate Bronze Age; something that means little to the layperson, but is obviously a big deal to the expert. So it is understandable why 1627-1613 BC is considered too early by many archaeologists and historians. On the other hand, it definitely raises questions about the timing of New Kingdom involvement. It leaves the specialist and lay person alike with something of a dilemma; the job of trying to weigh up the pros and cons of each piece of dating evidence.

The eruption is potentially a key time-marker for the Eastern Mediterranean and areas beyond, with the dating of lake sediments one aspect in aligning the history of the various peoples of the region. This prospect emerges, because the Thera eruption was so massive, with estimates of up to 60 cubic kilometres of Dense Rock Equivalent (DRE) of ejecta and dust sweeping eastward around the world several times.

And not to diminish the complexity of the issues, there are still some who question the accuracy of Carbon 14 dating, either because of the potential for contamination, or because of calibration issues with the method - the latter mainly due to C_{14} atmospheric variations over the years.

For example, in regard to contamination, any carbon issuing from the eruption in the form of CO_2 or methane (C_2H_4) would have zero C_{14}, as that would have been converted to C_{12} long ago. If this carbon was absorbed by trees on the island through emissions prior to the main eruption, it could potentially provide an older C_{14} result than would otherwise be expected; this compared to trees growing on the mainland far away.

However to be an intractable problem, it would have to be extended carbon contamination over a significant portion of the life of a tree or branch, i.e. over many years. It means that matching tree ring patterns over several years with other tree ring samples from elsewhere, and dating multiple samples using the C_{14} method over the growth period, all becomes critical in confirming when the timber was buried.

This tree ring matching has occurred, but whether to a sufficient degree of correlation for an unambiguous result is the question. The more artefacts that can be dated, the more our confidence level in the data increases, and the period over which the volcano was active is likely to be constrained.

To be frank, the jury is still out. But after considerable personal research and meditation, my view is; that there is sufficient evidence to conclude that the Thera eruption began in this general period, i.e. prior to the end of the 17th century BC, but continued through the 16th century and perhaps the next.

The scientific dating, historical records and archaeological material available at writing, support this view.

Before we look at more detail from the ancient past, I will continue with the story of contemporary archaeological investigations. These began with the first known proposals, that suggested we may indeed be looking at the origins of the Mosaic Exodus in an Aegean drama.

Earlier on, historians and Bible students speculated that Mt Sinai itself was a volcano. This was prompted by the associations in the book of Exodus of smoke and fire on the mountain, and therefore attempts to identify a particular peak became a priority. The trouble was, that so many mountains in and around the Sinai Peninsula ticked that box; not because they particularly looked like volcanoes, or even extinct ones (complete with crater and vent) but because much of the mountain landscape is igneous and volcanic in origin.

Archaeological investigations on the island of Crete in the Eastern Mediterranean, began with various Italian and English investigators, but gained a big boost, when Sir Arthur Evans unearthed the famous throne room in the palace of Knossos in 1900. Knossos was the principal city of the Minoans (as he called them after a King Minos) and is located on Crete's northern coastline. It was highlighted by Homer in his Odyssey 19 and was included as one of the 90 cities of Greek legend.

As with other semi-mythical Greek accounts, it sparked interest in locating real places and potentially solid history, as had already been achieved at Troy.

These Minoans were the self-same people known as the Keftiu by the Egyptians and who traded with them, just a few hundred kilometres away under the rule of pharaohs Hatshepsut, Tuthmose III and their predecessors. A wall painting in the tomb of Senenmut (who was quite probably, the chief architect of Queen Hatshepsut's magnificent west bank temple at Deir el-Bahari, Thebes) depicts such an event.

As more places were unearthed along the northern coast, it became apparent that nearly all the towns and ports had been destroyed (and often burnt) in a great disaster. In fact so bad was the destruction, that it was first assumed that this event, or events had completely destroyed Minoan civilisation.

It was only later, that it became clear the Minoans had survived the result of an absolutely huge tsunami, but were so weakened by the catastrophe, that they were eventually swamped by immigrants from Greece. These immigrants are known to us today as the Mycenaean.

The American archaeologist, Harriet Boyd, was one of the first to recognise the all-pervasive nature of the destruction throughout the region and the likely sequence of events.

But western investigators were not particularly familiar with tsunami events in the early days, and not all were experienced in vulcanology either. It took a Greek archaeologist, Spyridon Marinatos, to recognise the abundant pumice in some of the archaeological sites and the movement of huge blocks of stone, by obviously massive forces.

He unlocked the key to the origins of this near total devastation and was also familiar with Santorini (Thera). He knew instinctively the probable origin of the disaster; a massive volcanic blast and huge eruption of ash and pumice, accompanied by absolutely terrifying, off-the-scale, tsunami waves.

A much smaller eruption and earthquake in 1926, had resulted in the destruction of two thousand houses on Santorini and many on Crete. Other Greek islands suffered various degrees of serious damage, as did Turkey and Egypt, with 600 houses being destroyed in Alexandria and Cairo alone.

The 'smoking gun' was very much revealed. But who was the first to associate this event with the biblical Exodus?

An Englishman, called John G. Bennett was in Athens and experienced a small eruption in 1925. He took the opportunity to investigate Santorini with a boatload of other sightseers. The floating pumice, sulphur, the rocks being thrown high into the air, boiling fish, noise and general mayhem, reminded him of the Exodus account and the possible volcanic origins discussed before his time. In earlier days it was assumed a mountain in the Sinai region was the culprit. Here however, he was observing something on a distant island in the Aegean Sea.

In 1947-48 immediately after World War II, a Swedish expedition conducted core sampling on a vessel called the Albatross. This was followed a decade later by an American investigation from Columbia University in 1956 and 1958. This survey was on the RV Vema, and headed by geologists Dragoslav Ninkovich and Bruce Heezen. Every new survey of the seafloor, provided evermore core data and confirmed the existence of two distinct layers of volcanic ash surrounding Thera.

The area was ovoid shaped, with deposition extending to the south east of the island between Crete and the Turkish south coast. Later it was determined much more had ended up in an easterly direction, covering parts of Turkey and depositing in lakes where it could be analysed and mapped.

The lower ash level was dated to 25,000 years ago and so was not of immediate interest. But the properties of the upper layer, exactly matched the ash that overlaid the Minoan settlement on Thera and was subsequently mapped by Marinatos.

Then in 1980 came the Mt St. Helens volcanic eruption in Washington State of the United States. This was an explosive event that cost many lives, with a resulting swath of destruction and poisoning of agricultural land and waterways downwind and to the east. It covered much of the northern United States and the southern portions of the prairie states of Canada.

The subsequent study of the interaction of the poisonous and sulphurous ash and acid rain, drew direct parallels with the 'plagues' that were inflicted on Egypt and the sequence of events described. In fact it had been noted previously, that environmental disasters which were *not* initiated by volcanic action, but by some other misadventure, had resulted in similar sequences of 'plagues' and the poisoning and subsequent degradation of lakes, streams and farmland. This then flows to the death of fish and a peak in insect populations, flies, frogs and so on, in a rolling and predictable chain of events.

So in 1981, Dr Hans Goedicke of John Hopkins University gave a lecture which compared the St. Helens and Thera eruptions, and suggested that the Thera eruption – on a much more massive scale than the St Helens one - was responsible for the Exodus saga. The New York Times printed the story and the idea went viral (as they say these days) and the rest is history.

There have been other huge volcanic eruptions in relatively recent times, and one of them was Krakatau in 1883 in the Indonesian island chain. Krakatau sits in the Sunda Straits between Java and Sumatra and the eruption wreaked absolute havoc, with a large tsunami doing a deal of damage. Unbelievably, the explosion was even heard as far away as Australia.

There was also the massive eruption of Tambora in 1915 east of Krakatau, and other explosions of large magnitude have occurred every few hundred years or so.

One of those places is Rabaul on the island of New Britain. The harbour at Rabaul has been created by a caldera flooding with ocean water, and is routinely monitored by New Guinea scientists at the local observatory. It is also a place where Australian and other overseas geoscientists assist with investigations from time to time, by monitoring the active volcanoes surrounding Simpson Harbour. This is one of the more active volcanic spots on the planet, and so seismic sensors continually measure earth movement and volcanic eruptions. In fact a few years ago, heat flow measurements were conducted as well, to better define the molten lava body under the harbour waters.

It becomes obvious that large volcanic eruptions are not unusual in certain parts of the world, and the granddaddy event of them all was discussed earlier; the Sumatran eruption of Toba circa 70,000 years ago.

But the Mediterranean is far from immune from this activity, and so back in Greece with his initial investigations on Crete completed - and scientific paper recording the results of his work well on the way - Spyridon Marinatos was eager to investigate Santorini itself and so turned his attention in that direction. The chance finding by an American visitor, of a fossilised and partially burnt head of an African monkey on Santorini, added extra impetus to the enterprise, and had Marinatos Thera bound with a magnetometer and other equipment in May 1967. His first excavations there were about to begin.

Of course, it was known that the island had been inhabited before the 3,600 year old eruption, because under the enormous pile of pumice and ash which overlaid a distinctive 4 to 5 metre layer of pumice from that event, the remains of a stone house had been revealed. This was in a Thera quarry.

However the thickness and instability of the overburden prevented any excavation, and Spyridon was forced to look for other places for his first archaeological attempt. Local landholders then described to him interesting voids in the grassed ash fields at Akrotiri to the south - and noting that some watering troughs used by animals were in fact of ancient Minoan origin - he turned his attention there.

I will not go into the excavations in detail, but suffice to say one of the most astounding and to some extent, unexpected finds of archaeology was about to be made.

Even today only a small portion of the original village has been excavated. But what has been revealed, would have made the Romans 1,500 years later, think twice about the significance of their own achievements. Under the cover of prefabricated roofing, a Bronze Age Minoan town contemporary with the Egyptian Middle Kingdom and the Hyksos Second Intermediate Period can be viewed.

Given the restrictions imposed by the island site and nature of the volcanic soils, the general amenity speaks of a well off middle class, living off the back of a trading society. The two and three storied homes, the stone walls and the timber construction of doorways and windows, the stairways and the water closets with clay pipes feeding a sewage system below paved streets, says it all. The inhabitant's extra cash was used to plaster the houses with many brightly coloured walls and artistic frescoes. They depict animals from the outside world, women in the fashion of the time and pictures of ports in the eastern Mediterranean and seascapes of home.

A fresco in the 'West House'(perhaps a captain's home) and referred to as the 'Ship Procession' is packed with details of port life and arriving vessels, crewed mainly by Nubian sailors and their ship's masters. Spectators, including women, watch from the vantage of their splendid harbour-side homes. The fresco may have once gone right around the four sides of the room. They depict a regular, roughly rectangular trading route; east along the southern coast of Asia Minor, south along the Levant coast, then to Egypt, westward to Libya, back north to Crete and the return home.

Because the early stages of the calamity gave some warning, there are no (or few) human remains to be found. However the obvious panic and subsequent partial collapse of buildings, means that there remained a jumbled pile of household goods, plaster, lamps, animal bones and tableware strewn throughout.

Other signs, such as a cracked stairway confirm that earthquake was the first warning to the residents, and so there was apparently time to embark in their boats. Ample evidence has also been found, that workmen returned after the initial eruption to implement repairs, but had to evacuate in turn, as the volcano began to rumble again.

The eruption has been classified into four stages according to the geological evidence. The first phase of the eruption, Minoan A, deposited pumice into the ocean. This was followed by Minoan B and C phases; with pyroclastic flows and lava fountain activity, tsunami generation and an ash column at least 40 kilometres high penetrating into the stratosphere. Other estimates suggest it may have been twice this altitude.

The pyroclastic flows would have topped the rim of the volcano and run down into the sea on the gentler outward slopes of Santorini. The final Minoan D phase resulted in surge deposits, lahars and ash falls and a caldera collapse producing a mega tsunami of almost unimaginable proportions. There is an ancient legend to the effect that pumice was so ubiquitous, that for hundreds of kilometres there was no sea to be seen anywhere.

This sequence is important, because it gives possible clues to the timing of events in the biblical narrative which might otherwise seem unusual. How many people ultimately escaped to safety and survived the whole episode (including the final tsunami) is a matter of conjecture. The huge waves caused by the Minoan D explosion are estimated to have been 35 to 150 metres high, when they crashed into Crete.

The height of a tsunami at any particular place, very much depends on the bathymetry of the bottom and whether an embayment or a headland. So where the waves are funnelled and concentrated into a convex beach or river estuary, it can be many times higher than might otherwise be expected. How high the tsunami was when it reached the eastern Nile delta (where the coast curves north toward Gaza) could be roughly computer modelled, given sufficient bathymetric data in that area.

Other obvious geographical considerations, such as the focussing and refraction effects from the Anatolian coast, Rhodes and the island arc to Crete (and the other major Mediterranean coastlines) would all need to be taken into account.

For now we can only guess that it was daunting in the extreme, certainly horrific, with wave heights immensely greater than the Indian Ocean Boxing Day tsunami of 2004.

In any eruption scenario, weather disruption also plays a part as the atmosphere attempts to clear itself. Atmospheric dust, electrical storms, violent wind and torrential rain are all potential by-products. At the same time, inrushing air at low altitude replaces the volcanic debris and gas which has been blasted into the upper atmosphere.

Avaris and Goshen

There is ample evidence for Semitic speaking people from the Levant living in Egypt around the time of Abraham, Isaac and Jacob. Perhaps their stories as portrayed in the Bible are stylised to a greater degree than we might think, but it is where our biblical history meets the road, as it were, and becomes embedded with other written evidence that cannot be dismissed. The inscriptions and monuments of ancient Egypt are fortunately the most enduring of all. More ubiquitous monumental, hereditary and linguistic evidence from archaeological and historic sources is impossible to find.

Even though the Second Intermediate Period was chaotic compared to the Middle Kingdom and the famous New Kingdom to follow, it is easily shown that the Canaanites and Hyksos who emigrated in, or invaded around this time, coincide with biblical history very nicely.

The definition of 'Hyksos' is surely apt and applicable in a general way to Abraham's extended pastoral family. Nevertheless it has been suggested by some pundits, that the 'kings' were not Habiru pastoralists themselves, but other leaders of northern Levant origin, such as the Hurrians. If we want to be picky in regard to our definition of Hyksos, then Abraham certainly was an Asiatic and Apiru and qualifies anyway. But he also came from Harran and the Hurrian heartland.

So from the time of Tao II (Seqenenre) who was the father of Kamose, the last pharaoh of the 17th Dynasty, and his brother, Ahmose (the first pharaoh of the 18th) we hear about their problems with the incoming Asiatics.

The Turin Canon and the Papyrus Sallier both give substance to Second Intermediate Period rulers, as do other inscriptions. There are also several records describing battles and contact between the Theban rulers and the Hyksos.

Tao died in battle with the invaders in a military campaign on the Nile. His son Kamose had more success and his battle victory is recorded on two stelae. Finally, brother Ahmose I drove out the Hyksos, according to the account of a namesake who served in his army. Whether this was overstating the case is not clear. Certainly there were many Asiatics still in the land three hundred years later.

In fact DNA analysis of the pharaohs starting from Amenhotep III, indicates they were partly descended from Anatolian and Northern Levant people. It remains to be seen whether earlier 18th Dynasty pharaohs such as Thutmose III, Hatshepsut and Amenhotep II were also.

We know from the Bible, that the city associated with the land of Goshen where the Hebrews (and Hyksos) lived, was the City of Ramses. The area is in the eastern Nile delta, centred on the most easterly Nile distributor of the day, called the Pelusiac. As hinted at earlier, because the name 'Ramses' is mentioned in the Bible account, it meant that for years people had no reason to believe anything other, than that Ramses II was the pharaoh of the Exodus. In fact some artefacts from his time were found in the locality of Tell el-Dab'a.

However much more Ramesside material was known from a location called Tanis, on the next Nile distributor to the west of the Pelusiac. Consequently, in the modern era Tanis was originally recognised as the old capital of Ramses.

But then inscriptions were found from a twenty first dynasty pharaoh called Psusennes I, who reigned from 1047-1001 BC (over two hundred years after Ramses). The inscriptions indicated he had moved the Ramses II monuments to the new site at Tanis. The reason Psusennes I undertook this considerable operation, was not to confuse archaeologists and historians in the 20th century, but because the Pelusiac branch had silted up.

Tanis therefore became the new delta capital, where Psusennes (originally a priest), became pharaoh.

And this makes sense of some of the food laws we find in the Old Testament, such as the ban on eating shellfish. As the Pelusiac waterway silted up and became increasingly contaminated by animal and human waste, any bottom feeders would have been polluted and dangerous to eat.

The modern archaeological story then threw up a rather shocking discovery. It was found that the site of the original Ramses city at Tell el-Dab'a was built on an even older urban site. We now know it by the name of Avaris (Auris) and it has been excavated over decades by the Austrian archaeologist Manfred Bietak. The site is mainly farmland today, but aerial and satellite imagery show that it was once quite extensive, extending over 5 square kilometres. This is all quite extraordinary, because it has proven to be the original city existing in the SIP and at the general time of the Exodus.

We now understand the use of the name 'Ramses', as the Exodus city in Goshen, is an example of anachronistic use. It is a consequence of a scribe updating the description of the town, with the name current at time of copying. This is simply so that every contemporary reader will know which town is being referred to (and its location) regardless of their knowledge of its previous history or earlier names. It should also be kept in mind that the alteration would have occurred centuries or more later.

In fact there is really nothing unusual about this; it is something a writer (in fact anyone) does without thinking.

Of course it would have been better to have the old name in brackets, but ancient Hebrew only used a couple of punctuation marks and well.... hindsight is twenty, twenty vision.

Unfortunately anachronisms have not been recognised by some critics of the biblical text. They then insinuate that the history outlined has been 'stolen', or concocted at a later date.

In fact, anachronisms actually authenticate the ancient historicity of the narrative, once independent evidence is forthcoming. They also give us clues of the literary history; that is, when things were edited and copied and where additions occurred.

Now Bietak's work has revealed nine separate strata at Avaris, from the earliest Stratum H, starting in 1800 BC to Stratum D/2, finishing in 1529 BC. The SIP is just 230 years long within this period, starting a little later from 1769 BC (about half way through the Egyptian 13th Dynasty) and again finishing in 1529 BC. His Stratum G/1-3 then ends and another layer, Stratum F begins at 1709 BC; a date he aligns with the beginning of the 14th Dynasty and the rule of its first king, Nehesi.

However, some of the SIP dynasties appear to overlap and although there are several papyri for the period, the data is nevertheless scarce and discussions continue in regard to the dynasties, their existence in reality, and their possible numbering.

David Rohl, another Egyptologist, has analysed Bietak's strata record at Avaris and divides them into three units:

o   The First Asiatic Settlement from 1800 to 1709 BC
o   The Second Asiatic Settlement from 1709 to 1619 BC
o   The Greater Hyksos dynasty starting at 1619 BC (Stratum E/1) until the end of occupation in 1529 BC

He suggests that the First Asiatic Settlement were the proto Israelites settling in the later 12th Dynasty and early 13th (presumably from Canaan) and that they were happy to be peacefully integrated into Egyptian society.

They were then followed by another wave of Hyksos / Asiatics who were more culturally challenged and unsettled, until finally, during the 'Greater Hyksos' period, there were waves of very aggressive hostile tribes who conducted outright military action against the Egyptian hierarchy.

His Greater Hyksos' period is more generally referred to as the 15th Dynasty, with at least five kings starting with Salitis and finishing with Khamudi around 1544 BC. It should be noted that these end dates for the Hyksos 15th Dynasty overlap the beginning of the Thebes 18th Dynasty and reign of Ahmose I.

It adds further weight to the concept of an Exodus period, where the warlike Hyksos give the native Theban Egyptians a very hard time.

So a picture is gradually emerging, but it is a jigsaw of many parts. We have the possibility of an early 1627 BC Thera eruption date and timing of the Exodus that requires a pre New Kingdom event within the 17th Egyptian Dynasty, conventionally dated from about 1650 to 1550 BC. If we incorporate the recently revised dating for the Thera eruption of 1613 BC (14 years later) derived from the North American sequence of tree ring data, we end up with the 1627–1613 BC window determined earlier. This straddles the Greater Hyksos dynasty start date of 1619 BC proposed by Rohl above.

No worries mate, hand me a beer, it's all sorted - except (as noted) dates for kings and pharaohs in this period are scarce.

It does suggest that the later Hyksos probably came from the warlike Amalekite tribes and that the Habiru / Israelites, aka Joseph and family arrived early from Canaan. Comfortingly we know that the later 17th Egyptian Dynasty pharaohs, Tao II the Brave and Kamose definitely did engage the Hyksos. What we do not know is if their predecessors did too, and whether that would align with the eruption more readily.

Unfortunately little is known about the length of reigns of the 17th Dynasty kings, but there are at least two who did not last long. One of course was Tao II who was hacked to death no later than an estimated 1554 BC and there is an earlier pharaoh Sobekemsaf I who is thought to have reigned just 3 years. Indeed even more likely is pharaoh Sekhemrewadjkhau, who may have preceded Sobekemsaf I despite the fact he is referred to as Sobekemsaf II; this due to the many uncertainties about the succession of these kings.

One scenario then (and I stress only one) is that a native 17th Dynasty pharaoh attacked the Habiru-Asiatics around 1619 BC, just as the volcanic eruption started to devastate the land in the manner recorded in Exodus. His army was then destroyed in the tsunami on a beachhead, or coastal lagoon of the Mediterranean and the Hebrews under Moses escaped south-east back into the northern Sinai. The 'Greater' Hyksos (undoubtedly including the Amalekites who Moses fought at Rephidim) then occupied the void left behind to form the 15th Dynasty under Salitis.

I seem to keep coming back to the Amalekites and they have not entirely gone away. One of the most loved Egyptian football teams, Zamalek, still does battle on the football field today.

The Moses mystery

Now we have already encountered one of the prime sources for Egyptian history in the Egyptian priest Manetho. He lived during the 3rd century BC in the Ptolemaic period. His Egyptian name was Tjebnutjer and he wrote the Aegyptiaca, i.e. History of Egypt.

There are only excerpts of his work extant, but these are quoted by several historians of yesteryear; Scholia, Plato, Africanus, Eusebius and particularly by Flavius Josephus.

Josephus was a renegade Jewish general of some rank and influence in the Galilee region in the first century. Josephus swapped sides to the Romans, when he realised that there was no point in continuing the Jewish rebellion in the first century - a move that did not endear him to his fellow Jews. He partially salvaged his reputation in so far as history buffs are concerned, when he accompanied Roman General Titus from Alexandria to the sacking of Jerusalem in AD 70.

There he was instrumental in saving as many scrolls and manuscripts as he could from the Temple library.

It is therefore via these conduits and Manetho's work, that we get the commonly used Egyptian dynastic periods; the ones used down to our current time. But as mentioned previously, support for the dynasties also comes from other sources such as the Turin Cannon, and these supplement the scarcity of data from the SIP / Hyksos period.

It is sometimes stated that there is no non-biblical source, which corroborates the Moses name. Now while there is some truth in this assertion in terms of a particular identity match - and if we disregard Manetho's testimony - it should be noted that the name in the form we recognise it, is only used of pharaohs in the late Middle Kingdom period through to the early New Kingdom 18th Dynasty.

Of the hundreds of pharaohs known throughout the thousands of years of Egyptian history, we only see the name rendered in a familiar way from 1650 to 1390 BC, a mere 260 years. They are Dudimose I and II of the 13th Dynasty, Kamose and Ahmose I, brothers at the turn of the 17th and 18th Dynasties, followed by Thutmose I, II, III and IV in the early 18th Dynasty; eight non Hyksos pharaohs before and after the very period that the Canaanites and Hyksos occupied the eastern delta and much of the Nile valley.

This is another indication that we are looking at the right time for the Hebrew visitation to Egypt, and shows that the name Moses (Mose or Moshe) was popular then and only used by Theban pharaohs during this limited period.

So what does Tjebnutjer (Manetho) have to say about Moses and his history, keeping in mind that he wrote his book over one thousand years later in Greek from previous Egyptian records?

Well he elaborates on Moses' achievements as a young man, where he is both admired and distrusted by the royal family and elite supporters. This distrust arose because of his obvious Hebrew Asiatic descent, and an Egyptian prophecy that predicted: nothing good would ever come from a river-rescued male of foreign ancestry. What we learn in addition to the biblical narrative, revolves around his time as a young general in pharaoh's army, fighting against the 'Ethiopians'. They had invaded and terrorized the Egyptians prior to the Exodus and the subsequent Mosaic events in Sinai, the Arabah and Dead Sea region.

Some of the place names mentioned by Josephus also seem to be from Nubia (Kush) in today's North Sudan. This suggests that the 'Ethiopians' were more generally recognizable as Nubians or Kushites in other references. From other correspondence, we also know that the Hyksos kings were communicating and conspiring with the Nubians, via the desert oases to the west of the Nile and thereby bypassing the Theban kingdom.

Although we might detect a ring of truth in the story and one that could easily dovetail into the biblical account, other parts of the narrative - like noting there were only seven generations from Abraham to Moses over 400 years - do not seem to add up. That is unless generation is meant to refer to a dynasty rather than a strict father-son succession - or there is some other historical issue being overlooked. There are also other chronological problems in relation to the pharaohs, such as a familial association between Ramses (19th Dynasty) and Sesostris / Senusret (12th Dynasty) which does not align with our knowledge of the Egyptian Dynasties. However, the preserved material is fragmentary in nature, and the original intent may simply have implied that Senusret was an ancestor of Ramses.

Regardless, it seems highly plausible that the young Moses had a royal leadership role and a military one early in his life.

Some have endeavoured to use new chronologies to explain seeming inconsistencies and / or vague similarities in Egyptian and Levantine history. But as our dating techniques and cross checking have improved, there is less room to move, and other explanations need to be sought. In any case, this revised (and earlier) dating emerging for the Exodus may resolve some of these difficulties.

It has also been questioned (and with some validity) why the Egyptians would persist in using the Sothic (heliacal rising of the Sirius dog star) system of chronology for thousands of years without apparent correction. Because the Sothic year was based on a strict 365 day year with no leap years applied, it meant that summer and winter solstices swapped about every 750 years.

The Egyptian calendar was effectively reversed to the seasons in that time and was therefore known as the 'wandering' or 'vague' calendar.

This would obviously be intolerable from a long term seasonal and crop planting perspective, and would have been out of kilter with the annual Nile flood each July. But perhaps the Sothic system as the entrenched method of accounting for the years, was not in fact what happened in everyday life.

In fact it could not have been, and may have been analogous to the Jewish sacred year (after Moses and the Exodus) and the civil year (which followed the Babylonian tradition) except that with the Egyptians, one calendar rotated slowly with respect to the other. Furthermore, some scholars now question aspects of Sothic dating; for example whether in fact it was corrected far more often, than the few known texts available today suggest.

I am reflecting on Sothic dating specifically, because the original dating of much of the Egyptian dynastic period was partially founded on dates established using that system.

However the chronology no longer rests on Sothic dating exclusively; Ahmose I and his immediate family are absolutely tied to the Hyksos and the start of the New Kingdom and 18th Dynasty through modern dating methods. This goes for subsequent New Kingdom pharaohs as well, with abundant material for increasingly accurate radiocarbon dating and an expanding catalogue of other archaeological and genealogical evidence.

The fact that the Egyptian Dynasties used today follow the Manetho lead and generally accord with other king lists, also suggests that the chronology is sound. It is generally agreed that writers such as Josephus and his contemporaries had many more manuscripts than we have today, and a New Testament reference in Acts 7 reinforces the idea that Moses had a very privileged Egyptian upbringing and that much was expected of him. Acts 7:20-22

"At that time Moses was born and he was no ordinary child. For three months he was cared for in his father's house. When he was placed outside, Pharaoh's daughter took him and brought him up as her own son. Moses was educated in all wisdom of the Egyptians and was powerful in speech and action." (NIV)

This short passage accords with both the Exodus and the extra-biblical descriptions of Moses impressive physique and intellect - and the fact that he was instructed in the Egyptian ways. It means that we should not be surprised when we see the attributes of the Hebrew God reflect many of the characteristics of Egyptian thought; an intellectual understanding to complement the personal spiritual experiences of the Abrahamic line.

Now that we are focussing on an earlier period for the Exodus, we would not necessarily expect the Egyptians of Upper Egypt (Thebes and Aswan) to know much about what was happening in Hyksos territory. But we do!

The evidence suggests that the Egyptians and their supporters were living virtually side by side with the Hyksos - Asiatics at the time, even though their administrative centre and capital was in Thebes. The hearsay from traders, spies and travellers would also keep them informed, together with experience of actual weather disruption caused by the Thera eruption and its global effects. The overall picture supports the view, that the Thebans had not yet been driven away from the delta by the last wave of aggressive Amalekite-Hyksos.

The famous, but fragmentary hieratic Ipuwer Papyrus from the 19th Dynasty (now held in the Netherlands) seems to point to some sort of calamity or state of decay in Egypt from earlier times. Ipuwer was a name typical of the period 1850 – 1450 BC and some analysts therefore believe, that the nature of the papyrus suggests it is a copy of a document from the late Middle Kingdom around the 12th Dynasty.

It refers to the 'river is blood' as does the Bible, but is essentially a treatise on political ethics and a lament for the state of current affairs. The text is not easily linked to the Exodus event, although it has been popular to try.

However in 1947 fragments of a stele were recovered from the Karnak Temple in Luxor (old Thebes) along with other material. The stele was later reconstructed, and the first attempts to decipher it were thought to show that it was just another military battle in which an Egyptian pharaoh was involved.

Quite recently in 2014, when the text was retranslated, it was realised that the imagery could be describing a natural calamity. It is now known as the Ahmose Tempest Stele as a result. It speaks of an apocalyptic storm, destruction by water, tremendous noise and darkness for days; all things associated with the Mosaic description in the Exodus - and indeed, with weather caused by violent volcanic eruptions.

Significantly, it was inscribed at the behest of the self-same pharaoh, Ahmose I, first ruler of the 18th Dynasty and the New

Kingdom. This follows closely after the proposed beginning of our Thera eruption and the date of his reign is now supported to within seven years of the archaeological date, by the latest Oxford University AMS C_{14} dating of short lived organic material.

This analysis of short-lived flora, includes the dating of seeds and other short life carbonaceous artefacts, from pots and tombs associated with the pharaoh.

There are also other very interesting inscriptions from the Kamose / Ahmose period. In 1954 Dr. Labib Habachi of the Egyptian Department of Antiquities found an account at Karnak, where the Theban Kamose fought an apparently successful series of skirmishes with the Hyksos, reaching almost to Avaris.

In addition, there is the inscription on the tomb of a ship's master (also called Ahmose) who fought under Ahmose the king, at the last stand of the Hyksos:

'*When the town of Avaris was besieged,*' it begins '*...I was appointed to the ship 'Appearing in Memphis'.*' It then goes on to describe the battle in the canals and matches other accounts of a city on the verge of collapse. The fighting then ends with a treaty and the Manetho description leads on from this episode. He states that no fewer than 240,000 Hyksos households with all their possessions, then left Egypt and travelled to 'Syria'.

The Bible offers two possible dating sequences from the First Temple period back to the Exodus; in 1 Kings 6:1, 480 years is quoted in the Masoretic text, and 440 years in the Greek Septuagint version. It has been suggested one may be at the beginning of the Exodus and the other after the 40 years in the desert. In verse 1 it also states that the 480th year was the 4th year of Solomon's reign in the (second) month of Ziv.

Edwin R. Thiele in his work 'The Mysterious Numbers of the Hebrew Kings', is one writer whose work has been used as a yardstick for biblical chronology. His study is based on comparisons between the regnal years of the kings of the Kingdom of Judah and those of the Kingdom of Israel, as well as synchronisms with Assyrian chronology. These include:

o The battle of Qarqar in 853 BC, the same year Ahab died
o When Jehu paid tribute to Shalmaneser III in 841 BC

- The last year of both Shalmaneser IV and Hoshea in 723 BC
- The 14th year (701 BC) when Sennacherib attacked Judah
- And 597 BC when Nebuchadnezzar attacked Jerusalem

In addition, at least one astro-archaeological observation has been used for confirmation and Thiele's work has been refined by Leslie McFall and others more recently.

So there now seems to be a general consensus, if not total agreement on the result. It means that the start of the Divided Kingdom and Rehoboam's reign in Jerusalem is fixed with some certainty at 931 BC. Based on this and further biblical references, it is adduced that his father, Solomon, died late in 931 BC to early 930 BC and David died late 972 BC or 971 BC when he was 70 years old.

Solomon took 13 years to finish the Temple, which then links us with the commemoration which is believed by many to have occurred in the mid-10th century BC; around 957-955 BC. Thirteen years prior to 957-955 BC brings us back to 970-968 BC (the fourth year of Solomon's reign) and together with 480 years to the Exodus, provides a date of 1450-1448 BC.

This date falls to the time of Thutmose III, but after the Queen Hatshepsut co-regnal period from 1473-1458 BC (Thutmose III being very young when he came to the throne).

The next pharaoh, Amenhotep II is also potentially a candidate in 1408 BC, if we use 440 years instead. He reigned from 1425-1400 BC and like Thutmose III, was also a pharaoh with a fearsome reputation.

Flavius Josephus reports a time of 592 years (or alternately 612 years) for the period. This is from an unknown source; perhaps Manetho. This would push the Exodus back to 1562-1560 BC (or 1582-1580 BC); an early to mid-16th century BC date closer to our proposed 1627-1613 BC Thera eruption. There are a few more clues in the books written in the time of Joshua and the Judges, but nothing that absolutely links back to Egypt.

Now 40 days or 40 years came to symbolise a time of trial to the Hebrews, whether it was the Flood, or in the wilderness or wherever. Given that 12 tribes by 40 years equals 480, another possibility is that the time may be symbolic and therefore indicative only.

Because of the apparent conflicting evidence, some commentators have assumed that either two events have been incorporated into one episode in the Bible, or the timeframe has been compressed overall. For example we know that Asiatics were labourers in Egypt right down to Ramses III's time, and so if the Exodus occurred earlier, it would mean that not all the Hebrews left at the time of the eruption. There are in fact hypotheses that are more extraordinary than this and frankly hard to believe, but the general idea of an initial dramatic evacuation followed by émigrés in a steady or intermittent stream over forty years, or indeed a century and a half, seems quite a plausible proposition.

In fact the largest migration may have happened under either the Ahmose or Thutmose watch, if the 17th dynasty pharaohs prove to be too early.

The Bible does not directly indicate this, but this is in fact what we have seen happen on nearly every occasion within more recent emigrations, starting with the colonising period from AD 1500 out of Europe and continuing today into that self-same continent. It is just what happens.

Not all migrants would necessarily go home to the Levant either. Some would try something new around the Mediterranean, Western Europe or further afield.

Let's face it, there were no doubt many Minoans, early Phoenician and other seafarers ready to shift people as far as possible from the Eastern Mediterranean. In fact cities and structures which have been investigated in the Western Mediterranean as possible sites for Atlantis, are often built on a circular plan with one point of entry. Where the cities are sea ports, the waterway tends to have a single opening.

This leads me to wonder whether these places were built by refugees from Thera, and if so, is it possible they were attempting to recreate the shape of their island home and its unique circular shape?

In summary; there is considerable disparity between the early Thera dates and the biblical dating and there are several possible scenarios which could explain this anomaly. They are not necessarily mutually exclusive and could include elements of the following:

- that the sequence of the eruption and therefore the plagues took a lot longer than we might imagine
- the whole story has been conflated into one account to expedite the telling
- the eruption sequence leading up to the tsunami, occurred decades later than the current scientific evidence suggests

One outcome is simply a long drawn out Exodus from Egypt, with our biblical Moses leading the earliest cohort out of the land at the time the largest eruption occurred. This was followed by a stream of refugees culminating in a Thutmose inspired evacuation at the very end.

Alternatively, the account recorded in the Ahmose Stele could have been of the earlier eruption that caused the evacuation of Thera by the Minoan inhabitants, and not the 'big bang' which destroyed the island and resulted in the biggest and most destructive tsunami. In this scenario, the eruption would have been on a scale approximating the eruption of 1926; but not the mega-blast that finished the sequence.

This would mean that the numbers that Josephus quotes would correspond to the earlier Ahmose event, and that our Moses was from the later period when Queen Hatshepsut and Thutmose III were in power - or the decades immediately prior.

This is an interesting prospect, because one or two historians have suggested that the associate of Queen Hatshepsut mentioned earlier, Senenmut, was actually the biblical Moses. He was the likely architect of her temple and the one who had a picture of the Keftiu (Minoan) traders painted on the walls of his tomb – a tomb that was never used. His interest in the Keftiu could also be significant, and perhaps this is in part because they had already been forced to evacuate the island of Thera.

The Hatshepsut Mortuary Temple at Deir el-Bahari on the west bank at Thebes (Luxor) also has distinctive architecture for Egypt and quite unlike the Luxor or Karnak Temples - or indeed that of Edfu further upstream. It has no-nonsense crisp, straight lines, more reminiscent of a modern 20th century building, and one with timeless appeal. It seems almost out of place and unexpected in its west bank setting and arguably suggests a foreign influence.

Hatshepsut is also noted for a famous trading expedition to the 'Land of Punt'. This is a term for the southeast of Arabia (Yemen today) and the adjacent Indian Ocean coast of Africa (Somalia). From there exotic plants and animals were brought back to decorate the Theban landscape. The painted pictures of this mission are still to be seen today on the walls of the temple.

The whole episode hints at a special influence. Perhaps someone who had knowledge of that part of the world and its fabulous natural treasures? An Asiatic / Hebrew taken in by Egyptian royalty and treated as a prince?

Senenmut had tombs prepared for the afterlife, as was the custom in Egypt, and one is actually in the Hatshepsut mortuary precinct. If he was indeed the biblical Moses, then the tomb would have been prepared before the actual Exodus and the catastrophic eruption and tsunami event, but probably after the evacuation of Thera and the beginning of the plague sequence.

The artefacts and statues of Queen Hatshepsut and her reign were also defaced and trashed, and her name removed from memory by a subsequent pharaoh. This was not uncommon where a royal was considered to have been an embarrassment or traitor to the nation. It has therefore been suggested, that the reason for this was that Hatshepsut was the young princess who rescued Moses (a Hebrew) from out of the river as recorded in the Bible. She would therefore have been blamed for the nightmare that eventuated and the loss of so much slave labour.

If Moses was born around 1520 BC, then the biblical events of Hebrew slavery would have been initiated from ca. 1480 BC onward and the actual Exodus would line up with the events surrounding the Hatshepsut – Thutmose transition of power in Moses later life. The fact that Senenmut's tombs show no sign of ever having been used, supports this intriguing theory.

I sense out of all this, a case for a far longer sequence of eruptive events at Thera, and a stretching of the plagues over a considerable period. However it is far from settled in my mind.

The Egyptian record is tantalising, but patchy and the scientific record very new and the interpretation evolving.

Addendum: *During 2018 a new scientific study was concluded which suggests a somewhat later date for the eruption; namely 1600 to 1525 BC. If the new study proves to be significant and align with the other evidence more readily, then some of the commentary here will require adjustment. For now the new study is briefly addressed in Appendix 6.*

But we would also expect the events to be recorded by the Asiatics, and of course they did; at least one particular group who had a penchant for writing and recording their travels. And because they were pastoralists and always on the move, they did not carry their records on clay tablets, or inscribe them on walls, but used parchment and papyrus exclusively; the latter material commonly used for at least the previous one thousand years.

They eventually called their record the Torah, and as it was added to, it became the Tanakh, or the Old Testament in the Christian Bible. It is as simple as that.

And so the Egyptian pharaoh 'who knew not Joseph' and was responsible for the Hebrew enslavement, could well have been a pharaoh from the Theban line, during one of the most poorly documented periods of Egyptian history.

He could also have lived a little later during the early Eighteenth Dynasty – someone like Ahmose I.

But then there is the other possibility we have just discussed, he could also have been Thutmose III, the 'Napoleon of Egypt'; one of the most active and aggressive of all the pharaohs of the Egyptian New Kingdom.

An early Exodus also means an extended period of up to 150 years from the Exodus to the Davidic Kingdom; but that is a problem for later. Right now there is more to investigate in regard to the biblical description of the plagues, the column of fire by night and the column of smoke by day. Also the fate of pharaoh's army and the ubiquitous cloud that covered the land.

The sequence of events is described graphically and in detail in Exodus chapters 7 to 12.

CHAPTER 7

Israel

'Then back to Cairo on our very last day
Stopping on by, the internet café
Dusk at the restaurant on the river in town
Singing 'and I miss you, like deserts miss rain'.'

The Exodus narrative

I have touched on the similarities of the Exodus sequence of events (the plagues) compared with those noted by the Mt St Helens eruption of 1980; this, despite the obvious contrast of the island centric location of the Thera eruption and its almost immeasurably greater impact with tsunamis, floating pumice and deep sea sedimentation.

The preamble to the biblical story is a promise of God to Moses, that he has heard the 'groanings' of the Israelites under slavery and he will bring them 'from under the yoke of the Egyptians'. Moses gives this message to the Israelites, but they are so without hope, they dismiss the likelihood of escape.

And then the plagues begin. On each occasion Moses gives Pharaoh warning, although he feels personally unequipped to do so. The first plague is one of blood; and the mighty Nile is turned to blood (or at least the colour of blood) as were the pools of water and containers everywhere.

The fish in the river die and the water becomes so putrid, that it is no longer drinkable.

Even so pharaoh is not moved and another plague is sent. This time a plague of frogs as Moses again extends his hand over the Nile delta. The picture is almost comical as recorded, with frogs invading homes everywhere and en-masse. The reality would have been very different.

Eventually this plague passes and the next is commanded as pharaoh again hardens his heart. This time the plague is one of gnats by the billions, followed by another of flies. As each plague peaks and passes, pharaoh gets false hope until another plague begins.

The next time the blight is one on livestock. We are not given details of the affliction, but somehow none of the Hebrew cattle were affected. This may have been because their cattle were banned from the choice (but now polluted) watering holes along the eastern distributaries in the delta.

In the following event, Moses takes soot from a furnace and tosses it in the air. The plague commences and boils break out on men and animals. This has been associated with the corrosive effects of acid rain and the pollution of water bodies and the atmosphere by volcanic ash.

The next plague is one of hail accompanied with thunder and lightning. It is the worst storm ever experienced in Egypt. But by this time some of pharaoh's officials were doing what Moses required and brought their animals inside and under cover to protect them. A plague of locusts follows, blown in by the wind. They devastate the remaining crops, fruit on the trees and grass on the ground. Nothing green remains.

The ninth plague is one of total darkness which covers the land for three days. Still Pharaoh resists and so the final plague is implemented.

Moses then rises and gives a speech, 'this is what the Lord says 'About midnight I will go throughout Egypt and every firstborn son in Egypt will die, from the firstborn son of Pharaoh, who sits on the throne, to the firstborn son of the slave girl, who is at her hand mill, and all the firstborn of the cattle as well'.'

In total, the ten plagues are a raw, dramatic and brutal account, perhaps embellished in some of the detail, but is it history? Definitely!

When compared to the Mount St. Helens and other eruptions and environmental disasters, the scenario that engulfed the region around Thera (and the observed sequences of plagues) becomes self-explanatory; a 'natural' disaster, but so much vaster (orders of scale larger) and dramatically overwhelming.

However the last plague of the death of the 'first born', demands more consideration.

One explanation proposed; is that there is a translation issue here and that the death of the 'firstborn' is likely to be due to a typo where choicest, or chosen - בָּחִיר / bachir was mistranslated as eldest - בְּכוֹר / bekor. This may have occurred at any time between the time of the event and the finalising of the square script Hebrew. Perhaps in Egyptian hieroglyphs or demotic script, or later as paleo-Hebrew - or indeed the change to the square script sometime around the Babylonian exile.

It has also been posited, that the eldest child may have had specific chores or privileges (such as getting the choicest servings of grain) which might have made the eldest especially susceptible to a disease arising from pollution. An invisible-to-the-eye ergot bacterium has been suggested as a possible culprit.

Another suggestion is that the Egyptians were so stressed after nine plagues, that they were tempted to sacrifice their most precious possessions; their eldest children. Certainly there is evidence that child sacrifice was practiced by the Canaanites, Minoans and probably the Greeks, and this is one of the saddest of all ancient pre-Christian traditions. However I am unaware of any such practice amongst the Egyptians. If it did occur, it could arguably suggest a pharaoh of Hyksos origin.

Of course, this plague could simply be one which is miraculous in an entirely unrelated way to our understanding, more than good timing and foreknowledge, but orchestrated by God directly, or through his Angels. Take your pick.

When the 'seven times' prophetic timeline is outlined in the following sections, it will become apparent that sometimes human activity such as war and siege are linked directly to events such as earthquake, by a precise prophetic number of years. Another example we will study, prophetically links a military attack with a timeline milestone thousands of years later - as does the death of an individual by a different (but easily identifiable prophetic number) to the same milestone.

These examples do not appear to be simply a matter of foreknowledge, but manipulation and planning in a way we cannot fathom at present - and may never know.

Our Omnipotent Creator has his ways, and we are told they are surely higher than our ways.

However the plagues are not the end of the story. The date for departure was set before the final plague, the Passover meal planned and the animals slaughtered on the 10th day after the nearest New Moon to the spring equinox. The doorways of the Hebrew homes and tents were then wiped with blood, so that the 'angel of death' would 'passover' their households. The meal was prepared on the second Sabbath and eaten in the evening after dark, at the beginning of the 15th day.

Obviously, everything had been readied for a secret evacuation in the night. Personal goods, carts of belongings and equipment, animals, dried and salted food, the unleavened bread ('damper' of Australian heritage) free of yeast, prepared and wrapped so it would keep for as long as possible.

The children were briefed for a quiet and sudden getaway in the middle of the night!

Two weeks after the New Moon of course is the Full Moon and the 15th as close as you can get to the middle of the lunar month. There should be ample light at night to escape into the desert, providing there was no cloud. Unfortunately there *was* 'cloud'; cloud which could obscure the moon. Cloud composed of fine volcanic ash and choking dust, from one of the largest mega-eruptions of historic times.

However it would be controlled by the wind, and that night I suggest the God of Moses allowed the prevailing wind to blow the ash away. Importantly, the sequence of events that unfolded in far-away Thera - and has been revealed by geology today - becomes significant. The final cataclysmic Minoan D explosion that caused the mega tsunami, and forced most of the volcanic material into the stratosphere, had yet to occur.

For believers, its times like these, where you have done everything possible to prepare, that you ask a short blessing and go for it; when the rest can only be left in the hands of the Lord.

Even folk who do not have a strong faith, but are nevertheless people of action, understand the importance of making a decision and getting on with it, in critical, interactive situations. This is no time to vacillate.

The scripture then records the start of the journey, albeit there are some aspects of their route which are less than clear.

Exodus 13:17-22

*17 When Pharaoh let the people go, God did not lead them on the road through the Philistine country, though that was shorter. For God said, "If they face war, they might change their minds and return to Egypt." 18 So God led the people around by the desert road toward the Red Sea.[a] The Israelites went up out of Egypt ready for battle.
19 Moses took the bones of Joseph with him because Joseph had made the Israelites swear an oath. He had said, "God will surely come to your aid, and then you must carry my bones up with you from this place."[b]
20 After leaving Sukkoth they camped at Etham on the edge of the desert. 21 By day the LORD went ahead of them in a pillar of cloud to guide them on their way and by night in a pillar of fire to give them light, so that they could travel by day or night. 22 Neither the pillar of cloud by day nor the pillar of fire by night left its place in front of the people.* (NIV)

It is apparent that from the beginning, there was no intention to travel to Canaan via the Mediterranean coast. They were to avoid the Philistine country; an anachronistic reference, because the Philistines were planted there by the Egyptians at a later date. But the intent is clear, and that was to avoid the heavily used coastal road north along the Mediterranean via Gaza; a highway known to the Egyptians as the Way of Horus.

Most scholars suggest that 'yam suph' in the Hebrew means 'sea of reeds' rather than the placename Red Sea, although in this case, the Gulf of Suez (a branch of the Red Sea) was more or less on the planned inland Sinai route.

Exodus 14:1-9; 19-20

'Then the LORD said to Moses, 2 "Tell the Israelites to turn back and encamp near Pi Hahiroth, between Migdol and the sea. They are to encamp by the sea, directly opposite Baal Zephon. 3 Pharaoh will think, 'The Israelites are wandering around the land in confusion, hemmed in by the desert.' 4 And I will harden Pharaoh's heart, and he will pursue them. But I will gain glory for myself through Pharaoh and all his army, and the Egyptians will know that I am the LORD." So the Israelites did this.

⁵ When the king of Egypt was told that the people had fled, Pharaoh and his officials changed their minds about them and said, "What have we done? We have let the Israelites go and have lost their services!" ⁶ So he had his chariot made ready and took his army with him. ⁷ He took six hundred of the best chariots, along with all the other chariots of Egypt, with officers over all of them. ⁸ The LORD hardened the heart of Pharaoh king of Egypt, so that he pursued the Israelites, who were marching out boldly. ⁹ The Egyptians—all Pharaoh's horses and chariots, horsemen[a] and troops—pursued the Israelites and overtook them as they camped by the sea near Pi Hahiroth, opposite Baal Zephon'. (NIV)

The next thing we learn, is that the Israelites were to turn back and camp near a place by the name of Pi Hahiroth; somewhere between Migdol and the sea. This is a change of plan – at least temporarily. After they camp on the edge of the Etham desert it seems they turn north instead. They are to head for the Mediterranean after all, and on this bearing the 'column of smoke / column of fire' was almost certainly in front of them as they head to the Nile delta and the coast. In this north-westerly direction they will also encounter reed seas or papyrus swamps.

Deltas also tend to prograde (grow outward) quite quickly, compared to most sedimentary deposition, and the eastern delta 3,500 years ago may well have been many kilometres inland, compared to today. It means that a coastal lagoon like today's Lake Manzala would fit the bill, and probably extended much further south three and a half thousand years ago.

In the remote chance that any archaeological remains of the journey are still preserved; including, for example the fate of pharaoh's army, we can safely assume that such remains would be buried under tens of metres of sediment and some kilometres inland from the present coastline.

Now Migdol is a name for a fort and one suggestion is that the location was the fortress of Zile, a stronghold on the delta – desert boundary. Hahiroth is described as opposite Baal Zephon. And this is where Pharaoh's army finally catches them.

Then something very interesting happens:

¹⁹ Then the angel of God, who had been travelling in front of Israel's army, withdrew and went behind them. The pillar of cloud also moved from in front and stood behind them, [20] *coming between the armies of Egypt and Israel. Throughout the night the cloud brought darkness to the one side and light to the other side; so neither went near the other all night long.* (NIV)

Now here is something which seems inexplicable, but when one follows the journey the Israelites have taken thus far, it becomes more understandable. Just as the army catches them (verses 19 and 20) we are told that the 'angel of God' who had been travelling before them on their route north (with the edge of the Etham desert on their right and to the east) is now behind them, i.e. the 'pillar of cloud' moved from in front to behind and shielded them from Pharaoh.

Well that's one way of describing it. But another way of looking at this is the most pragmatic and does not discount the protection of God. The Israelites had turned, and were now travelling on a near 180 degree reverse course to the one they had taken from Etham. They were travelling northward, or perhaps slightly west of north to the Mediterranean initially, but now they were heading east or south-east and sub-parallel to the coast. It was not the cloud that had changed direction, but they. The cloud had always been coming from the north-west.

The perception of the eyewitness of course, was different. He, or she, was describing the sequence of events relative to their direction of travel. Given we are looking at thousands of people (and potentially more according to the Bible and Josephus) it is possible that the eyewitness who recorded this event, may have been somewhere in the middle of this vast moving throng, and nowhere near the vanguard with the leadership.

They were all fixated on this strange apparition-like, column of smoke by day and column of fire by night, that was low on the horizon and beneath the atmospheric dust. Without a compass and under a leaden sky, most of the crowd would have had little idea which way they were going at any particular time. In addition a later priest is likely to have interpreted the events in a more spiritual light, or simply not understood the circumstances causing the peculiar relative movement of the column of smoke.

To me, this tiny detail indicates that the original account was written by someone on the move and not simply by a static observer somewhere on the eastern Mediterranean coast. Other scribes have almost certainly added to, and commented on the account, but were not the originators. They have interpreted the cloud as an 'avatar' of God moving from the front to behind, because to God believing non-participants, this seemed ultimately the truth.

Essentially, I am inferring that seemingly inconsequential minutiae like this stamps authenticity on the account.

Exodus 14:21-31

21 Then Moses stretched out his hand over the sea, and all that night the Lord drove the sea back with a strong east wind and turned it into dry land. The waters were divided, 22 and the Israelites went through the sea on dry ground, with a wall of water on their right and on their left.
23 The Egyptians pursued them, and all Pharaoh's horses and chariots and horsemen followed them into the sea. 24 During the last watch of the night the Lord looked down from the pillar of fire and cloud at the Egyptian army and threw it into confusion. 25 He jammed[b] the wheels of their chariots so that they had difficulty driving. And the Egyptians said, "Let's get away from the Israelites! The Lord is fighting for them against Egypt."
26 Then the Lord said to Moses, "Stretch out your hand over the sea so that the waters may flow back over the Egyptians and their chariots and horsemen." 27 Moses stretched out his hand over the sea, and at daybreak the sea went back to its place. The Egyptians were fleeing toward[c] it, and the Lord swept them into the sea. 28 The water flowed back and covered the chariots and horsemen—the entire army of Pharaoh that had followed the Israelites into the sea. Not one of them survived.
29 But the Israelites went through the sea on dry ground, with a wall of water on their right and on their left. 30 That day the Lord saved Israel from the hands of the Egyptians, and Israel saw the Egyptians lying dead on the shore. 31 And when the Israelites saw the mighty hand of the Lord displayed against the Egyptians, the people feared the Lord and put their trust in him and in Moses his servant. (NIV)

Another observation to be made here, is in regard to pharaoh's army being washed away. If they were negotiating a coastal inlet or embayment of the Mediterranean, or alternatively were many kilometres inland near a papyrus swamp vulnerable to a tsunami, I think it likely, that beyond the water body was some higher ground, which acted as a reflector for the monumentally large waves.

There would also be no quick way around the inlet, or inland water body for the vast crowd. Either the high ground or cliff face caused a barrier for them, or the watery obstacle was too large to skirt round with an army thundering on their heels.

Now water receding unnaturally and dramatically, without warning is the first sign of a tsunami. This well-known and now well described effect, allowed the Hebrews to cross where once sea had previously been. So they crossed, and then the tsunami crashed back for the first time as Pharaoh's chariots and foot soldiers followed in pursuit.

Where did they end up?

*"The Egyptians were fleeing toward[c] it, and the L*ORD *swept them into the sea. 28 The water flowed back and covered the chariots and horsemen—the entire army of Pharaoh that had followed the Israelites into the sea. Not one of them survived.*
*29 But the Israelites went through the sea on dry ground, with a wall of water on their right and on their left. 30 That day the L*ORD *saved Israel from the hands of the Egyptians, and Israel saw the Egyptians lying dead on the shore."* (NIV)

They were dragged into the sea, but then the water flowed back so that the Egyptians were also washed up and lying dead on the shore. Once you understand the classic action of a tsunami wave, you could hardly get a better description of the deadly backward and forward motion of the huge waves, and the extraordinary drawdown as the water returned to the sea.

Unfortunately the movie makers have focussed on verse 29. A good description given the tsunami wave may have been tens of metres high and the wave reflected from high ground as the next wave hit. However depicted unrealistically, owing to a lack of appreciation of the nature of the wave action.

In short, this corner of the Mediterranean, with its curving coastline turning from a west-east direction to a south-north one, may well have resulted in a funnelling of sea water that shaped one of the highest series of tsunami waves ever experienced. The waves must also have had a long (perhaps huge) wavelength between crests. It is hard to imagine that the waves were as high as those at Crete, and around the Aegean, where anyone near the shoreline was unlikely to live to tell the tale. Nevertheless it all very much depends on coastal geometry.

Another part of the description which raises questions is: 'that night the LORD drove the sea back with a strong east wind and turned it into dry land'.

Obviously the east wind is not going to influence a mighty tsunami in any material sense. However it should be noted, that in shallow lakes in the area, a strong wind can cause 'wind set-down' and move a substantial amount of water causing it to bank up on the opposite shore. The writer, or more probably a later copyist without any experience of a tsunami, has assumed that the strong east wind (which was noted by eyewitnesses) must have been the cause.

In fact, the east wind will be an effect due to a combination of the prevailing wind pattern, together with the induced effects of the volcanic action at Thera blasting ash, dust and gas into the stratosphere. The low altitude wind then would race in to fill the void left by the rising mass over the volcano.

It will therefore be coming approximately from the east in this eastern Nile delta and Sinai location. Elsewhere it would have been coming from other points of the compass, depending on their position relative to Thera. At the volcano itself the wind would be spiralling in violently from all directions.

So this whole event and the positioning of the Hebrews at this time, was of course, just a very happy coincidence?

Given my own feeble experiences in life as a believer, I think it highly likely that a 'word of knowledge' or similar indication guided Moses. But we are also guided by circumstances and that would have been a significant part of it.

Originally it seems he planned the direct desert route across the Sinai, but then decided on the detour to the coast. Pharaoh's surveillance troops noted the change in direction and interpreted this as the 'desert hemmed them in'. Moving toward the main coastal road had its own dramas (with the hostile forces behind and the unknown ahead) as they negotiated the beach / swamp terrain. They were trapped and it was time to commend their predicament to the Lord.

It is worth noting that it was usual for writers of history to put words in the mouths of their heroes, and so what is attributed to Moses and recorded, may not have been exactly what was said. I accept that this is likely. Playwrights still do this in dramatic presentations as a part of their trade; it is their job.

The dialogue is not designed to deceive, but enhance the impact of events, with the likely exchanges that would have taken place at the time.

In 'Commentary on Exodus' Thomas Dozeman analyses the Exodus narrative using modern techniques; this in order to unravel the provenance of the sources which make up the book that has come down to us today. For his analysis on the initial flight out of the Nile delta, Dozeman uses two sources; one he calls the 'p source' and the other simply the 'non p source'.

Now there are many propositions and hypotheses that can be put forward; but what I would be looking for and have suggested above, is that there is an original eyewitness account, with later commentary by translators and copyists designed to put the scene in context for the faithful.

Indeed, providing more background as to the when, why and wherefore, as well as expanding on the spiritual lessons that can be learnt from the ordeal.

One thing is for sure, the complex phenomena surrounding this extraordinary experience have been recorded so precisely and sequentially, that it would not be possible to invent such an intricate, fictional narrative. It is absolutely amazing in my opinion, that such a detailed scientific description exists from so long ago. Simply put, it is a graphic and vivid eyewitness account, provided by a trained observer.

In any case, once the event was over, it became obvious that they had been saved that day, by an act of Providence that was so unusual and unique, it was bound not to be forgotten by those that experienced it, or - as it happens - by the next one hundred generations, over three thousand five hundred years.

Return to the land

In the previous section, I determined that the Israelites turned north shortly after they left Succoth and their initial departure from Goshen (Avaris) and that this occurred when they reached the edge of the Etham desert. To pharaoh they appeared hemmed in by the desert, and so their final encounter with his army occurred within kilometres of the Mediterranean coast. It was also roughly in the direction of Thera, with the column of smoke by day and column of fire by night in front of them.

Arriving at this point on the trail, it becomes obvious that there are many possible routes through or around the Sinai for the journey to Ezion Geber. Because it was located due east at the head of the Gulf of Aqaba, the route initially backtracking to the south from the coast and then through the middle of the Sinai, might not seem intuitive at first. But on the other hand, with pharaoh's army now out of the picture, they may have felt safe enough to retrace their steps and take the most direct path.

This route across the Sinai Peninsula and away from the Mediterranean was almost certainly intended from the start, and before their detour via the coast. As suggested, the detour was probably prompted by a divine word of knowledge, but also reinforced because Moses feared that the Sinai Caravan Way was too obvious, and the route which pharaoh would immediately assume they had taken. Nevertheless the route to Ezion Geber and the famous King's Highway - and eventually Canaan - was the least of two evils given the dangers of the Gaza road. Unfortunately it was still a popular route used to enter Egypt by all sorts of traders and travellers - and worryingly - militia.

We have seen how the Amalekites seem to fit the Egyptian description of the military inclined Hyksos; the ones who were encountered toward the end of the SIP and the Avaris settlement. Bands of militia were still around in the 16th century BC, despite the military successes of Kamose and Ahmose. Manetho describes these people as particularly unsavoury, because they and other Asiatics were a menace to the Egyptians and used the ancient Sinai Caravan Way to the Gulf of Suez to enter Egypt.

Undoubtedly, the Israelites were trying to avoid such traffic, but nevertheless needed watering holes and springs, and would have followed main roads as much as a large contingent allowed. Both the Bible and Josephus (quoting the Egyptian Manetho) subscribe to a very large multitude of over a million.

I am fairly sanguine about this, in the sense that I think it entirely possible that the exodus of Hebrews and their associates at the time, may have been considerably less and even down to tens of thousands. In this scenario, the million and more would be the total number of émigrés over the decades, following the initial dramatic departure.

Of course I do not know this for sure, but as I hinted at earlier, I do think it worthwhile comparing with other mass migration stories, including more recent ones such as the refugee crisis out of Syria. Imagining realistic possibilities based on precedent and the likely decision making coming into play, will only facilitate our quest for the truth of the matter.

There is a detailed record of the Exodus track in the book of Numbers. The sceptics will discard the record completely, but even though there are one or two obvious anachronisms, they help support the idea of a reliable ancient version which has been subsequently edited. The problem is; that of well over a score of names of staging posts, finding modern counterparts is nigh on impossible. It suggests that Davidic era scribes and later copyists / translators from the Babylon exile, had lost contact with the place names and locations along the track.

For the most part, this is probably because they were Hebraic names given by the Mosaic mob en-route for recording sake. They may have had little reference to existing Egyptian locations anyway - if indeed there were any for most of their stops.

There are many other references to the Exodus event and places along the route throughout the Bible, but while they give support to its historicity, there are also enough differences in the various accounts, to cause difficulty in reconciling ancient place names with later ones. This is why the route taken is such a point of contention.

The Numbers 33:1-11 description starts with their movements over those first days up to the Mediterranean and the 'reed' sea tsunami:

'Here are the stages in the journey of the Israelites when they came out of Egypt by divisions under the leadership of Moses and Aaron. ² At the LORD's command Moses recorded the stages in their journey. This is their journey by stages:

³ The Israelites set out from Rameses on the fifteenth day of the first month, the day after the Passover. They marched out defiantly in full view of all the Egyptians, ⁴ who were burying all their firstborn, whom the LORD had struck down among them; for the LORD had brought judgment on their gods.

⁵ The Israelites left Rameses and camped at Sukkoth. ⁶ They left Sukkoth and camped at Etham, on the edge of the desert. ⁷ They left Etham, turned back to Pi Hahiroth, to the east of Baal Zephon, and camped near Migdol. ⁸ 'They left Pi Hahiroth[a] and passed through the sea into the desert, and when they had travelled for three days in the Desert of Etham, they camped at Marah.' (NIV)

'They marched out defiantly in full view of the Egyptians...' hardly accords with a secret departure at night. But if we ignore this apparent contradiction (and example of discord in some of the texts) it seems, after the close call of the tsunami, they more or less immediately left the Mediterranean coast near Migdol.

Perhaps after pharaoh's army was destroyed, there was an opportunity to stride out 'defiantly' close to Goshen, as they passed by on their way to the head of the Sinai Caravan Way.

In any event, they were now back into the desert on a roughly south-east reciprocal course to their route from Sukkoth to the coast. The difference seems to be, that this time they were travelling further east through the Etham Desert.

⁹ 'They left Marah and went to Elim, where there were twelve springs and seventy palm trees, and they camped there.
¹⁰ They left Elim and camped by the Red Sea.[b]
¹¹ They left the Red Sea and camped in the Desert of Sin.' (NIV)

Now this mention of the Red Sea in verse 10 is interesting. Is it referring to a papyrus swamp (yam suph) or are they now actually at the head of the Gulf of Suez and on the Red Sea coast we know today?

The tip of the gulf is believed to have been further inland (north-west) than today and this extension remains as isolated, salty inland lakes - some of which have now been intersected by the Suez Canal. So both may apply and we are looking at a Suez Gulf coastal lagoon.

From here the record indicates 24 consecutive legs of the journey - whether one day at a time, or many is not explained. At the end of the period they arrive at Ezion Geber; the port at the head of the Gulf of Aqaba (Eilat).

As suggested, the place names seem largely archaic Hebrew, but names like Ezion Geber are known from the time of Solomon and may have been revised at that later period.

¹² They left the Desert of Sin and camped at Dophkah.¹³ They left Dophkah and camped at Alush. ¹⁴ They left Alush and camped at Rephidim, where there was no water for the people to drink. ¹⁵ They left Rephidim and camped in the Desert of Sinai. (NIV)

Eight legs after leaving the coast at Pi Hahiroth, they were into the Sinai Desert. We note one leg entailed three days travelling, so it is possible that over a month had passed, when they arrived in the desert near a mountain called Sinai. In fact Exodus 16:1 tells us they were six weeks into their journey at the Desert of Sin and were therefore likely to arrive at Sinai, by the end of the second month:

'The whole Israelite community set out from Elim and came to the Desert of Sin, which is between Elim and Sinai, on the fifteenth day of the second month after they had come out of Egypt.' (NIV)

With such a large contingent, there is only so far they could have travelled and the idea of going via the south-east Sinai Peninsula in the vicinity of Egypt's Mt Sinai and Mt. Catherine, would not have been an inviting prospect. It certainly would be possible, because the distances are quite modest. But with the obvious choice of taking the direct route along the Sinai Caravan Way and getting the journey done and dusted quickly, the long way round would not be enthusiastically received.

My view is that the journey via the southern Peninsula is only possible, if we consider the Mediterranean Sea tsunami and 'crossing' scenario I favour and have described.

The idea that they crossed the Gulf of Aqaba at its southern entrance at the Straits of Tiran (or at the much hyped Nuweiba 'land bridge') and made it to Jabal al Lawz in Arabia as many suggest, is implausible and in my view not worth consideration. And just to recognise this fanciful indulgence for what it is, this has to all fit into the first three or so stages out of Avaris, with pharaoh and the Egyptian army hot on their heels.

That pharaoh would take his army this far away from the eastern delta, seems unlikely in the extreme, given the unstable situation which would have remained. With Amalekite militia attempting to pour into Egypt, it makes no sense at all.

To compound the situation, no east wind would make one iota of difference to sea levels at the Tiran Straits or Nuweiba; either in two hundred metres, or eight hundred metres water depth respectively. But then, I have already admitted to going nearly half way around the earth to check these things out; and that was after spending more than a few hours with satellite imagery, studying this unlikely possibility.

So who am I kidding? Still a visit does concentrate the mind. Yes, I can vouch for the scenic route along the Suez Gulf coast, the amazing mountains of the south-east of the Peninsula and the laid back Sawa Beach resort north of Nuweiba - but that was by mini coach. While the views to Saudi Arabia and Jabal al Lawz care of my own eyes (and a 300 mm telephoto lens) were as close as I came to that part of the world.

So a Gulf of Aqaba 'red sea' crossing appears impossible given the time constraints, logistics and external threats. But given a 'yam suph' crossing near the head of the Pelusiac deltaic distributary, it would be entirely possible to take practically any route on the Sinai Peninsula inside a few weeks.

The Hebrews' next hurdle was the attack of the Amalekites at Rephidim. This apparently occurred before they moved into the Sinai proper.

While at Rephidim, we learn that Moses was able to command the battle from the vantage point of higher terrain. We are also told, that there was no water there, and this would have exacerbated the strained relationship between the two groups.

Now some have suggested that Rephidim is in the southern Peninsula region and associated with Wadi Feiran. To me this seems unlikely, given that the Amalekites would have surely taken the direct route, and as armed militia, had little reason to feel threatened by other travellers. They may have already heard that pharaoh's army had been destroyed.

There is a place called Galala Mountain relatively close to Suez, which is a plateau now being urbanised and could have featured in the Exodus saga. If not this particular site, then somewhere on the main range which extends up along the Suez Gulf coast from the south-east corner, where Jebel Katrine and Jebel Sinai are located. This range with its famous ancient mines, also stands out in satellite imagery and potentially provides a biblical Rephidim and of course, Mount Sinai.

So moving beyond Rephidim into the Sinai desert, we come to a mountain, or high place where the action of the biblical Mt Sinai took place. The Hebrews obviously spent some time here, so there must have been enough water and some provisioning available. If they headed eastward along the Caravan Way, the location is not likely to be too far off the ancient trail, but provide some privacy and protection.

On the other hand if they decided the Sinai Caravan Way was too dangerous, with further Amalekite traffic on the way, they may have headed south-east along the Suez Gulf.

Mount Sinai is also where they meet Moses' father-in-law Jethro and his small Midianite party. It suggests communication and prior arrangement and a reason not to deviate too far off the planned track. In any case Jethro and his family were used to travel in desert conditions, and no doubt were familiar with the watering holes and routes available to them from Ezion Geber to Egypt. They would also have been aware of alternate routes.

Moses' father-in-law gave him great advice on all manner of things, in terms of the law and management of the people. On top of that, he would have known the lie of the land as well as Moses himself. So if the biblical Mt Sinai is much closer to the head of the Red Sea / Gulf of Suez region than not, and close to the Egyptian end of the Caravan Way, then Jethro would have been in a position to advise Moses and Aaron on the direct route and whether further Amalekite bands were a threat.

On the other hand, if Mt Sinai is south, closer to the traditional site in the Southern Peninsula, it would mean that Moses had already decided to go that way and had arranged with Jethro to meet him there before they left Goshen. Of course there could also be combination scenarios; with for example, Jethro taking the Caravan Way directly from Ezion Geber to Egypt and then directing the Israelites south on a safer route - with most of their time being spent at Sinai itself.

I have a slight preference for the direct Sinai Caravan route, with deviation as required to oases and places of refuge for a large contingent. On the other hand, the traditional Egyptian site of Jebel Sinai and Jebel Catherine, can be accessed from the Wadi Feiran, a 130 kilometre valley with a 5 kilometre oasis. This oasis would have been an ideal place for an extended stay and away from direct traffic from Asia to Egypt. The book of Numbers lists another twenty legs for the Hebrew journey from Sinai and therefore suggests a considerable way to go, until they arrive at Ezion Geber at the head of the Gulf of Aqaba / Eilat.

This unfortunately provides few clues, with both the direct caravan route and the journey from the south-east Sinai location being much the same in length. It has to be admitted however, my musings in this regard are largely based on the account in Numbers and the many stops noted in that book.

If this is not accurate, or I have missed something, then the Mt. Sinai of the Bible may be much closer to the Negev or Arabah. In this case, the oasis at Ein el Qudeirat could be where the Hebrews camped. Either way, a location such as this could well have been a waypoint on either route.

Jethro would have briefed Moses and Aaron on alternate oases and diversions in case of emergency, with a plan to hold and take stock once in Midian territory around Ezion Geber and the Arabah. So the score of stops defined in Numbers after the extended Mt Sinai break, may have been one night wonders with the occasional longer rest at an oasis. On arriving at Ezion Geber, Jethro would have had some political clout, and could guarantee some empathy and accommodation for the Israelites; this, without too much rancour amongst the locals.

After a few months, or a year or two, everyone's patience would have been wearing thin. But then that's life and sometimes it is better not to know the future.

The journey from the port of Ezion Geber was then northward up the King's Highway via Kadesh - and then on to Mount Hor where Aaron died. Hor is another name for mountain and so does not help us in terms of location, but it was obviously further north beyond Kadesh. Reliving Moses' troubles of 3,500 years ago, we find it is most likely from Kadesh Barnea and Edomite territory, that spies are sent out to Canaan to check out the inhabitants, their military strength and preparedness.

On one hand there are glowing reports of a land 'flowing with milk and honey', on the other, of giants and insurmountable obstacles. The imagery is still used today from the pulpit, as a lesson to us 'to keep on, keeping on'.

There has been a long history of individuals attempting to identify Kadesh, and those who have investigated the archaeological sites in the area include Charles Woolley of Ur fame and the amazing T. E. Lawrence (Lawrence of Arabia). Lawrence wrote his memoir 'Seven Pillars of Wisdom' about his involvement with the Arab revolt against the Ottoman Turks in World War 1. That saga was also set in this part of the world.

Both Woolley and Lawrence suggested Ein el Qudeirat was the Kadesh site; the large Oasis in the southern Negev mentioned above.

Other analysts say Kadesh was located at Beidha, a site in the Petra area, a few kilometres to the north of the town. Some of the early church fathers, such as Eusebius and Jerome, thought that Kadesh Barnea was at Petra, or close by and associated with another place in the area called Cades, the 'Fountain of Judgement'; referring to the judgement of the people.

Centuries earlier, Abraham seems to have had a connection through the placename known as En-mishpat, or the 'Spring of Judgement'. This reference found in Genesis 14:7, is the episode of the Abrahamic story we also encountered in the Ebla tablets of Northern Syria. It ties with the Cades description:

> *'Then they turned back and went to En-mishpat (that is, Kadesh) and they conquered the whole territory of the Amalekites, as well as the Amorites who were living in Hazazon-tamar'.* (NIV)

In Petra's more recent archaeological history, there was an Egyptian grave identified with typical Hebrew style grave goods. If the account is correct; the grave goods included a walking staff engraved, or painted with hieroglyphs.

Neither should it surprise. This area described as the 'backside of the desert' from an Egyptian perspective, is close to Egypt's border as recognised from the earliest Dynastic times.

It is also a place where you can imagine Moses striking a lime covered supersaturated rock face with his staff, and water rushing out; perhaps in the Siq, the Outer Siq or somewhere in the district. Of course it was said he did it in anger and maybe it was a party trick as well. Remember, according to the Exodus narrative, Moses almost certainly had been here before.

Another account, this time from the 1960s, is the existence of a large stone-column sculpture of a snake in the Petra vicinity. It was subsequently destroyed by the locals, because some hippies were seen to venerate the statue - or so it was interpreted by the Bedouin. Again, the story is reminiscent of the 'brazen serpent being lifted up in the wilderness' as per the timeless biblical account.

Summing up: it is obvious that nothing definitive has been agreed upon by scholars, but I have a preference for Petra and am comfortable joining with the church fathers on that one.

It is an impressive and unusual place. It is one of those outdoor spaces where people who otherwise would not darken a church door, will nevertheless acknowledge a sense of being, and oneness with nature.

The Nabataean, who built the impressive town in the gorge we visit today, saw its potential and its unique atmospherics. It was once complete with dams and water conduits, colonnaded streets, squares and horticultural gardens. They also built an altar and standing columns (similar to the description of the two columns in the Jerusalem Temple) for worship. This site is high up, on what is now known as the Attuf Ridge.

So the Israel contingent would have spent some time at the Gulf and in Midianite territory south of Kadesh, and then after arrangements were made, moved on to the Kadesh Barnea region for the rest of the 40 years. From there they camped at Mount Hor, where Aaron passed away and was buried.

The King's Highway runs along the eastern side of the Arabah north-south depression, the Dead Sea and Jordan River. This deep north-south valley is all part of the Rift, which eventually morphs into the Beqaa Valley in Lebanon and the East Anatolian Fault Line further north. In terms of ethnic groupings, the highway extended northward through the territory of Midian, Edom, Moab, Ammon and Amor in what is Jordan today, and then beyond to Syria and onward toward Mesopotamia.

So surprise, surprise, we find that somewhere on one of the legs of their journey, there was a 40 year stretch that is otherwise skipped over in the Numbers 33 narrative.

Fortunately Deuteronomy 2 tells us that the Israelites were held up for 38 years between Kadesh Barnea and Zered Creek further north. This is in the region of Moab (Jordan today) and the text gives two reasons for the delay, and there may have been other issues as well.

Firstly, Israel was warned not to harm the Edomites or Moabites, because this territory was assigned to them by God. They were cousins from way back and therefore told to go gently.

Secondly, the narrative suggests that it was not the Lord's intent to allow the Hebrew slaves who left Egypt, to enter Canaan anyway; a penalty for unfaithfulness and lack of application is implied.

In addition, there are other episodes in regard to the Sinai and wilderness narrative - such as the slaying of the Midianites referred to in Numbers 31 - which are somewhat disconcerting.

This last event is not easily reconciled with the concept, that the Exodus Hebrews were aided by the Midianites en route to Canaan without fuss. It seems to indicate some sort of falling out and double cross; some issue which arose after the Israelites had largely passed through their territory and were adjacent to the Jordan. It suggests that time had taken its toll and Egypt's escaping slaves were no longer the friends of the Midianites.

We are also told, that some time later, an Amorite king would not let them pass through his territory either, and that they were compelled to destroy that opposition as well.

From a practical point of view, it could be an indication that the Israelites had nowhere near the numbers initially, to fight pitched battles with determined adversaries. However over the 40 years their growing numbers and desperation overcame their fear, as further refugees arrived in the King's Highway region.

So it is certainly difficult to understand exactly how the period unfolded, but attitudes and relationships were bound to change for the worse, as they wore out their welcome.

This reminds me of the 'bright sparks' on social media who poke fun at poor Moses and depict him as some sort of loser, going around in circles in the desert. They either impose a pathetic figure of a man on a map of the Sinai and Jordan, or paint a picture describing the direct route on the main road from Suez to Ezion Geber and then on up to Amman and the border crossing into Israel. They then show the estimated travel times by motor car and forced march to add to the ridicule.

Travel time by car is only 12 - 14 hours driving and split roughly in half on the two legs. A few hours extra would be required for comfort stops and more for border crossings.

Mind you, that is providing you had all the papers and were the right ethnicity. No big deal for an American, Australian or any other traveller driving on good roads on a major continent. Walking is another thing. But a disciplined hike by fit individuals could arguably cover the journey in a couple of weeks. A month or two would not be enough for some of us.

But the point has to be made, that from 1948 when the State of Israel was created, until 1994 when the Allenby–King Hussein border crossing was reopened, no one could cross the Jordan–Israel border legally. Even now only tourists and Palestinians from the West Bank are allowed through. That is a 46 year wait because of the geopolitical situation.

A similar scenario is shaping up right now for Syrian refugees living in camps in northern Jordan. Some of them have been there for 6 years and potentially face a decade or two before they get to move on, or go home. It is too often a commonplace occurrence, and as many will be aware, some Palestinian refugees have now been stuck in Lebanese camps since 1948.

So yes, it is a bit of a laugh at Moses expense, but realistically the situation is not uncommon for many refugees today in the region - or around the world.

It is also so much harder when moving a whole population; children and old folk, frail and infirm, animals by the hundreds or thousands, carts, tents, equipment and utensils, food and water, a mobile worship centre - and the list goes on.

Now attempts were made early on by archaeologists, to determine the places in Jordan that the Israelites went through. Unfortunately excavations to date have not revealed anything of interest from the period.

However there is one ancient city, in the West Bank area, that is thought to date back to as early as 8,000 years ago. It is one of the oldest inhabited cities currently known. It is also the endpoint of the Exodus saga before entering Canaan.

It is the almost mythical city of Jericho in the Jordan Valley. Numbers 33:47-53

[47]{} They left Almon Diblathaim and camped in the mountains of Abarim, near Nebo.
[48]{} They left the mountains of Abarim and camped on the plains of Moab by the Jordan across from Jericho. [49]{} There on the plains of Moab they camped along the Jordan from Beth Jeshimoth to Abel Shittim.
[50]{} On the plains of Moab by the Jordan across from Jericho the LORD said to Moses, [51]{} "Speak to the Israelites and say to them: 'When you cross the Jordan into Canaan, [52]{} drive out all the inhabitants of the land before you. Destroy all their carved images and their cast idols, and demolish all their high places. [53]{} Take possession of the land and settle in it, for I have given you the land to possess. (NIV)

The modern archaeology story of the 20th century, vividly illustrates the difficulties encountered when interpreting the archaeological record. This is a story of the enlightenment of modern science and discovery, but also suppositions and popular biases, which can affect both ordinary people and otherwise objective researchers.

Firstly the Bible was not correctly interpreted. The record that has come down to us is now recognised by some scholars, as the product of an original narrative, but one that has been subsequently copied and interpreted according to the need of the contemporary readership of the copyist.

To that end, I have mentioned anachronisms frequently and examples such as the City of Ramses and the name Ezion Geber, and even Israel (in respect of the Mosaic Hebrews) are likely examples. They are names that were apparently in use in the First Temple period ca. 1000-800 BC, but not necessarily 600 years earlier. So when the archaeologists of a couple of institutes were working at Jericho in the early to mid-twentieth century, they understood they were looking for signs of an attack around 1250 BC when Ramses II was pharaoh.

Now I have already described, how two hundred years or so later, Pharaoh Psusennes I (1047-1001 BC) moved the monumental architecture of Ramses from the location of the old city of Avaris (Qantir / Tell al Dab'a) to Tanis on the next Nile branch to the west. It is therefore extremely unlikely, that the use of the name Ramses was perpetuated for the old Goshen city, very far into the Davidic Kingdom.

Unfortunately the guys excavating the walls of Jericho knew nothing about Avaris. The city was not discovered until much later, and a lot of other pieces of the jigsaw were not available in the 1950s.

I imagine they were aware of some evidence for the Hyksos and Apiru expulsion at the end of the SIP and early 18th Dynasty, but there seems to have been no switched-on linguists or historians at the time of the digs (1920s to 1970s) with sufficient interest to question the status quo.

The rest is a story of Hollywood, popular opinion and the power of myth.

So when Kathleen Kenyon and her colleagues interpreted the stratigraphic sequence (and being awed at the overall age of Jericho that it showed) a prominent burnt layer and collapsed walls dating around the mid 16th century BC mark, were interpreted as being far too early to attribute to Joshua's earthquake aided attack on the city.

King for a while

Back to my own personal journey, with Cairo and Egypt now well in the past. Not to be forgotten ever, but after a hectic time in Greece, Santorini Island and Turkey gaining firsthand experience of slightly more recent biblical locations, there was an itinerary to follow, and it was time to head for the Holy Land.

It was not quite plain sailing though, as my promised and paid for, limousine driver decided to take his cash and run – but without me. I knew he was not a local, but a blow-in from somewhere in the Middle East and I had wondered about his dependability. The reason I was concerned, was that on our previous trip together, he kept repeating how reliable he was and what a great reputation he had in the trade.

Hear that a few times on one fairly short ride and it sticks in the back of your mind. It makes me smile in retrospect; more amusing than annoying.

So while waiting in the foyer of the Mosaic Hotel, I gave him as much time as I could, until it was obvious he was not going to show up. It was then back to reception to grab another taxi, and fortunately I was soon away for the journey to the airport. It was a Sunday morning and the whole world seemed to be on the move at Ataturk International. Despite this, I was soon lined up for Turkish Airways flight TK786 for Tel Aviv.

These days you see a lot of airport security, but I had not experienced anything quite like this. We went through three separate body scanners as we stood in the queue, and this was before we took the bus out to board the jet. My passport was taken twice for inspection, which seemed a bit unusual. However it was only back in Oz later, that a scandal surfaced about Australian passports being photographed by Mossad (the Israeli Secret Service). It was then I began to seriously wonder, whether I had a namesake engaged in Israeli intelligence somewhere.

That episode is the sum total of my 007 adventures, except for... well, the time... but that is another story and no international incidents resulted.

So we arrived safely at Ben Gurion International in the early afternoon and I was soon at my Tel Aviv Hotel with time to take a relaxing afternoon walk; northward along the beach to begin with, then on through a park or two and back around the city. I had eight days to look forward to in Israel, followed by a further three days in Jordan and was full of anticipation.

The effects of the Icelandic volcanic eruption had died down and had not been a problem flying around the Aegean, but at the time there were also attempts by activists to run an Israeli maritime blockade on Gaza. Not surprisingly, there was a nervous guide or two who were intent on reassuring us not to worry too much about the TV news - and the rockets that might fly.

Basically the advice was, 'turn the box off and get outside and enjoy your visit'. So I did.

The following day, the coach headed south east through fields of thriving crops of one sort or another, including canola (if I remember correctly) and then on to dryer country further south. Australian eucalypts occasionally lined the road and eventually, we turned just before Be'ersheva and headed more easterly to Arad for a coffee break. The country was hilly here and we drove through forestry plantations, before plunging abruptly down toward the Arabah and Dead Sea. The aridity and heat hit like a sledgehammer.

At one point, I dragged out my handheld GPS and camera and recorded the altitude as we drove north on the lake shore. I am not sure 'altitude' is the right word; it read three hundred and eighty metres below sea level!

And so the day passed leisurely at the lakeside and in the water. A brief visit to a cosmetics establishment followed and then the impressive cable car ride up to Masada and its incredible history. Finally north again, back along the lakeside and past occasional groves of date palms in the arid setting, before finally heading back westerly and the climb up to Jerusalem.

If you come from a large country like the USA, Canada, Russia or Australia, it is about now you realise how tiny Israel actually is, and how it would be impossible to defend the place without military control of the West Bank. This is especially so, as the western boundary of that territory goes within a dozen kilometres or so of the Mediterranean coast, just north of Tel Aviv. In a day spent mostly out of the bus on foot; sightseeing, eating, swimming, relaxing, shopping and in a cable car, we had travelled from the coast, south to the fringe of the Negev Desert, around Hebron and old Judea, through the ancient town of Arad, along the Dead Sea to Masada, past Jericho and back west to Jerusalem. It is a pocket handkerchief sized country.

In fact, in the state of South Australia we have a cattle station called Anna Creek ('ranch' to the Americans) which is a little larger than the whole of Israel.

So ended our first day and I was booked into the Jerusalem Gate; a large hotel with a shopping mall below. It is a busy place with tourists, locals and more than a few black attired Orthodox Jews, wandering in and out for meals and gatherings of one sort or another.

The next day was occupied sightseeing from the Mount of Olives, visiting Gethsemane in the Kidron Valley and touring the Old City. It included a pleasant stroll around the southern wall, contemplative time at the Western Wall, old Roman ruins, the King David memorial and finally; walking the Via Dolorosa through the bazaars to the Church of the Holy Sepulchre.

At one point, I was totally entranced by an unexpected treat on a corner. Two musicians were playing a haunting Israeli melody, with school children gathered around. And there was one thing more; on the bus we had passed the Jerusalem Military Cemetery at Mt. Scopus on our way to the Mount of Olives. So I elected to walk up there from the Old City and then make my own way home. It was a brief, but poignant visit.

Here were the well-kept gravestones of 19 and 20 year old's who had died in 1917 liberating Palestine (as it was known) from the Ottoman Turks. They were all British, but from every part of the Empire including Australia. I was well aware of the Australian Light Horse role at Be'ersheva and would have liked some time there, but Mt Scopus was probably the place to remember the campaign and the fallen.... Lest We Forget.

There was one more rather unexpected surprise, for those of us who chose to have some entertainment later that evening; a delightful amateur, but very skilled musical presentation at the old YMCA auditorium.

It really finished off a memorable day in Jerusalem.

Next morning was a more expansive tour of the city and surrounds, including the Holocaust Memorial, the wonderful Shrine of the Book (the Dead Sea Scroll exhibition) followed later on by a drive to Bethlehem. It was then through the wall; swapping our Israeli guide for a Palestinian one, in order to see the Church of the Nativity. We toured the town and soaked up the atmosphere of this special place. It is not always possible to do this as a tourist, but we were fortunate.

There was also a sense of emptiness; the local Israeli population are barred from doing this trip. It is off limits to them and the wall is an ugly reminder of the threat of terrorism.

That evening back in Jerusalem, I chose to walk down the hill into the city centre. There was a pop concert in one of the squares and I enjoyed that for a while, before moving on a little further looking for a coffee shop.

I found a well presented one on a corner. It was reminiscent of your typical British styling, with lots of lovely dark timber work in the interior decor, furniture and fittings. The timber-topped counter was unusually long, given the size of the place, with a gentleman serving behind who appeared to be the owner.

I asked for a coffee and chose a cheesecake from the display beneath the counter. He then asked me something which I did not quite catch, so I asked him to repeat it.

'Your name please... so I can call you when your coffee is ready.'

"Oh" I said, "David". Not giving a second's thought to my reply.

He looked straight up, and as it turned out, changed his mind and walked right around the back of the counter and then turned toward me. Given he had just asked for my name, so I could come and collect my coffee, I was taken aback.

He then led me to the very best seat in the cafe overlooking the courtyard outside and told me to relax and enjoy; waited until I was comfortable and returned shortly after with my perfectly presented order. This was the second time on my trip, and in more decades than I could remember that I truly appreciated the name my parents had given me. I was in no doubt; this was the City of David. I felt like a King.

Note: *I know I should not encourage others to use an alias in order to get exceptional service in town, but it pays to have the right name when you come to Jerusalem.*

The following day we were back on the bus, picking everyone up from their hotels again (including the King David) and off down the road to Jericho once more. This time we were headed north along the Jordan Valley (all West Bank country) and gradually the landscape changed from near desert to green farmland, as we headed for Beit She'an.

This is another truly ancient city with a tell going back to distant Canaanite times, but with a colourful history since then.

It is also the town Thutmose III made his provincial outpost back around 1450 BC and where some of our biblical characters lived and died. Later the town was given the name Scythopolis by the Greeks - and just as we were becoming totally immersed in a historical reverie - the crack of a military jet breaking the sound barrier jolted us back to the present; a frequent occurrence judging by the lack of interest on the part of the locals.

Later that day we visited Nazareth and the famous church site, and further up the road managed the precipitous climb to Safed; another ancient fortress town perched precariously on a mountain. Today it is a picturesque art and tourist centre, which also has an interesting Kabbalist connection.

After looking through the galleries, taking in the history and grabbing another coffee, it was back to the Jordan Valley again and north toward Lebanon. Our accommodation at Kibbutz Kfar Giladi proved to be a comfortable and well-appointed retreat, with spacious public areas. A large under-cover swimming pool and extended recreational facilities were a feature and a lovely place to relax. Close to the border in a picturesque location, it was our home for the next couple of evenings.

The following day was a most pleasurable one for an outdoors boy; circling right around Galilee and following in the steps of Jesus. The Sermon on the Mount location was our first stop, followed by a fishing boat ride to the sound of music, by local lads with a flair for history and the theatrical. It was then a short drive to Capharnaum (Capernaum) the home of Mary and where Jesus also spent time. The old basalt buildings from the 1st century can still be viewed, with the central focus a home that was apparently that of the disciple Peter's extended family.

The limestone synagogue was built in the mid fourth century, by Greek speaking returnees from the north and their local brethren. The whole site is utterly steeped in history from a New Testament and late Roman point of view.

We then stopped by the Jordan for lunch and watched Coptic water baptisms being conducted, before heading east to the Golan Heights. It was then onward past military vehicles hidden in valleys and gullies, and on up to a lookout under a military outpost overlooking Syria. After some time surveying the scene to the east, it was back in the bus driving through picturesque country and the Druze villages near Mount Hermon.

Further on we descended the steep and winding road back to the valley, past the ancient remains of Dan and finally to Kfar Giladi once more - another beautiful day.

The last day's drive was south and then westward across to the coast, to the delightful old Crusader port city of Acre (Akko) where this old mapmaker got lost for a while. The labyrinth of historic passageways and fort structures was too much for me, although that embarrassment was forgotten in the beautiful maritime setting that is the Akko waterfront.

Then down to Haifa and Mount Carmel of Prophet Elijah and Old Testament fame, with its rather spectacular views of the port, Baha'i gardens and terraces.

Further south we stopped by Caesarea Maritima, the old Roman port with its amphitheatre by the sea and famous plaque authorised by Pontius Pilate of New Testament infamy. Finally it was back to Tel Aviv, with just enough time for a brief visit to Yafo, the biblical Joppa (Jaffa) in the early evening.

Yafo is situated immediately south and within easy walking distance of the city along the beach.

It was the old port for Jerusalem which also featured in the Jonah story. These days it is a relaxed centre for artists and their studios and a lovely historic place to wander around.

The following day a coach arrived for the next part of the adventure; this time headed for Jordan via the main road to Jerusalem. Once more we did the rounds of the hotels there and picked up more travellers. It was then back down the winding road past the Bedouin camps on the hillsides, Jericho and Beit HaArava for the third and last time.

Everybody else on board was returning to Israel, but I was scheduled to fly out of Amman's Queen Alia International Airport to Britain and the last leg of my holiday.

We crossed at the Allenby / King Hussein Bridge and border gate; a rather lengthy process through security on both sides. The coach ride then continued up the winding valley on the east bank, skirting Amman and heading north to the ancient site of Jerash. Founded by Alexander the Great as a holiday and retirement location for his troops back in the 4th century BC, this is another city with an interesting history.

Quite a remarkable place, restored after the terrible earthquake of AD 749 which also hit other Greco-Roman cities throughout the Levant, including Beit She'an which I had visited only a few days earlier. This area was all part of the biblical Greek Decapolis (10 cities) of Jordan, Syria and northern Israel which were established from the 3rd century BC onward.

There is no doubt that Jerash must have been very impressive in its heyday. The amphitheatre is still functional and we were treated to an instrumental ensemble of a drummer and - wait for it - bagpipe player. I was later to learn that the bagpipes are actually indigenous to this part of the world, probably including Egypt, and the connection with the Scots who perfected the modern instrument is pretty obviously a story of migration (or travel) persistence and plain bloody mindedness.

The educational side was a demonstration of the acoustics of the auditorium; the 'sweet spot' in the centre, allowing the orator or entertainer to speak in a near normal voice and be heard anywhere around the amphitheatre.

Our destination for a couple of nights was Amman where we spent some time touring the older historic areas. We also passed new high rise buildings, attesting to the robust modern service industries which underpin their economy.

They are not blessed with oil reserves like most of their neighbours and the refugee population has always been a large one - but they seem forward thinking nonetheless. Unfortunately since my visit, their economy has been stressed by a vastly increased refugee population and is now in crisis.

The next day our target was well to the south at the most famous of Jordan's tourist sites and indeed a world famous destination – Petra. Travelling south the countryside soon became arid and flat, but after several rest stops on the way we arrived at the hilly and quite mountainous Shara Mountains.

The Siq; the narrow ravine barely wide enough for a horse and cart is remarkable and of course, the keyhole view of 'The Treasury' at the southern end, truly legendary.

The water works and reticulation that allowed this city to function is very interesting from an engineering perspective, and a scaled up version of similar storage and conduits seen at Masada near the Dead Sea. The Nabataean architecture is unique, but with influences from everywhere: including Egypt and Persia, and it would not surprise if Greek trained artisans were influential in its design and construction.

As the Outer Siq broadens, some of the older cliff-carved buildings and structures still remain, with sepulchres and tombs featuring strongly. However they are not restored in the same manner as the inner ones and eventually the Siq opens out even further to the Wadi Musa, an area that has a Byzantine era building with mosaic floors; a feature of that era and prolific in Jordan. There were many other points of interest beside.

It is also an area large enough for a whole nation to camp in a vastly older time period. But the Wadi Musa name meant little to me at the time. After all, it seems every second town in Egypt, the Sinai, Arabia and the southern Levant has a Mt. Sinai or a Wadi Moses; and the idea that I might be looking at a real location for a significant portion of the Exodus account, was not a serious consideration at the time.

After a solid day exploring, we reluctantly made the journey back to Amman with a relaxing meal stop on the way. The following day our destination was the pretty town of Madaba, where we visited the tourist shops and the famous church there.

The church has a Byzantine era mosaic map of the Middle East and Mediterranean embedded in the floor; one which was fortuitously revealed some years ago during restoration work. It provides an insight into the lengthy Byzantine period.

Finally we made it to Mount Nebo where Moses is said to have seen the Promised Land. At 120 years old, he did not live long enough to travel 'over Jordan' with the now Joshua led Hebrews, but I think Moses would have been near enough to have satisfied himself that his faithful efforts were about to bear fruit. Ours, on the other hand, was not a clear day; the volcanic dust of the Icelandic eruption stubbornly persisted. Still, a glint of sunlight reflecting from Jerusalem sparkled in the haze.

I was fortunate as well. Although I was parting company with everyone on the coach and they were travelling back to Israel, I was provided with a chauffeur driven limousine to Queen Alia International airport and my flight to the UK. No offence to the Israelis, but I had never travelled on a royal airline before and my five hour flight to Heath Row on Royal Jordanian capped off my time as a Davidic 'king' in the Near East.

David and Solomon

The history of Israel (commencing with the Kingdom of David and Solomon and the Divided Kingdom that followed) is well documented in terms of the kings, priesthood, prophets, places and events. So too the Second Temple period, leading up to the events of the Christian era and the AD 70 sacking of Jerusalem by the Romans. The remains of many villages and towns from 2,000 years ago have now been rediscovered as well, augmenting our knowledge from the scriptures, other Jewish texts and secular sources of the period.

However the archaeology of the pre-Babylonian exile / Kingdom period is rather sketchy and therefore, open to multiple interpretations. The points of view espoused, depend very much on the background of the individual researchers and writers involved. Nevertheless, over the last century or two, biblical sites, characters and events that have never previously been substantiated outside of the Bible, have been confirmed. However a tug of war has developed between the 'minimalists', who see little evidence for a separate Israel presence in Canaan prior to the exiles and the conservatives who see maximum support for a traditional biblical view. Of course as would be expected, there is a variety of opinion between the extremes.

Because of this situation, I will focus on and summarise items of interest from the Israel Kingdom period, and share some thoughts on the implications of an early entry to the land.

It is a moot point that we talk about 'Jacob's other name being Israel' or the 'Israelite Exodus out of Egypt', when the Israel name (or something very much like it) may not have been used until around the time of pharaohs Ramses II and Merneptah. The key consideration is that the people we understand by that name, have a written history which is totally compatible within the wider context of immigrants from Anatolia / Syria into Egypt in the early second millennium BC. People, who along with others, made their way to the eastern Nile delta only to subsequently return to Canaan during a once off, historic natural disaster recorded in minute detail in the Bible.

Since we are re-evaluating the evidence for a Hebrew return to the land, and one as early as the first half of the 16th century BC, there is the obvious problem of accounting for something like a hundred and fifty years from the beginning of the 18th Egyptian Dynasty to their return to the Jordan Valley and Judean and Samarian hill country.

However this early return to Canaan is tempered somewhat, by the prospect of an extended Exodus into the mid-15th century BC (and the Thutmose III era) when the pharaoh initiated his campaign into Canaan and beyond. In addition, it is worth remembering that the Hebrew were mainly shepherds and tent dwellers, and lived their lives around animals and their upkeep. When a Bedouin moves his tents, no matter how substantial and lavish inside, there is not much to show for it in the landscape after he has pulled up his tent pegs and moved on. There is also no real reason why it would be easy to distinguish remains from an Israelite tribe, from that of other Canaanite pastoralists.

On the other hand, there were also Canaanites who were well and truly city dwellers, and had been so since the third millennium BC. They lived in moderate to large cities like

Hazor, Ai and an early Jerusalem and would have been far more impacted by Egyptian incursions up the Gaza coastal route.

In the Bible we have the stories of the Judges period found in that book and also the book of Joshua, but generally little archaeology that can definitely be called Israelite to confirm the picture. However the excavations at Jericho, as well as other locations on the east bank of the Jordan Valley in the region of the Dead Sea, support the early biblical narrative and the ruins of Sodom are believed to have been found at Tall el-Hammam - a site which appears to have been abandoned around 1540 BC.

In the main, the biblical narrative is focussed on the Tribes themselves and their immediate neighbours, and we get a picture of the concerns of the Israelite leadership of the day and their aspirations. The Egyptian presence would obviously be a constant, but background concern, and would have suppressed the activities of all Canaanites. Keeping their heads down and maintaining a low profile while the Egyptian, Assyrian and Hittite Empires raced into battle along the coastal highway was a no-brainer. When centres such as Beit She'an became the local centre of Egyptian power, it did not leave a lot of wriggle room for city dwellers or the hill tribes to flex their muscles.

It seems a loose association of Judges and tribal chiefs was the best the Israelites could do for the time being.

Following Tuthmose, Amenhotep III was one of the pharaohs known to leave his mark in the land and this is described below in terms of excavations at Gezer. Later, Rameses II (The Great) the self-same pharaoh who built his reputation through warfare and monument building from the Sudan to Anatolia, reaches a stalemate with the Hittites further north in 1275 BC and becomes a force in the land.

In Pharaoh Merneptah's reign (1213-1203 BC), a stele written in Egyptian hieroglyphs, identifies a number of Canaanite groups. One name in particular may well be the Israelite people. The Merneptah Stele, also known as the Israel Stele was discovered in 1898 at a site on the west bank at Thebes (Luxor) by Flinders Petrie; the early British archaeologist. It has been regarded as his greatest discovery.

Now if we are looking at a time centuries before 1203 BC as the date of the Exodus, then the stele was inscribed at a time when the Israelite tribes were no longer directly under the Egyptian yoke. They needed to stay out of the way of Egyptian power though, and therefore could not become the significant national identity of later years under David and Solomon.

During this late pre-kingdom period, we can therefore surmise from both biblical history and archaeology, that there must have existed a practical coexistence with the other Canaanites and with Egypt; this despite religious differences and cultural taboos. In addition, the story of Ruth indicates that first and foremost Israel was a faith and ideal driven people, rather than an ethnic cohort. Intermarriage was not out of the question if respect and worship were acknowledged by the foreign party to Yahuah, the God of Israel. At least this seems to have been the official position based on the biblical record.

The Merneptah Stele was first translated by Wilhelm Spiegelberg, and in Petrie's 1897 publication 'Six Temples at Thebes' it is described as 'engraved on the rough back of a stele of Amenhotep III'. It had obviously been removed from his temple, and placed back-outward against a wall in the forecourt of Merneptah's own temple precinct.

On the stele, the Egyptian god Amun is shown giving a sword to the king for his efforts, thereby glorifying Merneptah's victories over his Libyan enemies and their Sea People allies.

In the context of our story here, the final two lines mention a campaign in Canaan where Merneptah says he defeated and destroyed Ashkelon, Gezer, Yanoam and Israel. While the hieroglyph grouping for 'Israel' has been disputed by some as being too generic to apply to particular itinerant tribes, the association with Canaan cannot be dismissed, with the Israelites being seen as part of that company of people living in the hill areas.

This Ramses II / Merneptah time in history (13th century BC) is significant in terms of other archaeological discoveries in the Holy Land. In the 1950s eminent Israeli archaeologist Dr Yigael Yadin excavated the site at Tell el-Qedah north of the Sea of Galilee. This was the old city of Hazor and here he found evidence of a violent destruction of the city ca. 1240 BC which he attributed to the assault described in Joshua 11; an assault which apparently occurred immediately after the capture of Ai.

Both Hazor and Ai are ancient sites going back to the Early Bronze Age, with Hazor also being the largest city in Canaan at its height. Although Hazor and Ai may well have been attacked by Israelites, we now know that the Hazor attack was well after the initial destruction and capture of Jericho. So if Joshua was the hero of the early Exodus years, then it appears that a namesake, later Israelite leader, or someone else altogether has to be responsible for the 13th century BC military action. It could have been anyone, even an Egyptian pharaoh such as Ramses II.

This is a problem compounded by the possibility of an early Exodus, as seems likely from current evidence, but it clears up a lot of misconceptions that archaeologists were working under through much of the 20th century; i.e. that Rameses II was the pharaoh of the Exodus. It also explains why there was no kingdom in the days of the Judges. Why they kept to the high country, Jordan Valley and east of the Jordan and therefore out of harm's way from those on the coast. And above all, why the Israelites remained mobile and semi nomadic, preferring the guerrilla lifestyle rather than opposing the empires directly.

Even if Joshua was part of the Thutmose III saga (and the 40 years in the wilderness should be associated with that time in the mid-15th century) a direct association is closer, but still not close enough. It is also possible that Joshua's attack on Hazor was an earlier event to that of the destruction revealed by archaeology. That would also provide a solution of sorts.

However it is difficult not to come to the conclusion, that some of the Bible account may be condensed for the period (a conflated timeline) with too much being ascribed to Joshua.

The famous Sea People arrived in the Eastern Mediterranean around the time of Merneptah's reign. They were on the move because of the onset of drought, earthquake and war which began a 'dark age' in the Eastern Mediterranean that lasted more than 300 years. They came from the Mediterranean islands with Crete being a likely home - or perhaps Anatolia - and with a Libyan confederate attacked Egypt. They therefore may have been the Keftiu of earlier times, and / or the Caphtorites mentioned in the Old Testament.

In fact Keftiu (Egyptian) and Caphtor (Hebrew Semitic) are probably the same people and again the similarity of the consonants suggests this. In any event, the result of this campaign meant they ended up in the Southern Levant and Gaza region (and later in northern Israel) where they were known as the Peleset (Philistines). They become the famous biblical adversaries of Israel in the Davidic era, and Philistine cities of the 10th century have now been excavated, and accord closely with what we read in biblical descriptions.

How on earth extreme minimalists discount the biblical record in this regard, is a little perplexing to say the least. Nevertheless a King of all Israel was only a dream for years, and any nation so formed was always going to incorporate the existing Canaanites if it was successful. That is simply what happens when one dynasty becomes dominant and the rest are incorporated into the new regime. Consequently a fragment of Mycenaean, or Phoenician pottery unearthed in recent archaeological investigations, is unlikely to tell us whether the owner was of Jacob's line or from one of his neighbours.

Eventually though, the Kingdom of Israel had its glory days and reached its height of power and influence with the reign of Solomon. To a great extent the height of its religious and spiritual power too, given that Solomon implemented his father's vision for a permanent Temple in Jerusalem. The site chosen was on the mountain site at Jerusalem, where Abraham was tempted to offer Isaac as a sacrifice to God.

Little has been found archaeologically in regard to the First Temple, but some small clay seals have been recovered from a Western Wall dig and other artefacts from nearby sites are believed to be Temple related. Water tunnels and cisterns similar to those described in the Bible have been discovered and one or two specifically identified. At least one has been dated to the First Temple period, and the location of the original City of David is believed to lie south of the current wall (Figure 3).

In regard to the famous Solomon's Mines, the location has now been excavated north of the Gulf of Aqaba in the Timna

Valley. This was Edomite territory and archaeological investigations at the site, show that the mining was a joint Edom and Israel effort. Solomon apparently had considerable control of the operations; even supplying food from Israel. This included fish from the Mediterranean and northern Israel rivers.

The artefacts discovered also indicate that the copper smelting artisans were treated as elites, when compared to their comrades who had more mundane support jobs.

At Dan in the far north of Israel, an inscription on basalt was unearthed in 1993-94. On it a Damascus King boasts of a victory against Omri the king of Israel and his ally the king of the 'House of David'. It is dated to the 9th century BC.

The Mesha Stele is a stele also set up in the 9th century by a Moabite king. It recounts how Moab had been subject to Israel, but was eventually able to throw off the Israelite yoke.

A similar story about the battle of Qarqar in 853 BC (the same year Ahab died) is also recorded by the Assyrians on the Khurk Stele. It refers to 'Ahab the Israelite' as a member of a coalition of twelve involved in the battle. Ahab apparently contributed 2,000 chariots and 10,000 foot soldiers to the engagement, where the Assyrian king Shalmaneser III claims victory.

Victory or no, the battle seems to have set Assyrian military ambitions back some years when viewed in historical context.

Even so the 9th century was not a great one for Israel, with trouble on several fronts. It perhaps indicates the beginning of the steady decline following the Solomon era and the troubles which lay ahead.

The archaeology of Israel and the Davidic Kingdom era has revealed all sorts of associations recorded in the Bible. How the Canaanites / Israelites used to have their own personal 'arks' for votive purposes; little boxes in which they housed their favourite deity. One presumes a devout Israelite who disdained the Canaanite idols had a box without an idol; perhaps a tiny copy of the Law, or just an empty vessel to show he worshipped the unseen God.

In other locations, square or rectangular topped altars have been found with four horns at each corner. These are typical of the descriptions of altars used by the Israelites in the Bible.

Because of idolatry amongst the Israelites, sometimes these horned altars were used for Canaanite / Babylonian worship. Amos, a prophet of the Ten Tribes describes a case where the horns were smashed by a king bent on destroying idol worship.

Amos 3:14:

"On the day I punish Israel for her sins, I will destroy the altars of Bethel; the horns of the altar will be cut off and fall to the ground."
(NIV)

The Bible also describes an altar at Dan and archaeologists have unearthed the remains of horned altars at Megiddo and south at Be'ersheva as well. These are from the 10th to 8th centuries BC and quite large at nine feet wide. They are decorated with squiggle decoration, that distinguishes them from plainer altars found at Arad and elsewhere.

Y. Aharoni, the archaeologist who excavated the site at Arad, suggested that this altar may have been dismantled when sacrifice was centralised at Jerusalem. An interesting thought, but accurate dating of its final use is not available.

A few years ago, a copy of the Epic of Gilgamesh was also found in Israel dating back to the 8th century BC. It shows that the Flood story was known before the exile period, although that of course should not come as a surprise. The Ten Tribes were never far away from Assyrian influence and geographically adjacent to the Euphrates and the behavioural sphere of Mesopotamia. From scripture alone, we know this was a constant gripe of both the prophets of Israel and the prophets of Judah during the Divided Kingdom days.

The Near East is the most direct connecting point between three continents and the battle space of the ancient empires. It means everything has been destroyed many times over.

Whatever evidence remains of ancient architecture, inscriptions and artefacts in a city like Jerusalem, is unsurprisingly buried under multiple structures erected since, together with immense quantities of fill from multiple demolitions. In addition, much of the current Old City was erected during the Islamic period extending from the early 7th century until the 20th. This understandably has reduced the very early archaeological evidence available. The city therefore has been built up like a 'tell' over the millennia and no doubt hides many secrets below.

One day not too far away, I anticipate much more material will be found; probably astounding things.

This reminds me of an interesting incident. The old Eastern Gate from the Second Temple period was reportedly seen and photographed a few years ago when an archaeology student accidently fell into an Islamic tomb. The ancient Gate is apparently directly under the walled up one we see today.

The latter was built by the Muslims when the current walls were constructed in the Islamic period, but was then subsequently blocked up in 1541 by Sultan Suleiman the Magnificent (obviously no introvert) to prevent the Messiah, or some other 'prince of the covenant' from entering on his return. This episode as per prophetic indications in the book of Ezekiel.

History and prophecy go hand in hand in the Holy Land!

Earthquake is the obvious agency that will reveal such secrets and that will certainly occur, according to the visions of the Jewish prophets and their end-time predictions. But this will require considerable destruction within the Old City and on the Temple Mount. It will be a topic of further discussion.

Outside Jerusalem, at the sites of smaller Canaanite cities and towns, excavations have been more successful and in some cases have yielded surprising results. For example, at the biblical city of Gezer located between Tel Aviv and Jerusalem, archaeological investigations have been ongoing and the Davidic Kingdom period city well defined. However recently a Late Bronze Age (14th century BC) Canaanite city was found lurking beneath the known ruins. It was discovered by a team lead by Dr. Steven Ortiz and Dr. Samuel Wolff.

This hidden, pre-Davidic city was destroyed during the Egyptian 18th Dynasty period and the new Gezer subsequently built over the top. It was apparently a strategic stopover between Egypt and Mesopotamia.

Pottery vessels, cylinder seals, a large scarab with the cartouche of Amenhotep III, together with some Philistine traded goods, attest to the city's existence and Egyptian New Kingdom influence. In addition, the Tel el-Amarna letters found in Egypt describe the wars between the Canaanite city states of the time and either these, or an Egyptian campaign, probably resulted in the 14th century city's demise.

Now before we study the prophets, I would like to turn to one of the big issues of legitimate concern. It is the lack of any paleo-Hebrew or Canaanite scriptures from the Bible to be unearthed in Israel from the pre-exilic period. This, as opposed to everyday inscriptions in the paleo scripts, some of which have been found for the corresponding period.

The Hebrews used papyrus and leather parchment exclusively it seems and by necessity as itinerant pastoralists. No doubt there was considerable pride in that achievement, even when they settled down in Jerusalem. Scratch pads of wax-coated wooden tablets were also used and reused for notes. But there was a catch – the scrolls were not as robust as inscriptions on rock, deteriorated rapidly by comparison to tablets in a library, and therefore had to be copied regularly.

So the relatively short history in the land and the nomadic nature of that association up until the Kingdom of David, obviously continued to constrain their writing to perishable and easily carried materials. This was in contrast to the large and powerful empires, who had the time and money to have their more important papyrus texts inscribed on clay or stone as well.

One of the earliest Egyptian papyrus texts comes from the building of the Giza pyramids around 2550 BC. So we know the Egyptians also used texts and drawings on papyrus, and did so prolifically. Amongst many other things it allowed them to easily carry plans and instructions around the country for their monumental architecture. Egyptologist Kenneth Kitchen estimated that about 99% of the papyri dating from 3000 BC through to the 4th century BC has been lost or destroyed.

So this is also a problem for our quest for Israelite or Phoenician documents pre-exile. And although the square script was introduced during, or as a consequence of the exiles, the early paleo script remained in use for centuries in some parts. Fragments were found amongst the Dead Sea Scrolls dating from the first century BC and other examples date even later. This suggests it was prized, as might well be imagined, amongst religious and conservative groups.

It would seem an obvious priority at the Babylonian sacking of Jerusalem, for the priests to hide and make safe treasured scrolls from earlier times. After all, before the final siege in 588 BC there had been earlier interventions in Jerusalem by the Babylonians. Even kings were deposed and substituted for others and Jewish elites including prophets, taken into captivity. So it was not for want of previous warnings, that they knew they had to act decisively in order to secure their heritage.

To take care of everyday secular dealings would have been one thing for Israelite society, but the preservation of sacred texts and artefacts, both hereditary and holy, would have been something else altogether. They would have protected them with their lives and hidden them carefully, using at least one secret location and preplanned arrangements; no personal sacrifice of priests and officials would have been too much.

They obviously managed to copy these early texts; otherwise we would never have ended up with the square script Hebrew Bible with the Abrahamic, Exodus and other narratives.

If they hid the old scrolls written in paleo-Hebrew, it would most likely have been in Jerusalem and Judea, but it could have been anywhere, and localities such as the Dead Sea area, Mount Nebo and Petra have been suggested. Even Egypt is a possibility. The question is, were the originals stashed away and could they still exist today?

It would be a fabulous discovery if something remains to be found and one lives in hope. But as comforting and as interesting as it would be to have pre-exilic biblical texts in paleo-Hebrew, it is by no means essential. We have what we need in the later Hebrew and Greek translations.

Now, I have presented a very condensed version of the Israel archaeological heritage, in order to move on to the prophetic visions of the writers toward the end of the kingdom period.

Despite this, I hope that the summary above gives some continuity that adds to the overall picture of Israel and the subsequent prophetic interpretation to follow. The significant time in question is the brief few years at the end of the Divided

Kingdom. This was when the Kingdom of Judah (or David) remained based at Jerusalem and the Kingdom of Israel (the Ten Tribes) had already been taken captive by the Assyrians and removed from northern Israel and Samaria.

The visions of the Jewish prophets of the Babylonian exile and the timeline of events they reveal - along with the New Testament prophecies which describe so vividly end time events and national identities - will be the centrepiece of this work. It is what is most important for us today.

The essential relationship between the Judaic visionaries and their Creator God will also be laid bare and some of the pain, torment and sacrifices they made will be touched upon.

In the end, the written history of the people (while fascinating) is not why we are focussed on Jerusalem and Israel. It is about a practical faith and spiritual growth, which can be experienced and proven by anyone who has a will to do so. Once we understand that the prophets had very real spiritual experiences, we can begin to appreciate the visions, inspiration and their prophetic predictions.

The Ten Tribes and their prophets will be on the periphery, having been taken captive 130 years earlier and moved en-masse to Assyria. From there they were forcibly spread around the country, while Judah on the other hand, only had their elite removed to Babylon and that for a mere seventy years.

The dispersion of the northern tribes may be the reason, that their prophets and spokespersons do not feature in defining the chronology of future events. Perhaps it was always meant to be the royal role of Judah and their Jerusalem based elite anyway.

For whatever reason, at the time of the exile it was the prophets of Judah who provide us with timelines and milestones to show how history will play out. The prophecies are Jerusalem centric, but worldwide in scope.

The successive empires which would rise and fall, the final empire called Babylon the Great and the apocalyptic events of the 'end of the age', all come from them. It is a theme reinforced with much more detail in the New Testament.

The prophecies reveal the numbers and the plan of God.

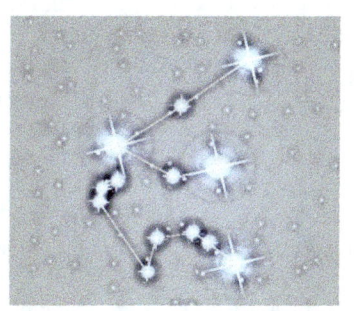

VISIONS OF ELOHIM

CHAPTER 8

A Prophet or Two

'This is what the Lord says, he who made the earth, the Lord who formed it and established it...'Call to me and I will answer you and tell you great and unsearchable things you do not know'.'
The Prophet JEREMIAH

To whom it may concern

If you have persevered with the journey I have outlined 'In Those Days' and you are a believer and strong Bible advocate, you may be shocked at this point and dismayed at the naturalistic explanations given within the context of the Creation, Flood, Babel and Exodus narratives.

How can there be any pretence at defending the basic tenants of the Judeo-Christian faith?

That perhaps I have shown that the stories outlined are based on some sort of actual history – but what of miracle? In other words, the mystery of ancient biblical events may seem to have been stripped significantly of their magic.

What I believe the previous chapters ultimately illustrate, is exactly what we find the Bible teaches in so many places. That God is master of all His creation and uses every aspect of it to bless, test and discipline His people. This includes what we call 'nature'; the natural world in all its diversity, with its wonders and disasters of every kind.

The miracle from our point of view is often in the timing, as well as the power, extent and drama of the event.

The miracle is also in the special protection through all things afforded the 'chosen'. Those souls who go about their daily lives in an unselfish and faithful manner; little works of God, however humble and practical they might seem to be to the casual observer.

The world of the supernatural and miracle are first and foremost the things we do not understand. They are things beyond our ken. We think we have some understanding these days of the natural world that our predecessors did not - and that is certainly the case. To them the more dramatic and unusual things which they experienced; such as mighty floods, volcanic eruptions, earthquake, landslides, meteorites and strange astronomical events - all appeared miraculous.

Essentially this has not changed in our day. What appears to have happened is that the dividing line between the normal and the super normal has moved on a little, as our understanding has increased. The Virgin birth of Jesus of Nazareth is a miracle by most measures and therefore scoffed at by many. However since the advent of in-vitro fertilisation (IVF) mankind is in a position to do something similar. It is therefore entirely possible, that one day we will know how the birth of Jesus occurred and it will be a mystery no more. In the meantime, IVF shows us that it is conceivable (excuse the pun) and that our attempts are probably only rudimentary, compared to what is possible with superior power and knowledge.

It is also worth remembering, that 200 or so years ago, most would have smiled knowingly at the idea of transmitting information through apparent nothingness – 'you are a little crazy methinks'. I am talking about radio and television of course, and the transmission in real time of conversations, data and pictures (including videos) GPS coordinates and so much more.

Many of the properties of light were understood, but no one appreciated in any material sense, that visible light was only a fraction of something we now call the electromagnetic spectrum. Neither, that it only required the vacuum of space through which to propagate. An enormously wide frequency range is available; from very low ground hugging transmissions, through the radio bands to very-high frequencies, microwaves (including radar), infrared up to the visible spectrum and on to the ultraviolet, X-rays and extremely high frequency gamma rays.

This was all a totally hidden world just a few short years ago, and if those before our time had been subject to the experience of television, or an X-ray check-up in an out-of-time, back to the future experience, would have been amazed at what it was all about. Today we are subject to these things and much more, and yet most have no idea of how these things operate. We just seem to accept things without understanding and it is amazing how blasé we become in our ignorance.

I am not saying that we will ever know everything there is to know. Perhaps some things will always be miracle to us.

But I hope the above illustrates, that we should not be afraid that by illuminating a small portion of the natural world in the Flood, or the dramatic volcanic activity of the Exodus, that this in any way diminishes the work of our Creator. Not one iota. It is his world to use as he wishes

Now the general principles articulated above will be emphasised later when we seriously search the prophets. Once we see what was prophesied in regard to future natural disasters and what they represent, it will become obvious that in the context of Israel and Bible prophecy, that they are usually depicted as acts or judgements of God. So what is good for predicted history, must also apply now and in the past. It also means that while outcomes may appear random to us, there are other factors at play which are ultimately foreknown and pre-planned.

This is not at all intuitive for a technocrat and explorer, I have to say. At first sight it smacks of superstition.

But I think both the world of Newtonian physics and the random nature of the quantum-mechanics underpinning it all, gives us a glimpse of how everything (including the future) could be manipulated by an almighty Power and Intellect, and yet still not contravene the physical laws of the universe.

What I hope will emerge from this, is that we can associate more closely with our biblical heroes like Abraham, Moses and David, because we can see that our world (minus our industrial age technology) is little different to the world of their day.

They had no more advantage in believing in an invisible God than we do. Like us they had doubts, sometimes deep doubts, and they too were sometimes self-absorbed and willfully disobedient when it came to spiritual and moral issues. But the faithful soldiered on because they saw prayers answered, had moments when they were truly blessed and from time to time had dramatic experiences, which seemed truly God ordained.

So if I have stripped away some of our Sunday / Sabbath school preconceptions and dragged us into the real world, please bear with me. I think the time has come at the very end of the age, for a reality check and for the truth, however brutal.

But note how similar it is to our own experience today. Freed from any superstitious notions of scripture, we can now identify special works of God in our lives and more easily acknowledge his overall presence in the here and now - and not just thousands of years ago. It may even help us to digest the meaty bits and interpret them correctly.

After all, most believers *can* identify personal miracles in their own lives, and this is where it matters. So with that in mind, let us consider some early examples in a little detail.

We are told that Jacob 'laid hands' on his sons and blessed them as he did so. For those who have had some experience in this regard, the idea that a very real personal blessing accompanied this act will not surprise. This is not about handing out discipline, or an approving pat on the head, but a blessing in faith before God. It appears to have been an ancient practice handed down since time immemorial.

Whether the act was also accompanied by a manifestation, i.e. the 'dwelling' or 'settling' of the Spirit of God is not really clear. If that in fact occurred, it was what is called a manifestation of the Shechinah 'glory' in Jewish spiritual literature.

When Moses begins having experiences of God at the 'burning' bush and on Sinai itself, it becomes apparent that the Lord God begins to show himself once more, through physical manifestations of his Spirit, and that Moses also appoints his priests and leaders with special blessings.

In Numbers 27:18, we find that Joshua the son of Nun is full of the spirit of wisdom and through the 'laying on of hands', Moses appoints him as his successor.

It would seem that in this case Joshua was already Holy Spirit blessed, because the 'spirit of wisdom' is very much the same as the 'holy wisdom' and is another term for the Spirit of God. He is the One who is behind the Shechinah manifestation and in fact, a very famous Christian church building is called exactly that and features in the New Testament.

A public blessing or dedication ceremony of later times, involves anointing the head of the person with oil. This is an act akin to water baptism and especially used at the inauguration of kings and priests. It can also be a personal act of faith for any believer who requires healing, or is preparing for a special task.

These types of event are acts of human commitment, whereas the Holy Spirit action is God's blessing on the individual.

Moving on to the united Davidic Kingdom and the dedication of the First Temple under Solomon, we see even clearer images of what occurred in terms of the Shechinah manifestation of God. This event is described in two places, 1 Kings 8:11 and 2 Chronicles 5:14. It follows the completion of the Temple and the Ark of the Covenant being brought into the sanctuary.

All the elders of Israel were gathered there and we are told the priests and the entire company of Israel were also assembled and celebrating the occasion. There was loud music and much excitement and it must have been a wonderfully uplifting experience. Now as much as I love my NIV version of the Bible, this is where it is useful to examine other translations from the Hebrew. Some versions depict this event in a significantly more literal way, not that many would pick it.

Often we read that the Temple was filled with a cloud, e.g. 'the priests could not perform their duties, because of the cloud and the glory of the Lord filled the Temple'. My JPS Tanakh says something similar 'the priests could not remain to perform their service...' or 'could not stay...'. Now these descriptions paint an impressive scene, but they do not do justice to the words. The literal meaning of the phrase is that the priests **'could not stand'** because of the presence of the Lord.

Anyone who has experienced the 'settling' or 'dwelling' will understand exactly what I am suggesting here. In more recent times it is called 'being slain in the Spirit', a 'touch from the Lord', 'baptised in the Spirit' or some such similar description. The problem is, that if you have never had the experience, it is nigh on impossible to know what the text is implying. It makes it very difficult for the translator.

In my opinion, it is undoubtedly the primary physical manifestation of God on the believer. The simple manifestation where people are laid prostrate, was derided by onlookers at the first Pentecost recorded in the New Testament. They said these people must be drunk. Sceptics today sometimes disparage as mass hysteria, fainting, hypnosis - or simply disregard as inconsequential. But to those who are so blessed, *the change is internal and sublime.*

In the New Testament we often read occasions where the Holy Spirit 'came upon them'. This again is about manifestation; it is a euphemism referring to the same experience. It is *not* to be mistaken for an event without a physical component.

When Jesus was baptised in the Jordan, we see the 'Spirit **came upon him** like a dove', i.e. gently, bringing him down.

We see it as a fairly common experience in the New Testament and sometimes accompanied with other signs referred to as 'gifts of the spirit'; messages of wisdom, knowledge, faith, healing, prophecy, discernment and tongues (other languages).

No doubt these are not the limit of God's Holy Spirit manifestations and sometimes a burning, heating or an unnatural sensation of complete joy and calm has been noted as well.

Many Pentecostals believe 'tongues' are paramount, because of the close association with the Holy Spirit in descriptions in the New Testament. But I suspect they do not associate phrases such as 'the Holy Spirit came upon them', repeated time and again in the NT, with the 'falling down' and it is obvious that many have not made the connection with the Shechinah of the Old Testament either.

In fact from a purely physical point of view, a water baptism is very much an analogue of the Holy Spirit baptism. People are invariably laid backward **in the very same manner**.

The story of Apollos and the Apostle Paul's arrival in Ephesus, is an example of the issues here and is found in Acts 19.

When Paul arrives in the city, he finds some disciples have not heard of the Holy Spirit and have never experienced the anointing themselves. He asks them what baptism they had already received and was told John's baptism (that of water and repentance). When Paul explains there is more and places his hands on them, 'the Holy Spirit came on them, and they spoke in tongues and prophesied'.

Now this is easy to misinterpret, if you do not know about the Shechinah of the First Temple period. Three things happen here – *not* two. The Holy Spirit comes on them *and* they speak in tongues *and* they prophecy. The last two are gifts and it is important to recognise we are looking at a collective experience here. In other words perhaps not all spoke in tongues and perhaps not all prophesied, but the key is that *the Holy Spirit came on them,* i.e. they experienced the *settling* of the *Shechinah.*

The Holy Spirit baptism is often expressed as 'the baptism of the Spirit evidenced by speaking in tongues', but to me this is unnecessarily restrictive and does not acknowledge the fundamental physical evidence – the 'falling down'. In Paul's list of spiritual gifts laid out above, it is obvious that 'tongues' is the least of the gifts. It can therefore be expected that tongues will often occur at an initial baptism, or anointing experience - or indeed, very soon after. And so the question arises, does the 'evidenced by speaking in tongues' definition really matter?

I think it does. And that those who minister should not short change, or confuse those seeking the anointing, by putting the cart before the horse in regard to the blessing.

I have always had concerns that the misconceptions here, could be skewing the understanding of many and limiting the work of God. In fact, I suspect there are people in the churches who have been told that they have been so blest, but are unsatisfied and are still waiting for the real deal.

In addition, the issue leads to the scripturally unsound doctrine that there are two expressions of tongues; a 'baptism evidenced by speaking in tongues' as well as a secondary 'tongues' that is a spiritual gift. *This teaching is not found in scripture.*

Now at the first Pentecost after Jesus' resurrection, the Spirit was manifest 'like the sound of a rushing wind' and also described like 'tongues of fire'. These are melodramatic images and really demonstrate what the recipients were feeling, not what was necessarily seen by observers. A close look at the text and how these images are expressed, shows that for the most part they are superlatives, similes and metaphors.

But manifestations there were, people could not stand and when they spoke, it was often in other tongues. The mob said they were drunk and laughed it away.

Further on in the New Testament, we find that the Apostles (including Paul) returned to Jerusalem from their missionary journeys to Asia Minor, Greece, Egypt and elsewhere and after much discussion, agreed that believing Gentiles were now true 'grafted in' Jews. This was their conclusion after seeing the Holy Spirit 'fall' on the Gentiles, in the very same manner as it had fallen on the Jews at the first Pentecost meeting.

It does not necessarily mean that every believer was so blessed either. Perhaps they were, perhaps not. But the blessing was ubiquitous enough amongst the Gentiles, to know that the ramifications could not be ignored. In other words, the Apostles saw in this manifestation and other associated miracles that God himself had chosen these people. It was not the work of men, so who were they – the Apostles - to deny the Gentiles a place as de-facto Jews with all that entailed? God's chosen people.

This was shocking and a concern. Surely this was only for the privileged Jew, such as the high priests at the First Temple!

The episode is recounted in Paul's letter to the Galatians and the book of Acts from chapter 20 onward. These passages are only easily understood by a few today; those within the church fellowships where the reality of the Holy Spirit anointing has been experienced. And despite a seeming drought of the Holy Spirit anointing in later years, we see that the early Roman Church had its 'confirmation'. It shows that in the early centuries the experience was a reality in that church as well.

After the Reformation, various preachers such as John and Charles Wesley in Britain, also saw these manifestations in a mighty way and finally in the early 20th century, revivals in Wales and America began an outpouring of the Spirit of God, probably unseen since the beginning. Is it now time for practising Jews to inherit their birthright; particularly in Israel itself?

All this demonstrates, that anyone who thinks that they can discredit the faith by nit-picking the written record, are doomed to bitter disappointment. Like an iceberg, it is the immense unseen that underpins the simple words in scripture. Yes, those that are not committed may be victims of shallow textual criticism, but the Spirit of God is the foundation of the faith and our very existence - and that will never be extinguished.

These Holy Spirit expressions of God's power and presence, also underpin the many prophetic utterances in the Bible and speak quietly, but firmly across time and space. The prophecies are not about foretelling for the most part, but a commentary on life, belief, morality and direction; both personal and national.

In retrospect, the great prophets of old are revered, respected and held in awe as the shining lights they were - at least to the extent that some people treat their pronouncements seriously. At the time though, one gets a sense that their contemporaries in general loathed them. At best they were the flavour of the month at one time and hated the next. Across society, the common man, the intelligentsia, the law makers, royalty and even priests commonly treated their warnings with contempt.

Jeremiah is an example. He was influential and sometimes in good standing with the royal family. But his life was one challenge after another and anytime he was outspoken against the powers that be, he ended up in jail or feared for his life.

Not a lot has changed in our time. Human beings do not like being told, even when it is in their self-interest to listen and act. Such is the fate of those who hold up warnings of immorality, decadence, greed and poor decision making within society - or indeed those who bring attention to external threats from foreign forces, oppressive ideologies or adverse environmental impacts. Their lot is ridicule and derision.

If spared, they can only watch on in dismay, as the consequences of poor decision making adversely impacts the society around them. 'I told you so' is no satisfaction when family and friends are harmed and national disaster threatens.

So foretelling was part of it and was usually in the context 'if you do not repent, if you do not change your ways, if you do not start to work together once more, then expect a calamitous outcome'. There are prophecies recounted right from the days of Abraham, where he and his sons and grandsons were promised things, mainly good, often grand, but also personal warnings, some with dire effects. But the grand, nationally focussed pronouncements were meant to shine a light on the future; to give hope and direction to the people of Israel and ultimately to the world for millennia to come.

To sum up; our God is the God of the physical world as well as the unseen, he created it and he uses it as he wishes. He works through the forces of nature, but he also works through very personal intercession in our daily lives - sometimes dramatically, but often subtle ways that we may miss if we are not tuned in. This showcases the protection that believers enjoy and the personal guidance we are assured is ours, even where there appears to be no amazing miracle that we can easily identify 'in the moment'. It is often in retrospect that the opportunities, guidance and protection with which we were blessed becomes apparent, and the path the Lord has chosen for us crystal clear.

This sadly is in contrast to those who are totally oblivious to the presence of God, or the unseen world around us.

Conventional wisdom

So now we move on to the amazing timeline of the Prophet Daniel and the Apostle John. I will show through the prophetic visions they were given, there is a link from the ancient past to our present day. The message is extremely urgent, because the timeline as described by Daniel is at an end.

Both Daniel and John were prophets of Judah and the timeline they outline is identical, the milestones and kingdom identities seamless. Everything is in total harmony. It ties all pertinent history together from ca. 956 BC until today and almost incomprehensibly, it reveals a bigger picture than we expect initially.

The timeline also helps us understand the place of the many specific prophetic events in scripture, how they fit historically and relate to the big picture. From the time of the Jewish exile, there has been just one overarching age to the present; the Age of the Gentiles, aka the Time of Jacob's Trouble. It incorporates one of the zodiacal ages completely, Pisces - and the latter quarter of the previous age – Aries; and this vast time span was prophesied by those who had 'visions of God'.

Daniel noted that his night visions included an introduction by an impressive human like figure (an image of a Son of Man) as it is described, who directs the Archangel Gabriel on exactly what to reveal, what the various visions meant, their timing and what to write down. But Daniel is not alone. His book is pivotal in understanding the timeline of trial and redemption, but it is repeated using other visions and ultimately expanded on in Revelation. It is also supported by much other fine detail from prophetic visions in both Testaments.

Of course the prophet Daniel, whose book has often fascinated, is also neglected in places of worship. Avoided is probably not too harsh an observation. The problem is; that there have been so many question marks around interpretation in so many quarters. Other vested interests have made deliberate attempts to ignore his book, lest it reveals too much.

Now to my knowledge there is no support from contemporary psychology, psychoanalysis or the neurosciences for anything more than normal dreams. Dreams or visions emanating from external sources other than our own experience or subconscious are viewed with scepticism - and I suspect, dismissed.

This is not surprising. These sciences as they exist today are quite young. However a lot has been learnt since the 1800s and many researchers (from Sigmund Freud to our contemporary scientists) have taken significant strides forward in related research; this, if documentaries and the popular media are anything to go by. Whether these sciences have come as far as say, geology, which originated a little earlier in the modern era, is something I cannot say. In addition there has always been the fear of 'quacks' in such a difficult sphere of research.

But with advances in neuroscience and the amazing technology behind various scanning techniques, we seem to be entering a new phase where almost anything is imaginable in the 21st century.

Intriguing experiments now include manipulating the mind, in order to consciously alter perceptions and physical actions. This is achieved using the external application of computer controlled sensors and inputs. The sense or limb is controlled in such a way, that the subject thinks it is their own doing.

It also appears that the function of the various parts of the human brain - and how those parts interact for both conscious and unconscious actions - is now understood to quite a remarkable degree. Other curious phenomena such as hypnosis have been known for many years and also reveal interesting attributes of the mind. But here is a thought: if these frontiers are now being explored by us today, how much more may our minds have been interacted with in the past and even today, by higher powers and more advanced intelligence?

Since we have a world underpinned by a sense of uncertainty at an underlying level, more akin to the 'spirit' of the Bible, it suggests to the layperson there could well exist connections between the human brain and higher intelligence. In fact, ultimately directly, or indirectly with the Divine.

So can people see the future, and is it true that only a few really take the Bible seriously in this respect? And of those that do, how many really believe there is a message which can be understood? Importantly for us: in the babble and clamour of so much diverse opinion today, is there *any* possibility of absolute clarity of the historical predictions of the prophets?

In Christian circles there are definitely more than a few who are devoted to discussion and study of the biblical prophets. It may not be so pronounced in Judaism and Islam, but nevertheless everyone has their teaching and theories.

Of course, in the latter religion the focus is on the Qur'an; a relatively small book, and so the Hadith explanatory texts and testimonials provide much of the detail.

My experience online, is that the so-called foretelling of the 'prophet' is more about stretching interpretations of particular verses to encompass modern discoveries. These are retro interpretations used to justify the text as prophecy. This usually involves convincing the faithful of some pre-knowledge of 20th and 21st century science by Muhammad. To date, I have only encountered simplistic attempts, which only fool the gullible.

However matters of religion and belief are not often viewed objectively, and so it is understandable that followers grab at anything they can. This stretching the envelope is also seen in Christian circles and accounts for much ill-conceived interpretation of prophecy. Specifically those prophecies given before the Babylonian exile by Jeremiah, Ezekiel et al, which warn that

Jerusalem was to be sacked, and which are too often willy-nilly repackaged (or repurposed) to describe supposed future events even now. Of course prophecies that have more than one fulfilment cannot be discounted; 'types' and forerunner examples are not uncommon. But when you realise that those who propose such future applications, are often oblivious to the primary fulfilment that occurred 2,600 years ago, you begin to appreciate the historic ignorance that abounds and the random hypothesising which is so easy to spout on that basis.

The difference in the Judeo-Christian scripture, as compared to the embroidering efforts of modern commentators of the Qur'an, is that the biblical prophecies and the circumstances surrounding them are usually explicitly described and in considerable detail. Where there are exceptions, the setting is usually clear in the context of the narrative. There is almost always an obvious claim to receiving information from Above, and the circumstances surrounding that described in some detail.

But Christians are also subject to every wind of interpretation. The amount of material in films, literature, videos and online blogs, or indeed from the pulpit, is immense. In our busy world too, few find time to study for themselves and are therefore the victims of way too much second and third hand opinion; opinion which is often contradictory. On the plus side there is much detail and a good deal of genuine interest and learning.

However it means there is a lot to wade through and so despite this upside, confusion generally reigns. It means there is a desperate need for a solid biblical timeline; something that will allow us to correctly connect the dots of the individual prophecies. It is something I hope to provide here.

I also believe that the Lord has his purposes and he reveals things when they are required - not when we would like to be indulged. And so the current age of mass communication, accessible data and general connectedness is a unique time, never experienced in the past and this facilitates our quest here.

There is a time and a season we are told in Ecclesiastes 3. It is in fact a new age of enlightenment, but only one for those who are not deceived by every imagination of men. A strong familiarity of the science underpinning our world, the lessons of history and a sound knowledge of fundamental biblical principles are essential.

The bane of biblical prophecy in the 20th and early 21st centuries is the scourge of 'futurism'. This interpretation is a literalist approach which reduces events occurring over millennia, down to a few short years. It is a scheme where most of the prophecies of the 'end time' have been bundled into the future. It is much easier to be a futurist interpreter with an open canvas and let your imagination run riot, than grapple with reality. It is easier to plonk everything into a speculative future world, than do some serious historical homework.

Although there are hints of futurism in earlier times, it was primarily introduced by the Catholic Church in the 16th century as a defence against Reformation teaching. This defence was a result of Protestant claims that the Roman Church was the Whore of Revelation 17, and therefore antichrist at its heart.

Earlier in the century (1516) the Roman Church attempted to stop the printing and sale of the Bible. This was an attempt to nullify Martin Luther's 95 Theses and his declaration against the Church. Their sale of indulgences in return for salvation, and a plethora of other autocratic decrees, were anathema to those who had actually read the New Testament and knew what it contained. Luther wanted Christendom to return to a Bible based faith, rather than a faith dictated to, and controlled by the Catholic intelligentsia. He knew without doubt that ultimately God's forgiveness could not be bought by subscription.

Luther also wanted free discussion of scripture. This was fair enough in principle, but of course it also allowed 'every wind of doctrine' and interpretation.

So the Roman Church attempted to outlaw renegade Reformation interpretations and opinion to counter this. Their efforts were to no avail given the situation of the time, so the Jesuit order was created and various Catholic scholars were tasked with looking for other ways of interpreting the prophets.

A Spaniard, Francisco Ribera (1537-1591) wrote a commentary, where he proposed that the first few chapters of the book of Daniel (Daniel 2 and 7 specifically) applied to the various empires starting with Babylon and ending with pagan Rome and the Ten Kingdoms. This early part of the timeline still followed the widely accepted, 'historicist' method. But then he consigned the last 3.5 'times' to a future period of just three and a half years, immediately prior to the second coming of Christ.

So was born the 'futurist' method of interpreting Bible prophecy. Today it has morphed into a version where all 'seven times' are declared to be covered in a brief, future seven year period. This teaching has all sorts of unfortunate consequences and perversions. They include:

- o A secret rapture of believers before the seven year period (at least by some interpreters)
- o A future antichrist figure negotiating a peace deal in Jerusalem at the three and a half year midpoint
- o Jesus' second coming at the end of the seven years
- o A further thousand year age after the second advent, where Satan is bound and released again
- o Any prophecy can be inserted, providing the interpretation is distorted and manipulated enough to fit the short timeframe
- o Options for the Whore City and 'mark of the beast' are limited only by the imagination

All this would have horrified Bible scholars and Christians from earlier periods, including those of the Crusader centuries.

The Crusaders for the most part appear to have been biblical historicists by understanding. It is also extremely likely that they believed the 1,000 years of Revelation 20 was contemporary with the church age, and therefore initiated the Crusades on the basis of Jesus' imminent return. The Crusades were then agreed on at the Council of Clermont in AD 1095, no doubt after decades of discussion and argument.

In hindsight it is obvious that the Crusader expositors got something wrong and misread the ancient wisdom and knowledge of the Apostles and Prophets. I will seek to address this later. Since then it has been pretty much downhill all the way, with even historicist interpreters losing sight of the truth.

Now the Catholic Church was casting around for any solution, because another scholar, Luis de Alcazar (1554-1630) also worked on the 'preterist' prophetic method. In its most basic form 'preterism' teaches that all prophecy was completed around AD 70-73 and the fall of Jerusalem to Roman General, Titus. There are passages in the New Testament which indicate a form of preterism was at work even in the first century.

Paul in writing to the Thessalonians attempts to counter this, by explaining that the Antichrist must come *before* Jesus returns. However it was in response to the Reformers of the 16th century, that the doctrine was articulated precisely.

The motives of the Papacy at the time seem pretty suspect, despite being hard pressed by a chorus of accusation from the dissenters. A blanket ban on individuals reading the Bible for themselves, was hardly going to help infuse knowledge and promote understanding of scripture. It smacked too much of control for control's sake and is the antithesis of the Christian way; free access to knowledge, freedom of thought and personal responsibility in regard to faith.

But they had a right to be concerned by misguided interpretations and new prophetic identifications which departed from previous wisdom. And unfortunately there seem to have been considerable agreement by Protestant leaders on their identification of Rome as the Whore City. It shows that freedom of thought, does not always guarantee correct interpretation.

So whether Francisco Ribera and the other Jesuits entrusted to look for alternate prophetic schemes, were also being devious and part of the conspiracy, is not for me to say. Some may have been genuinely convinced that something had to be wrong with the traditional prophetic interpretation; after all, it was being used to smear the Church. Others perhaps decided to fight fire with fire, regardless of the moral issues.

The futurist interpretation was later revived by John Nelson Darby (1800-1882) of the Plymouth Brethren and subsequently followed by others. Since the 1960's it has swept through many Protestant evangelical churches on the back of popular books by Hal Lindsey and a whole bandwagon of associates. How the Adversary will try any means whatsoever to deceive the elect!

Today the future has caught up with us, and many of the expected events forecast can be shown to have already occurred.

This is exposing futurism for what it is; an incorrect chronology of the biblical prophecies. Futurism encourages speculation of nearly all prophetic events without accountability. Even in futurist circles the confusion is being described as 'prophecy burnout' - a sure sign something is wrong.

The problem is manifold I suspect and more about spiritual and moral issues than not. They include *not* following the pattern of salvation and enlightenment outlined by the Apostles.

Fundamentalism seems to demand simplistic answers and so carelessly interpreting time periods in prophecy such as 'hours', 'days' and 'months' as literal periods of time, when the prophets taught otherwise, is one of the issues. Other more obscure terms such as 'times' are also misunderstood. Scripturally the problem revolves around the misinterpretation of Daniel 9:27.

I have already outlined some of the issues with the creation story in the section 'In Those Days'. Here I will address a propensity for many fundamentalist and evangelical Christians to overlook, or ignore the prophetic rules outlined in Leviticus, Numbers and Ezekiel, as well as the general lessons that can be learnt from the history of Israel. Instead they adopt simplistic understandings of the duration of a 'time', a 'time, times and half a time', 42 'months', 1260 'days' and ignore previous wisdom on national identities in prophecy, e.g. Daniel 2 and 7.

We are all guilty of this at times, because we naturally use common, contemporary definitions and assume they were applicable to prophecy in ancient times. It is a case of too much talking and not enough listening and understanding. In part, it is also laziness in studying the scripture. In any event, we are now being forced toward a better understanding of the historical sequence of world events and their alignment with the Bible.

Sadly, many supporters of futurism, and nearly all Reformation historicists are bent on attacking the Roman Church and identifying it with the Whore / Mystery Babylon tag of Revelation 17. This has happened despite Francisco Ribera's best efforts. As a consequence evangelical futurists (and tragically historicists) as well as the Catholic theologians of the past, all fail and the Evil One wins. The futurists who do not finger Rome, point to New York, Jerusalem or some other city of choice, with all manner of conspiracy theory and imaginable 'mark of the beast' interpretation attached. As intimated earlier, the impetus for much of this seems to have arisen after World War II and followed 'six day' creationism.

Sometimes the obvious meaning in prophecy, or the intended target location is hidden. As in the thinking behind the parables of Jesus, not everybody is meant to understand. On the other hand, even those that sincerely desire to know the truth, may have to be patient and wait for the clues to fall into place. Much greater minds than 'yours truly' have studied these things, but I believe the time has arrived, the details are coming together, and the Lord is speaking. .

If you have been a steadfast student of prophecy, you will have read many forecasts and attempts to harmonise current events with verses in the Bible. But consider; can you imagine a credible prophecy of Israel where the promised return to the land (eretz Israel) is not prominent? Or a prophecy where the World Wars and the Holocaust of 1942 do not feature somewhere in the narrative? And yet this is what has happened in popular circles, where Bible prophecy has been treated as a game, rather than the Plan of God.

There are some prophetic terms and descriptions that are not necessarily incompatible with the historicist approach, while others can be exposed as false or not applicable.

The various classifications and terms include Pre-tribulation, Post-tribulation, Pre-millennialism, Amillennialism and Dispensationalism. They are used to emphasise the order in which events occur; events such as the 'tribulation', the second coming of Christ, the rapture, and the millennial Messianic reign. They are terms particularly applicable to Revelation descriptions, but flow through to all prophecy.

Although the diligent student will find useful material in many books and articles where the author favours one or other of these scenarios, it goes without saying, that interpreting a prophetic vision of the future must be based on an understanding of past history. And that is what the historicist or historic method is all about; following the prophets of Judah and their God given guidelines. It is what I am about to describe here.

However I know the approach described here will unsettle most traditional historicists, because they invariably follow the Protestant Reformers. I was summarily dismissed from one prophecy group as soon as I suggested the Reformers were a little off track. It did not help that the Administrator was also a pastor and passionate 'six day' creationist. Not my day as they say, but we can all point to such times in our lives.

I will now discuss the 'preterist' interpretation of scripture in more detail. Preterism is a more insidious prophetic belief than most, because it denies that the prophets saw anything beyond AD 70-73 and therefore have no relevance thereafter. It has a small, but growing following and comes in a number of guises. It is described thus by one of its more strict proponents:

*"In short, a **Preterist** is the generally accepted term for those who believe that **all** biblical prophecy events have been completed, and were accomplished in the past - the first century to be exact. The event usually associated with the fulfilment of Bible prophecy is the Roman-Jewish War of AD 66-73, especially during the destruction of Jerusalem in AD 70."*

and

"Just for the record, only those who believe that the Scriptures teach that all the events related to Christ's Parousia (i.e. "Second Advent," the resurrection and judgment, etc.) are past, deserve the name "Preterist." All others are simply futurists of a sort."

Note: Full preterists claim that *every* prophecy in the Bible was fulfilled before the Jewish War of AD 66-73 and particularly the sacking of Jerusalem by Titus in AD 70. It therefore includes the Day of the Lord, usually called the Second coming of Christ. This is a huge call, because of several issues that become apparent once you examine them. There is also a form of preterism called 'partial preterism' which forlornly attempts to tackle these issues. The problems of full preterism include:

1. The second coming around AD 70
2. The fulfilment of many prophecies in the OT are not addressed and the timelines and instructions in Ezekiel and Daniel in regard to prophetic understanding are ignored

3. Many prophecies in Revelation could never have been fulfilled because of insufficient time. These include disease and plagues, wars and rumours of war, 'natural' disasters and celestial occurrences
4. An attempt is made to associate pronouncements like the 'Kingdom of God is near' with the Second Coming and not the work of Jesus, the Apostles and Saints in the power of the Holy Spirit. Other teaching in the gospels, that listeners around the Lord would not see death before the Kingdom of God had come, are in a similar vein
5. A conviction that the various mentions of 'generation' and its variations in the gospels are restricted to our common usage today (a generational age, such as 'in Jesus' day') without consideration of the obvious broader Greek usage which includes genealogy, i.e. lineage, ancestry, dynasty, bloodline or race - implying an ongoing hereditary sequence; e.g. Gentiles (nations)

Firstly, if Jesus came in the clouds in a way that every eye could see him around AD 70, then we have absolutely no record of this event. It seems inconceivable to me that it would not have been recorded somewhere; perhaps by Josephus or the Romans, the Greeks, and if not by official sources, certainly by his followers and the 1st century church.

The Apostle John who received the Revelation passed away in AD 98. The Revelation could have been received any time before this date, but according to one school of thought, was written as early as AD 70.

Regardless, John and many others were around to experience and record such a remarkable event and did not.

So this is the scenario painted by preterists, but without evidence. There is an account by the church fathers, that some Christians in Jerusalem heeded the warning prophecy by Jesus of impending destruction and headed for the town of Pella in today's northern Jordan. This seems quite plausible, as this was one of the cities of the Greek Decapolis to the east of the Jordan River - and safe enough away from Judea.

Points 2 – 3 will be addressed as a matter of course, with the unravelling of the lengthy biblical tribulation timeline. A key verse in regard to point 4 is the following. Matthew 16:27-28

*"For the Son of man **is going to come** in his Father's glory with his angels, and **then he will reward each person according to what they have done**. Truly I tell you, some who are **standing here**, will not taste death before they see the Son of man coming in his kingdom."* (NIV)

I suggest that Matthew 16:27-28 needs to be interpreted in the light of the comparable, more succinct and unambiguous verses in Mark and Luke, which record the same event. The emphasis in those verses, is that some of those standing there would not see death before they saw the 'kingdom of God'.

Elsewhere in the gospels Jesus teaches that the kingdom of God had arrived and indeed was within them, referring to his disciples and followers. The kingdom had arrived because Jesus had arrived, the promised Messiah endowed with the Holy Spirit and performing miracles of healing and renewal - and with the message of love, hope and goodwill. Moreover they were to see the crucifixion, resurrection and ascension of Messiah denoting the new order, and the promise of an immortal body to believers at their own resurrection. Mark 9:1

*[1] And he said to them, "Truly I tell you, **some who are standing here will not taste death before they see that the kingdom of God has come with power.**"* (NIV)

Luke 9:27

*[27] "Truly I tell you, **some who are standing here will not taste death before they see the kingdom of God.**"* (NIV)

The Matthew 16 reference tends to suggest an immediate second coming, but Mark and Luke's short statements directed at Jesus' work in the here and now, militate against a second coming fulfilment only, but the beginning of the overall kingdom ministry. Mark emphasises that the kingdom 'has come with power'. Surely the miracles of Jesus, which even some of the Pharisees, Roman soldiers and ruling class acknowledged, were absolute confirmation of 'power'? They heralded the fact that the kingdom had arrived, as Jesus stated many times throughout the gospels. Indeed, they were to see Jesus ascending into the heavens. The Kingdom had certainly come!

The disciples and all believers are the citizens of that kingdom and empowered by the Holy Spirit. It can therefore be safely concluded that Matthew 16:28 is referring to the advent of the kingdom and not its consummation when Jesus returns.

It is sad to relate, but I have met one or two in the Messianic tradition who will not accept that we are living under a second covenant. They are waiting until the second coming and the New Jerusalem. Fundamentally they are either ignorant of the power and reality of the Holy Spirit's presence in their own lives, or they are actively avoiding seeking that reality. So for them there is no renewal and no kingdom experience.

On the other hand, the 'fullness' of the Kingdom is at the Second Coming, when the promise is there for a 'resurrected' and decay free body, as Jesus received at his resurrection.

Interestingly, if Matthew is correct, many were raised from the dead when Jesus ascended to heaven after his resurrection. Matthew 27:50-54

"And when Jesus had cried out again in a loud voice, he gave up his spirit.[51] At that moment the curtain of the temple was torn in two from top to bottom. The earth shook, the rocks split [52] **and the tombs broke open. The bodies of many holy people who had died were raised to life. [53] They came out of the tombs after Jesus' resurrection and[e] went into the holy city and appeared to many** *people.[54] When the centurion and those with him who were guarding Jesus saw the earthquake and all that had happened, they were terrified, and exclaimed, "Surely he was the Son of God!"* (NIV)

This should also be taken into account. Presumably it speaks of the dead from the tombs outside the Old City in the Kidron Valley and the slopes of the Mount of Olives. So it is possible that there were a few who may have heard Jesus talk about the kingdom, but then passed away. This is not the scenario painted by Matthew, Mark and Luke earlier, but does beg the question, were they 'raised up' in mortal, or resurrection bodies?

These verses from Matthew are the only mention of this event, but I believe they should not be forgotten. If Jesus' own resurrection is the first resurrection - and the event described here is related in some way - then the resurrection that is mentioned in Matthew 24, Revelation 11 and elsewhere, is clearly yet to occur and becomes the second resurrection.

So certainly the Kingdom of God had come, redemption for all time was here, but had all prophecy been fulfilled this side of the new heaven and earth *and* the New Jerusalem?

That this is not the case will become self-evident as we study the prophecies and timing guidelines. Once we understand that a very particular interpretation of Matthew 16 and a few relative terms such as 'soon' and 'not far off' are ultimately the only support for the preterist view - and the interpretation otherwise fails to pass all other tests, including the great Daniel timeline - we can put it to bed as a failed hypothesis.

Point 5 revolves around two possible issues: The broader meaning of the Koine Greek usage of 'generation' and the various tenses and spelling which cover more than in English today; including genealogy, lineage or race.

Secondly, the possibility that a copyist has used the incorrect form of the word in Matthew 24:34. Matthew 24 Verse 34:

'Truly I tell you, this generation will certainly not pass away until all these things have happened.' (NIV)

Shortly, I will show that the historical content of Jesus' Matthew 24 Olivet address plays out over a very long period. It is chronological in nature and therefore Point 5 can be grouped with Points 2 and 3. For example, the 'abomination of desolation' described in the chapter at verse 15, will be shown not to have occurred until 600 years after Jesus uttered the words.

He also specifically states that when he returns we will see the 'sign of the Son of Man in the sky'. That comment places his second coming to *a time no earlier* than the 21st century.

Now it will come as no surprise, that considerable discernment and a fair degree of study is required, to determine the correspondence of history with the prophetic visions recorded by Isaiah, Jeremiah, Ezekiel and Daniel. If we include the Minor Prophets, Gospels, Epistles and Revelation and throw in some Torah to kick it all off, we *really* have our work cut out for us. We simply just cannot postulate anything we like about future events by letting our imaginations run wild, but must take a scholarly approach to known history.

It is also essential to know the circumstances of each prophet, and in respect of current and future events, make absolutely sure that what was prophesied, has not already taken place. If we are not very careful, this will lead to error in identifying people, nations, events and places. Of course there may be more than one fulfilment, but we need to identify the primary one.

In such instances one fulfilment will usually be over-arching and may involve thousands of years, while the other (if it exists) may refer to a specific event or experience. This is sometimes referred to as a 'type'. In addition there can be parallels between Israel's experience and that of their (and our) Messiah.

In all this we need something else as well; divine guidance and inspiration. We can never appreciate how all this hangs together, unless the Lord lifts the curtain from our eyes.

One thing that has absolutely stunned me in this exercise, is how key prophetic rules and pivotal identifications have been completely hidden from our view. It is truly amazing and at the same time immensely humbling. It shows again, that His ways are truly higher than our ways.

Returning to biblical history; if the reader remains unconvinced that Abraham even existed after my earlier section 'In Those Days', or has not had the opportunity to meditate upon that section, then bear with me. Our prophetic journey began unexpectedly with Genesis 1 and the patriarchs, but it was with Abraham that events turned personal and prophecies, specific. In fact where the Genesis 1 prophecy came from is uncertain. It may have been a prophecy from as late as the exiles, or as early as the patriarchs. Regardless of provenance, I proposed that the Genesis 1 account was so close to modern cosmological and geological theory, that it seems inspired one way or another and far ahead of its time when compared to other creation stories.

Then with Abraham, we learn of very personal accounts – recollections of family – that are focussed on everyday events, but which include special moments of divine blessing, promises, predictions and warnings for the future.

We eventually came to the Exodus account, which has now been collaborated by modern scientific observation in regard to the processes that occurred. Because of the magnitude of the Exodus saga and the manifestation of God at Mt. Sinai, Moses is regarded as the greatest of all human prophets by those in the Hebreo-Christian faith. Moses was tuned into the Almighty, while all those around were like ships tossed on a stormy sea. Jesus himself attested to that. Moses probably had the most spiritual and intimate experience with God at Sinai, that any ordinary human being ever had. For all that, he was not given detailed visions of future events, complete with timetables. It was the Babylon era Jewish prophets and their prophecies, that are undeniably the most significant and detailed of all.

It may have been foreshadowed by the significant Torah prophecy of Leviticus 26, but alternatively chapter 26 may in fact also date from the same period. Simply put, the inspiration and visions experienced and seen by the likes of Isaiah, Jeremiah, Ezekiel and Daniel, speak directly to us today, just as much as they spoke to their fellows in antiquity. Why?

Well, because it is a warning against rebellion, unfaithfulness and selfishness and begins an almost mind boggling age of judgement and correction. The immediate pre-exilic period which lead up to this time, gives us a sense that the dividing of the nation into Judah and the Ten Tribes shows a disharmony which is particularly abhorrent in the eyes of God. You cannot build a Kingdom of Heaven on earth, when the nation selected to have the foremost part in that enterprise is itself divided.

There was therefore a terrible price to pay and this comes in the form of disenfranchisement, exile, slavery and death.

For Judah and their associates, the Babylonian exile was arguably a less turbulent time than the initial fate that befell the Northern Tribes. After all, they were completely removed from their land. It was nevertheless a time of fear and distress unparalleled since the Exodus, and just as uncertain.

What they did not realise; was that it was the beginning of an age of tribulation, stress and judgement that would last for millennia. If they had known at the outset how long it was to last and how it would end, their hearts would surely have failed them. It was a time of fear and distress which ended catastrophically in the World Wars of the twentieth century, the carnage in Europe and the Jewish Holocaust.

As I will emphasise later, the Jewish Exile and the period of the World Wars parallel each other to a considerable degree when dissected. They also bookend the most significant 'age' recorded in the Judeo-Christian Bible.

So it is under extraordinary circumstances that God worked amongst his prophets. They were humbled, their lives in danger, they often had nowhere to turn and so they sought the Lord like no others. Daniel is a prominent example; when we read of his protection in the lion's den it seems surreal, almost to the point of the ridiculous. It certainly is amazing.

But could it be that God had set Daniel apart for a very special task from the beginning? That there was absolutely no way Daniel was ever going to succumb to the lions, or any other dire circumstance until his work was completed?

I believe it is so. Later in pagan Rome some of the more blood thirsty Caesars of the Claudian and Flavian Dynasties were to put the Christians of the first century (and several other groups) to death in a similar manner in the arenas. They were not all saved; but Daniel was. Plainly he was not only loved like other saints, he was also prepared by God for the special task of recording his 'night visions' and their interpretations.

It was absolutely vital, that this was completed for our future edification and enlightenment.

So we have many prophets whose utterances and predictions are spread across the pages of the Jewish Tanakh (OT). Portions are still read in synagogues today for solace and encouragement. Unfortunately over 2,000 years the verses read have been reduced and the book of Daniel (for example) consigned to the 'Writings' rather than the 'Prophets'. Isaiah 53 apparently has been virtually banned as censorship has increased.

The prophetic passages are read in Christian churches as well. But there again the interpretations of these verses and how they have applied down through the ages, are often glossed over. An easily digested take home message is provided for everyday living, but with little discussion or understanding of the historical or prophetic significance. Those that are interested in prophecy complain most pastors and preachers will not touch the subject. Objective onlookers cynically say 'I wonder why?'

Islamic students have a reverse interpretation of events where Abraham's chosen becomes Ishmael, Muhammad becomes *the* prophet, a personage called the Mahdi is a messianic-like leader of the end time and Babylon becomes a western conspiracy. The 'last days' scenario painted by Islamic teachers is confusing and convoluted and based on teaching that has evolved over time. Later, I will enlarge on the events that they suggest will unfold in the last days.

On the other hand, secularist sceptics consign any prophetic word to make-believe, not proven by science and ignored by psychology. Possibility thinking is not for them – even if the 'arrow of time' hit them on the bum.

So we need to rediscover the thread of truth; the timeline of history that holds the plan together. When we do that, we will not convince those that choose to believe otherwise, but they will have to dodge and weave more vigorously to avoid the obvious. For those who choose to believe, we will have illuminated more evidence of Divine foreknowledge and indeed Divine direction and guidance, both national and personal.

From here on, the focus will be on things that are new and special in regard to the 'seven times' biblical timeline. Things I believe I have been shown. It is not that this timeline is new to historicist interpreters, but rather it requires refinement in order to clarify the timeline and through that, the identification of the national players. It is also about the keys to interpretation, including the pivotal geographical importance of Jerusalem and the peoples associated with that city.

The prophecy I will begin with is encountered in Leviticus 26. In this chapter, we find dire warnings of the various afflictions Israel will suffer in the future, if they do not turn from their evil ways. It goes like this in Young's Literal Translation, which gives us a little bit of the Hebrew phraseology and flavour. Here is a portion of the chapter:

[13] *I [am] Jehovah your God, who have brought you out of the land of the Egyptians, from being their servants; and I break the bars of your*

yoke, and cause you to go erect. [14] *`And if ye do not hearken to Me, and do not all these commands;* [15] *and if at My statutes ye kick, and if My judgments your soul loathe, so as not to do all My commands -- to your breaking My covenant –*

[16] *I also do this to you, and I have appointed over you trouble, the consumption, and the burning fever, consuming eyes, and causing pain of soul; and your seed in vain ye have sowed, and your enemies have eaten it;* [17] *and I have set My face against you, and ye have been smitten before your enemies; and those hating you have ruled over you, and ye have fled, and there is none pursuing you.* [18] *`And if unto these ye hearken not to Me, -- then I have added to chastise you **seven times for your sins**;*

And:

[27] *`And if for this ye hearken not to Me, and have walked with Me in opposition,* [28] *then I have walked with you in the fury of opposition, and have chastised you, even I,* ***seven times for your sins.***

In conclusion:

[32] *and I have made desolate the land, and your enemies, who are dwelling in it, have been astonished at it.* [33] *And you **I scatter among nations, and have drawn out after you a sword, and your land hath been a desolation and your cities are a waste.*** (YLT)

In this chapter, the central thesis and warning is that Israel will be punished seven times for their sins. The question is what does this mean? Is it just a figure of speech for the severity of the coming trial and punishment? Or is it meant to be more specific? And seven times what? We can appreciate from the text that the punishment outlined is severe, that the nation will be dispersed and so on, but is it only a threat?

The chapter gives no clues as to the length of the punishment, but the passage does suggest, that 'time' is literally of the essence. If we then look at Ezekiel (particularly chapter 4) and even closer at the Book of Daniel, the idea that severity is also measured by the *length* of the punishment, becomes more and more likely - even inevitable. When we understand how long this time of judgement in fact is, the words of Leviticus 26 become particularly dire and devastating.

This 'seven times' scheme is an integral part of Daniel's timeline and also that of the Book of Revelation in the New

Testament. The full 'seven times' becomes the total period of chastisement, where self determination, national land rights, peace and prosperity are all forgone. However it is easy to confuse the two halves of the 'seven times' in some prophecies, and some historicist scholars only recognise one half the period. The result of this is confusion with the historical sequence.

So with this to guide us, it follows that two consecutive 'three and a half times' is the total punishment for Israel; a period with a half time interval like a very rugged game of football.

Now I have already explained, that most 20th and 21st century Protestant Christian commentators who teach futurism, see the 'time, times and half a time', 42 months, or 1,260 'days' of prophecy in the Bible as literal days, months or years. They ignore Ezekiel and the overarching historical reach of the book of Daniel completely. They also tend to disregard the 2,000 years of history from the 1st to 21th centuries, as if prophecy has nothing to say about the period. Others attempt to identify the animal kingdoms of Daniel with current or future empires at the expense of history - and the Daniel 9:27 summary of Israel's tribulation is taught as a dislocated seven years still future.

Some realise none of this is satisfactory and are simply perplexed at best, or give it all away. It is the 'futurist trap'.

It is a great pity, that those saints down through the ages who understood the correct scriptural principles, have now been ignored. Of course it was far from plain sailing through the centuries as I have readily acknowledged, but Godly, perceptive and intelligent men such as most of the Church fathers, the theologians of the Crusader period, devout biblical scholars until the 16th century and physicist Sir Isaac Newton at the turn of the 17th and 18th centuries - all understood the instructions of the prophets to a remarkable degree, rarely seen today.

The Godly legacy continued until the mid-twentieth century, despite false interpretations creeping back into some sectors of Christendom a century earlier. But then disaster; the trickle of false teaching developed into a tsunami that finally swamped the elect. Regardless of prophetic preferences, a curtain of blindness now prevails on a great deal of the prophetic word.

CHAPTER 9

Rules of Engagement

'In the thirtieth year, in the fourth month on the fifth day, while I was amongst the exiles by the Kebar River, the heavens were opened......and I saw, Visions of God'. The Prophet EZEKIEL

A day for a year

The Leviticus 26 passage referring to the 'seven times' of judgement of Israel and quoted in the previous chapter is our Lesson 1. I believe this chapter in Leviticus is also referred to in Daniel 9:27 as an overarching covenant or agreement between God and the nation. This is in addition to the obvious message of chapter 9 which is pointing to the coming of Messiah; the 'Lamb of God'.

Lesson 1: there is **to be seven 'times' of punishment / trial / refinement for the people of Israel.**

Israel was especially chosen to be a faithful and inspired race and an example to the rest of the world. However at the national level the warning was ignored, they failed miserably and were therefore subject to the penalties outlined in the Leviticus agreement. The judgment is on the people rather than the land, but necessarily the focus will be in the Holy Land itself.

We will return to the discussion of the 'seven times' and the 'time, times and half a time' in a moment, but the next and most basic lesson in regard to the application of the historicist (and biblical) method of prophecy, is something that commentators who favour this interpretation have known for years.

The rule is found in three or four passages in the Old Testament, including chapter 4 of the Book of Ezekiel where we have a prediction of the siege of Jerusalem.

"Now, son of man, take a block of clay, put it in front of you and draw the city of Jerusalem on it. ² Then lay siege to it: Erect siege works against it, build a ramp up to it, set up camps against it and put battering rams around it. ³ Then take an iron pan, place it as an iron wall between you and the city and turn your face toward it. It will be under siege, and you shall besiege it. This will be a sign to the people of Israel.

*⁴ "Then lie on your left side and put the sin of the people of Israel upon yourself.[a] You are to bear their sin for the number of days you lie on your side. ⁵ I have assigned **you the same number of days as the years** of their sin. So for **390 days** you will bear the sin of the people of Israel, i.e. the Ten Tribes.*

*⁶ "After you have finished this, lie down again, this time on your right side, and bear the sin of the people of Judah. I have assigned you **40 days**, **a day for each year**. ⁷ Turn your face toward the siege of Jerusalem and with bared arm prophesy against her. ⁸ I will tie you up with ropes so that you cannot turn from one side to the other until you have finished the days of your siege.* (NIV)

There are a number of points here that are worth discussing. But for the moment I will confine my comments to the simple instruction given in the vision to the prophet. It is the routine that Ezekiel should go through, to demonstrate the impending disaster to strike Judah. It was meant to make a statement that could not be ignored. Jerusalem was to come under a horrific siege and the population needed to believe it!

Now Ezekiel must have indeed gone through with this enactment exercise, because he was given detailed instructions on how to ration out his food. In fact he played out the whole thing in order to catch the attention of anyone who would listen.

No doubt Ezekiel also understood the principle of interpretation that was meant to be followed subsequently; that a 'day' in prophecy is NOT a literal day from evening to morning (as the Israelites measured their days) but a whole calendar year.

This rule is further reinforced with the 'seventy week' prophecy of Daniel 9.

While many Christian commentators would agree that the rule described in chapter 9 is that 'weeks', 'sevens' or 'seven days' refer to a period of seven years, i.e. each day represents one year, there are other interpretations found in Christian literature. This is a very unfortunate state of affairs and is seen in the footnotes of at least one English translation which appears to follow erroneous, quite probably Jewish teaching.

In the obvious interpretation, the prophecy covers a time span of 70 weeks and therefore 7 x 70 or 490 years. Daniel 9:24

"Seventy 'sevens'[c] are decreed for your people and your holy city to finish[d] transgression, to put an end to sin, to atone for wickedness, to bring in everlasting righteousness, to seal up vision and prophecy and to anoint the Most Holy Place." (NIV)

Even interpreters of the futurist school are aware of this prophetic rule. Unfortunately they sometimes interpret 'times' as days or years, and use it where it is not meant to be applied. This misunderstanding of prophetic terminology will be cleared up forthwith, but first I will formalise the second lesson:

Lesson 2: **One prophetic 'day' is equal to a defined period of time, which is most commonly one whole calendar year.**

Even the prophetic phraseology **'evening and morning'** denoting a 24 hour day actually means 'beginning and ending' or 'start and finish' - **of a year** or **other defined length of time**.

This 'day for a year' principle is a fundamental rule of biblical prophecy, although not the only one. It has been around for centuries – indeed millennia according to the testimony of Moses, Ezekiel, Daniel and others in scripture.

However the broader meaning is also important in some contexts.

It is true that *primarily* a prophetic day represents one calendar year, but the broader principle states that **'a 'day' stands for a defined length of time'**. If it is not a year, it is usually longer and may for example denote a 'time', or an age.

This rule could also be divided into subsets, or alternatively separated into more rules, and further discussion on the definition of prophetic periods of time will follow.

The year is how long?

The next period of prophetic time I will discuss is the actual length of the year. By that I mean the number of days in the year.

Now you might think this is trivial, or that I have gone completely bananas. But if you are aware of the different terminology used in Revelation chapters 11 and 12 to define a 'time, times and half a time', 42 months, or 1,260 days – all seemingly for a nominal period of 3.5 years - it becomes apparent that there is a problem. Not that I saw the problem until it was revealed to me six years ago.

It is a problem caused by Babylonian custom and their interest in round numbers such as 360, 60 and 12 based on their sexagesimal i.e. 60 based counting system. This is a discussion I had earlier, in regard to the Babylonian counting system in 'Cuneiform and Clay' under the heading of 'The numbers game'.

It is also seen in other Middle Eastern traditions. For example some elements of this system, such as the 360 day year convention, were also followed by Macedonian Greeks, Greek speaking Hebrews and others from the broader region. They were well versed in Mesopotamian and Babylonian tradition and most used the lunisolar year in some form or other.

However some communities subsequently moved to a direct solar year. This, under the influence of the Egyptian calendar and later Roman (Julian) reforms. This gained momentum after 31 BC as a way of honouring the Roman Caesar, Octavius Augustus. It means that the Babylonians and Greeks at one time saw the value of a nominal 360 day year. It was easily adjusted to match a true solar (tropical) year, by adding a big end-of-year party period over a five or six day holiday. And even if you were using a lunisolar calendar, you needed to keep the lunar months and solar years roughly aligned for the practical purposes of life - like sowing, planting and harvesting.

The consequences for biblical prophetic interpretation, is that it has been the fashion for most commentators of the historicist approach (at least to my knowledge) to use what they call a 'prophetic' year of 360 days. After all, it seems in step with the

Revelation definition and therefore, what could be possibly wrong with that?

As it turns out, rather a lot; enough to thwart many an interpretation for a very, very long time!

It is true that interesting timelines can be built using a 360 day year, and that they seem to be significant to some degree or other. They may even be designed to give us a taste of the truth.

But there have been other schemes proposed as well. In a later chapter I will address the Daniel 9 prophecy of the seventy 'weeks' in more detail. It is a timeline of seventy 'sevens' (490 years) which can also be viewed as ten times 49 years – or as 10 jubilees of 49 years. Some analysts have therefore followed that scheme to our present time, and linked in some dates which they feel significant.

I have some issues with a 49 year jubilee, as against the preferred 50 year one, because like the 360 day year, it too seems to be a product of the Babylonian exile. Nevertheless there is no point in ruling out the work of others arbitrarily, given they may be parallel strands of a more complex, prophetically linked timeline. Whatever the case, here I will be outlining the rules of prophetic interpretation that I believe have been shown to me.

Very significantly, they result in a timeline that terminates in the very near future and heralds the return of the Lord.

Others have sought to justify the 360 day interpretation by invoking the sidereal day, which is the day based on the earth's rotation on its axis in relation to the background stars. But a sidereal day is only a few minutes short of 24 hours and the sidereal year only one day short of 365.24 days. This occurs because the reference adopted is that of the background stars outside the solar system, rather than the sun. It means that a sidereal year adjustment is no solution at all.

Now there have been many analyses of these prophecies using the historicist method over the years. The most famous practitioner over the last 400 years is surely Sir Isaac Newton. He was an extraordinary gifted mathematician, astronomer and theologian and the father of modern physics. Hardly anything escaped his attention; classical mechanics (the laws of motion), gravity, light, heating and cooling were just some of the things that gripped his imagination and were subsequently studied and quantified by him. In fact 'Newtonian' physics is still the basis for our everyday lives. Albert Einstein's relativity theory has only supplanted classical theory in the sub atomic and cosmological worlds, where speed-of-light interactions, nuclear energy considerations and related gravitational effects are important.

Isaac Newton was also the inventor, or co-inventor of the calculus and the one who introduced the gold standard in monetary policy. This brilliant mind and dedicated Christian also studied the prophecies seriously. However he was so far ahead of his time, that he was not around to witness the historical events that we have seen in the 19th to 21st centuries. Indeed, he was not around for most of the 18th century either. These pivotal events include the return of the state of Israel, and the various milestones along the way for the returning Jews – 1917, 1922-3, 1947-8, 1967 and 1979 being some significant examples.

Nevertheless, Sir Isaac was familiar with the Daniel 12 prediction that travel in search of knowledge would increase greatly in the last days, and it would not surprise if he wondered whether his research, and those of his contemporaries, were to be the catalyst for those last days.

Amongst the expositors of the historicist method around in the late 19th and early 20th century period, was an Irish pastor by the name of Grattan Guinness. He was from the famous brewing family, and was one of those to turn his mind to the task of unravelling the prophecies. As a young man, this was my first encounter with the historicist approach to understanding Bible prophecy. This was because Guinness was one of the writers who heavily influenced my father.

In Guinness' time, leading up to the events of World War 1 and beyond (and particularly the significant events around 1917 and the liberation of Jerusalem) it meant that it was natural to believe that a 360 day year, and a 'seven times' timeline may describe a period from ca. 604 BC to 1917. In fact one of his predictions in the late 1800s, apparently pointed to 1917.

However in my later 20's, I came to an impasse with this view of prophecy and let my interest ride for over 30 years.

I felt strongly when comparing futurist interpretations (that tended to get all the play time on church film nights) that the historical approach was correct, although both interpretations seemed to portray an imminent end of the age. However futurists kept changing their identifications as various 'cold war' and post 'cold war' national and global scenarios played out.

This was obviously because they had no timeline on which to hang their predictions. In contrast, the historicist approach seemed to have the potential to provide an overarching continuity through the centuries; a 'master plan' of world history.

But there were problems here as well; there was often little consistency with the timeline amongst historicist commentators. A variety of different start and finish dates were offered, with milestones which had no bearing on Jerusalem at all. Others did not acknowledge that dual 'time, times and half a time' applied, and used a single 1,260 year timeline instead.

The Guinness 'seven times' timeline appealed, but even there, too many options were offered as possibilities. In general, when the timelines were applied to historical events, they did not seem to fit as neatly as one would like. Often it appeared the interpreter had worked backward, from a known historical event and arrived at a questionable start date.

Now none of this is necessarily a sin, and it is also understandable that one would work from the known to the unknown. In essence the historicist methodology seemed sound, but the parameters and milestones were not convincing. There must be more to it I thought.

So I was unconvinced and my career and family took over my overall attention. At least that was something I could progress, and interestingly as it happened, I had my first of many Holy Spirit anointing experiences which anchored my being. It meant that analysing the prophetic timeline was no longer a priority for me personally. My journey in life (and faith) could progress harmoniously and productively regardless.

In fact from a philosophical standpoint, I was intent mainly on confirming that the contemporary geology I was taught at high school was indeed correct - and I touched on this earlier when describing my first encounter with a strong and active 'six day' advocate. Given that I was a technical specialist / surveyor working in a geophysical and geological exploration world, this was a very obvious and compatible career focus.

So, roughly fifteen years before my retirement, I started to look at the historical basis for biblical events again. I have discussed the major ones in earlier chapters and have shared my visit to Egypt in some detail - as well as a little of my experience in the Eastern Mediterranean, Israel and Jordan.

> It was therefore only in late 2011 (well after I had returned home) that the prophecies of Daniel caught my attention at all.
>
> Although it was not directly related to my travels and previous historical interest, these experiences prepared me in part for the inspiration that followed. I say inspiration, because even though I was prepared in spirit, it all came as a surprise; a massive shock in fact. I had not taken up studying the timelines again at all, and was not even aware of thinking about them.
>
> I had not read anything from prophetic authors for decades and I belonged to no prophecy groups either; it was just not 'in-frame'. In fact on reflection, the Lord had me to himself.
>
> On the first occasion, *I woke up in the wee early hours with the sound of a voice, seemingly from nowhere; apparently from somewhere higher, more sublime.* **It was like the calm, steady voice that Samuel heard,** but insistent nonetheless. It was the first, but not the last time that I had an inspirational episode.
>
> No dramatic night vision, *just bolt upright in the middle of the night and a voice, clear as a bell, that startled me* and was sufficient to send me racing to my living room for a calculator, my Bible and on subsequent occasions to Mr. Google. It was a simple message, but it felt like I was being chastised for my ignorance, **'There are 365 and a part days in one solar year'.**

The strange thing is that I knew *exactly* what the message was about. Perhaps Grattan Guinness and those that went before were a little bit off course! Why did they always use a 360 day year? Sure, the Babylonians used a ceremonial or sacred 360 day solar year, but for record keeping they were not interested in the solar year at all, but instead counted using lunar months. And returning to thoughts about our more recent historicist prophetic commentators; the ones who had called the Babylonian 360 day year a 'prophetic' year: Interesting I decided!

> ***Everything bad and undesirable is epitomised by the very name Babylon, in both the New and Old Testaments!***

I was dumbfounded! Why had I never seen this before? *Even primary school kids know there are 365 and a bit days in a year.* It therefore becomes uncomfortably apparent, that first century Judeo-Christian writers knew something that historicist commentators of most of the Christian era have never realised - at least not up until now. **It is only a convention!**

Mind you, it is understandable why the historicist scholars believed they should use a 360 day year. They were blindsided by the way numbers were represented in the books of Genesis and Revelation, while not understanding the history!

It would certainly have persuaded commentators in the past that a literal 360 day year should be used. But it is written in Greek and the Greeks followed the Babylonians and Assyrians before them. The writer of Revelation (living as he did in his later years in Asia Minor) knew without doubt that 360 days were only representative of a year, so he just followed his vision and the usual convention. Given the Babylonians used a lunisolar based calendar, whereas the Hebrew Mosaic calendar as outlined in Leviticus shows every indication of being a solar based one, it means we cannot ignore the implications of the length of the solar year in Bible prophecy.

Reinforcing this view, Leviticus gives a template for all the days starting from the first day of the year, beginning at the day of the first new moon at the northern vernal equinox, all the way through to Passover preparation day (on the 14th day). This is followed by the Feast of Unleavened Bread (starting the following day and the Sabbath) for seven days to the next Sabbath.

On the day after that Sabbath, *if and only if,* the first fruits from the field had come in, then 7 weeks plus one day (to the fiftieth day or Pentecost) is to be counted off. This means that 2 weeks + 1 or more weeks + 7 weeks + 1 day were counted off, *without* regard to any further new moons after the one at the beginning of the year. In other words, there is no sign in Exodus or Leviticus that the days are reset by the moon in two and a half months and it is therefore highly likely, that new moons were not used to adjust the calendar for the remainder of the solar year either; but only used to mark the passing weeks.

In my opinion, this puts the Mosaic calendar in the solar basket - at least for the purposes of prophecy and redemption.

There are potential caveats in terms of provenance that I will address later, but the important thing is that it is consistent with the Egyptian model, where a 360 day year plus 5 extra holidays were used to make up their solar year.

So if the Mosaic calendar is solar based, where does the observation of the new moon at the equinox come into the calculation? My considered view is this: because the equinox was not easy to pick without a sundial, or well-known landmarks, the designated astronomer-priest measured the shadows and noted the equinox. The message was then broadcast to the plebeians and shepherds in the regions, announcing that at the next new moon they should start their calendar year. In other words, the new moon was primarily a synchronisation aid to begin the year. Thereafter, using the new moons (or full moons) to count the days off would only have been a secondary consideration.

So, despite this lunar connection to kick off the year, the Hebrew Sacred calendar appears to be solar based and derived from the Egyptian model to a significant extent. After all, by the time the Exodus occurred, Israel had lived in the Nile delta (including Egypt controlled Canaan) for 430 years. It means we can be pretty confident that the 365 day solar calendar was the basis for their timekeeping in the early days after the Exodus.

However once they left the Nile delta, there was a problem. For starters there was no annual flood to check their solar clock and they needed another solution. Using the first new moon after the equinox as the marker, also has the issue that its appearance varies by up to one complete lunar cycle when compared to a solar calendar. Compounding the problem was the Egyptian 365 day year (0.24 of a day short of an actual solar year). This all meant that adjustment was needed one way or another in order to synchronise the seasons.

Fortunately they found that checking their new moon observation against the new shoots of spring and ripening crops, solved this difficulty. Once the 'first-fruits' were presented to the priest, the year was well and truly established by using nature to confirm the seasons.

Now this might not suit us today in our tightly controlled, time critical technical world, but that is the way it was done before a more institutionalised year was adopted. In those times the system was as fit for purpose as ours is today. The seasons might not always be consistent, but the seasons and crops were paramount. It means the first seven weeks of the omer (a grain measure) was variable, depending on whether the crops had ripened or not. This only emphasises the point that they used a solar (season) controlled calendar, not a lunar one.

In addition it matters little if your solar reference is an equinox, solstice or something in between. Importantly, you do not need to insert an extra month here and there as the Babylonians did. It was also a solution that avoided the pitfall of the Egyptian Sothic year and a wandering calendar.

But then the Hebrews were pastoralists and farmers and often on the move, and had different concerns to the city dwellers. They were very much keen observers of nature and no doubt the heavens - and they were certainly not distracted by television, social media, streaming services and home delivered meals.

Now despite the fact that the history of the Exodus and giving of the Law is embedded in the calendar (together with prophetic implications for the future) we cannot discount the possibility that the origins of the Sacred Calendar could be much older again; even back to the Patriarchs of earlier times.

Returning again to the prophetic analysts of more recent years; it is certainly not my intention to denigrate the previous work of historicist interpreters. After all, it is on their diligent, primary scholarship that this interpretation is based and I am grateful that others took the first steps and that my own father brought the historicist method to my attention.

In fact the message here is largely about fine tuning and focussing the prophetic binoculars more precisely. For my part, I can only assume that any misconceptions which have occurred, have been used by the Lord to obscure the prophecies until they are needed; right now, in the 'last days' at the end of the age.

And so the timeline that has been revealed to me and I will annunciate here, is therefore based on this new historicist interpretation of Daniel and Revelation using actual solar years.

Here I was, grabbing my calculator and looking at dates. Out with the Babylonian and in with a 365.24 day solar year to calculate the 'times' of Leviticus and Daniel's magnificent prophetic timeline of world history. Don't get me wrong. Just because it is a simple insight, I take no credit. I was shown it. It was as if the Lord himself shook me by the shoulders and said 'Wake up sleepy head, this is not rocket science, get to work!'

Lesson 3 (our third rule): **a prophetic year is a solar calendar year of 365.24 actual days.**

But as they say in the domestic appliance advertisements, 'there's more....!' If a prophetic year is **365.24 days,** then this is what is **understood as a 'time' when converted to years.** It applies to all the timeline prophecies in both Hebrew and Greek. More broadly a **'time'** appears to be a 'multiple' of a **'day'** and can therefore refer to other periods as well. I will discuss this shortly in relation to all solar system parameters.

Note: Before moving on, there are issues of provenance that may be of interest to scholars of ancient calendar systems.

The Egyptian Civil calendar started with the annual Nile flood and / or the rising of Sirius around July in the Julian calendar. Or it did so until a strict 365 day calendar was instituted and it started wandering through the seasons. The year was also broken into three seasons of four months each, reflecting the inundation, planting and ripening seasons along the river.

In addition the Egyptians also had a very ancient lunar religious calendar that honoured their gods. We know the names of the days for each lunar month, but little else in terms of timing. It too may have influenced the Levitical calendar.

Regardless of the above, there is also another question. How do we reconcile the fact that the Mosaic calendar began near the spring equinox (as per the book of Leviticus) while the ancient Egyptian Civil calendar kicked off with a July Nile flood?

This could be a special case of anachronistic use of a later calendar. The Levitical narrative may have been updated by Jewish priests during, or after, the Babylonian exile. We use our current western calendar in a way that allows us to date things that occurred before it was introduced, i.e. proleptically.

We do this so that we have a continuous log of history using a calendar with which we are thoroughly familiar.

The AD chronological system was only instituted in the 6th century. This was long after the Julian calendar was introduced by Julius Caesar in 46 BC. So dating events and the reigns of monarchs before the 6th century using the AD-BC (or CE-BCE) system is a case of backward extrapolation. It is a case of using a calendar with which readers are familiar with, to date *all* events, while keeping everything in correct historic perspective.

And so we find that the Babylonian lunisolar calendar began at the northern spring equinox, unlike the later Macedonian Greek equivalent introduced into Asia by Alexander the Great. The Greeks too had a lunisolar calendar, but it began at the autumnal equinox and this is where the Jewish Civil calendar year derives from; beginning as it does in the northern autumn.

So whether from the Babylon of Daniel's day, or millennia earlier in Mesopotamia, there was a calendar which began with the month of Nisanu at the vernal (spring) equinox, which is the basis for the commencement of the Mosaic Sacred Year as well.

If this is the case, the Leviticus account is either using the later Mesopotamian calendar retrospectively (when it comes to the beginning of the Mosaic year) or they inherited the custom much earlier from their Patriarch ancestors; those who lived in the northern Levant-Mesopotamian region. Either way, its uptake by the Jews would have occurred before the use of the structured 19 year lunisolar system was instituted, somewhere between the 8th to 5th centuries BC. The Mosaic solar year then morphed into a lunisolar one, following the Babylonian lead.

The Babylonian day also began at sundown, whereas the Egyptian day began in the pre-dawn or dawn. If so, this is a further example of backward extrapolation, and despite being an apparent puzzle to us now, has the advantage of providing more information on the likely history of the biblical text.

Regardless of provenance and the vast amount of evidence that has simply been lost in the desert sands and the mists of time, there is no reason to suppose that the seasonal timing of the Exodus saga was anything but in the northern spring. I suggest the copyists would have made sure that detail was not lost!

I have added this information as background and out of historical and linguistic interest, rather than a matter of immediate prophetic concern. Further discussion of ancient calendars is provided in the Appendices.

Time, times and half a time

The lesson above is completely ignored, or overlooked by preterist and modern futurist expositors. That omission completely trashes any coherent understanding of biblical prophecy, or the 'seven times' timeline; the very period outlining the plan of God over the last 2,500 years.

For the most part, the sequence of nations and empires and their identity becomes hidden, disjointed or confused if we do not appreciate this. It means we are not as prepared for the Lord's coming as we could be, given a true understanding of the timing and warnings embedded in the prophecies.

Where 'time' is used as a pointer to a particular event or action, or an indeterminate length of time as in 'at that time', 'give him some time', 'the times and the seasons', 'therefore at that time' or 'at the same time'; then the word 'zeman' is used in the Hebrew according to Strong's and Brown-Driver-Briggs.

These sources say that 'zeman' or 'זְמָן' (original script) is Aramaic from an Old Persian origin (Strong's Hebrew 2166).

But the word for 'time' used to define a period of a year - as in a prophecy of 365.24 days is the word 'iddan', where 'iddan, (original word 'עִדָּן'), is of Syriac origin, and probably from the Assyrian loan word '*adannu*', meaning *fixed, appointed*, or *definite time;* again according to Strong's (Strong's Hebrew 5732).

This latter word is the one we are interested in for our prophetic purposes. Just as a day, or a year, is used to define the length of a particular period of time, so using the day for a year prophetic principle, one 'day' becomes one 'year', and one year (365.24 days) defines a 'time' of 365.24 years.

Based on this definition of a 'time', we can then work out the duration of the period referred to throughout the Bible as a 'time, times and half a time'. It becomes three and a half solar years (of days) meaning that it is (365.24 x 3) plus 365.24/2 (a half). This equals **1,278.34 years** or 1,278 years to a sufficient degree of accuracy for most prophetic purposes.

Lesson 4: **a prophetic 'time, times and half a time' is actually equal to 1,278.34 literal years** when converted from days to years. Therefore **42 months or 1,260 days** in Revelation 11 and 12 is actually code for **1,278.34 calendar years**.

Let us now consider some scriptural examples. In Daniel 7 we find one of Daniel's night visions where he sees four 'beasts' i.e. animals. Each animal depicts one of four successive kingdoms. They are world empires starting with Babylon, and the portion we are interested in here is the interpretation of the dream, starting at verse 15.

15 "I, Daniel, was troubled in spirit, and the visions that passed through my mind disturbed me. 16 I approached one of those standing there and asked him the meaning of all this.

"So he told me and gave me the interpretation of these things: 17 'The four great beasts are four kings that will rise from the earth. 18 But the holy people of the Most High will receive the kingdom and will possess it forever—yes, forever and ever.'

19 "Then I wanted to know the meaning of the fourth beast, which was different from all the others and most terrifying, with its iron teeth and bronze claws—the beast that crushed and devoured its victims and trampled underfoot whatever was left. 20 I also wanted to know about the ten horns on its head and about the other horn that came up, before which three of them fell—the horn that looked more imposing than the others and that had eyes and a mouth that spoke boastfully.

²¹ As I watched, this horn was waging war against the holy people and defeating them, ²² until the Ancient of Days came and pronounced judgment in favour of the holy people of the Most High, and the time came when they possessed the kingdom.

*²³ "He gave me this explanation: 'The fourth beast is a fourth kingdom that will appear on earth. It will be different from all the other kingdoms and will devour the whole earth, trampling it down and crushing it. ²⁴ The ten horns are ten kings who will come from this kingdom. After them another king will arise, different from the earlier ones; he will subdue three kings. ²⁵ He will speak against the Most High and oppress his holy people and try to change the set times and the laws. The holy people will be delivered into his hands **for a time, times and half a time**.[b]* (NIV)

I have presented the interpretation portion of the vision here, because it brings to us the concept of 'time, times and half a time' for the first time in a famous prophecy.

We will look into this in more detail later, but according to our rule we are looking at a period of 1,278 years. This will actually give us valuable information about the possible identity of this later king, who we are told will 'subdue three kings'. However it will mean little, until we have established which 1,278 years the verses refer to within the overall timeline.

Another of Daniel's prophetic visions is from Daniel 12 and I will provide an excerpt from this chapter as well:

*"At that time Michael, the great prince who protects your people, will arise. **There will be a time of distress such as has not happened from the beginning of nations until then.** But at that time your people—everyone whose name is found written in the book—will be delivered. ² Multitudes who sleep in the dust of the earth will awake: some to everlasting life, others to shame and everlasting contempt. ³ Those who are wise[a] will shine like the brightness of the heavens, and those who lead many to righteousness, like the stars forever and ever. ⁴ But you, Daniel, roll up and seal the words of the scroll until the time of the end. **Many will go here and there to increase knowledge.**"*

⁵ Then I, Daniel, looked, and there before me stood two others, one on this bank of the river and one on the opposite bank. ⁶ One of them said to the man clothed in linen, who was above the waters of the river, "How long will it be before these astonishing things are fulfilled?"

*⁷ The man clothed in linen, who was above the waters of the river, lifted his right hand and his left hand toward heaven, and I heard him swear by him who lives forever, saying, **"It will be for a time, times and half a time.**[b] When the power of the holy people has been finally broken, all these things will be completed."* (NIV)

Daniel chapter 12 is one of the most amazing prophetic utterances we will ever know. There are others which are just as incredible, but this one sums up the complete 'seven times' timeline and more. Here I am only focussing on the first part of the chapter, because it refers to the 'three and a half times'.

We will analyse the whole chapter later, because it is the only passage in the Bible which clearly refers to the whole 'seven times' and the significant years beyond. As such, I will show that it covers a period beyond 2017 and our current time, but not much further. It is therefore an urgent message to proclaim.

In Revelation in the Christian New Testament, a series of visions were given to the Apostle John on the Greek island of Patmos. These visions appear to have been received from about AD 50 to AD 70 (and the destruction of Jerusalem) with the possibility that some may have been received later - this, given John is believed to have passed away in AD 98.

This was again a terrible time of persecution and fear, where a few, whose steadfast faith did not desert them, continued to seek their God and his word. Revelation 11:1-6

*I was given a reed like a measuring rod and was told, "Go and measure the temple of God and the altar, with its worshipers. ² But exclude the outer court; do not measure it, because it has been given to the Gentiles. They will trample on the holy **city for 42 months**. ³ And I will appoint my two witnesses, and they will **prophesy for 1,260 days**, clothed in sackcloth."*

⁴ They are "the two olive trees" and the two lampstands, and "they stand before the Lord of the earth."[a] *⁵ If anyone tries to harm them, fire comes from their mouths and devours their enemies. This is how anyone who wants to harm them must die. ⁶ They have power to shut up the heavens so that it will not rain during the time they are prophesying; and they have power to turn the waters into blood and to strike the earth with every kind of plague as often as they want.* (NIV)

Again in Revelation 12, there is an extraordinary picture of Israel and the scattering that followed, firstly after AD 70 and the rebellion of AD 135 – but then followed by the 'desolation'.

If there are any chapters in the New Testament that Jews should appreciate (regardless of their view of Christian doctrine) these are they; a pure prophetic word from God describing Israel, its early existence and the further tribulation and diaspora to come beyond the 1st century.

The beautiful image presented here and its meaning will be addressed in detail in a later chapter. For now I will just quote the first few verses that expressly describe the woman who is Israel, the birth of her child, the Diaspora to come and how long it will take for this vision to unfold. It is an example where the 1,260 description for 3.5 times or 42 months is used.

It demonstrates why Christian interpreters have so easily assumed the vision is referring to actual years, rather than a literary / number convention. An excerpt from Revelation 12:

> *A great sign appeared in heaven: a woman clothed with the sun, with the moon under her feet and a crown of twelve stars on her head. 2 She was pregnant and cried out in pain as she was about to give birth. 3 Then another sign appeared in heaven: an enormous red dragon with seven heads and ten horns and seven crowns on its heads. 4 Its tail swept a third of the stars out of the sky and flung them to the earth. The dragon stood in front of the woman who was about to give birth, so that it might devour her child the moment he was born. 5 She gave birth to a son, a male child, who "will rule all the nations with an iron sceptre."[a] And her child was snatched up to God and to his throne. 6 The woman fled into the wilderness to a place prepared for her by God, where she might be taken care of **for 1,260 days**.* (NIV)

Because both these passages only refer to a single 3.5 'times', it behoves us to try to understand which one - the first or the second?

We could reasonably guess, but we will shortly move on and make absolutely sure how the total seven times of judgement unfolds. In the mean time, the addition of the two halves of the tribulation becomes a corollary to our 'time, times and half a time' definition. It is the full period of 'trouble' or 'tribulation' of Israel as described in Leviticus 26. So back to our counting.

It goes without saying; **seven times is twice 1,278.34 years and is therefore a total of 2,556.68 years.**

I will round down to 2,556 years in general discussion, but with some trepidation. In the rare cases we are looking at fractions of a year, this may have to be taken into account.

Before moving on, there is at least one instance in the book of Daniel where the term 'time' is used for a literal calendar year, and therefore is an exception to the rule. At least that appears to be the case, unless we define 'time' in a broader fashion. The example is found in Daniel 4:1-36 and is a personal prophecy about Nebuchadnezzar himself.

It is a vision of a tree, which is an illustration of the king in all his glory. However it describes how the tree will be cut down, such that only a stump remains and is a reflection of how his mighty kingdom will fall. Despite this calamity, he is assured that the stump will regrow, if he but acknowledges the Lord. The pertinent portion is Daniel 4:31-32:

> *³¹ Even as the words were on his lips, a voice came from heaven, "This is what is decreed for you, King Nebuchadnezzar: Your royal authority has been taken from you. ³² You will be driven away from people and will live with the wild animals; you will eat grass like the ox.* **Seven times will pass** *by for you until you acknowledge that the Most High is sovereign over all kingdoms on earth and gives them to anyone he wishes."* (NIV)

Now when we examine this passage, it should be noted that there are some important differences with the other passages in chapter seven and twelve talking about 'times' for a fixed period. Firstly there are a number of the early chapters of Daniel written in Aramaic, rather than Hebrew, which could call into question the translation and intended meaning. So Daniel 1:1-2:4a and 8:1-12:13 are in Hebrew, while 2:4b -7:28 are in biblical Aramaic. This mixture of Aramaic and Hebrew is also found in the book of Ezra and presumably for similar reasons.

Secondly, the Hebrew introduction of chapters 1:1-2:4a is an introduction written later; apparently at the time of compilation. Thirdly, the prophecy is talking about a personal prophecy (although none other than a king) and not about a nation or society and is therefore constrained to a human lifetime.

This is personal for both Nebuchadnezzar and Daniel, and recounts Daniel's early life living in a foreign capital and working at the behest of the king. By contrast, the later chapters refer to Israel nationally. So the context constrains the usage of 'seven times' in this prophecy anyway, but if the use of 'time' in prophecy simply means multiples of a discrete 'day', then this example is not an exception, but part of a broader definition.

However there is another lesson here which we have already encountered in our previous history of Israel; whether way back with the Flood, Babel or Exodus. We need to be cognisant that biblical history is not always presented in a way we might expect in a modern academic work, where war and strife has not unduly compromised the result. In fact this biblical narrative was influenced by the literary norms of the times; and the times were desperate and fractured throughout the Babylonian exile.

As alluded to earlier, the prophecy is far more important to the scribe than the accuracy of the surrounding events.

In this case, I suggest accurately recording the prophecies was everything, while the circumstances surrounding the prophet were not so vital. In any event, the historical details would have been difficult to establish by later compilers, who may have been rank amateurs anyway.

Most of the elite Jews did not return from Babylon and consequently, it is likely that the prophecy refers to something that happened to Nabonidus, a later king of the neo-Babylonian line. He was a devotee of the moon god Sin according to extra-biblical sources and was exiled to the Oasis of Taima (Tayma) in north-west Arabia. So the unfortunate circumstances may have been his fate. If so, it confirms that the book of Daniel has been edited and recompiled at a later date; possibly from more than one manuscript. This has been undertaken by someone who was unaware of who was who, and assumed the anecdote was about Nebuchadnezzar, the Babylonian king who Daniel knew. The prophecy is true enough, but the names mixed up.

Fortunately none of this detracts from the prophecies in any significant way. It indicates however, that the compilers had more difficulty reassembling the pieces of the puzzle in front of them, than we might suppose. These lessons also serve to clarify further the prophetic nature of the terms 'day' and 'time' and the importance of the particular word used. It also illustrates that context is critical and that there has been some muddying of the waters due to human error.

This should not surprise. This is real life. It illustrates that understanding the overall prophetic timeline is critical. Once established, it provides a solid framework in which to place individual prophecies and historic evidence, and clear up misconceptions and erroneous interpretations.

A new millennium

But there is another bombshell; a millennium is not a millennium! You have got to be kidding me mister! After all, a millennium is a thousand years isn't it?

Well, not always. The history books tell us that the First Crusade was initiated by Pope Urban II at the Council of Clermont in 1095 and lasted from 1096 to 1099. However it had probably been on the minds of church authorities and theologians for many decades. It seems earlier scholars were concerned that the end of the millennium would occur from AD 1000 onward, and I suggest that this was prompted by the content of Revelation 20 and the so-called 'millennial reign of Christ'.

The Crusade eventually happened after Pope Urban rallied the faithful and gained the support from influential western figures, including kings and military men.

Now, the Crusades were organised to provide safe passage for European pilgrims to Jerusalem and the Holy Land. This occurred after the Sultan of the day decided to deny access to the region. The idea that the end of the millennium heralded the end of the age, appears to have been very much a driver of the burgeoning pilgrimage, and in any event, the Muslims had taken the Holy Land by force and guile and it seemed appropriate to reverse this abomination.

So there were altruistic motives for the earlier Crusades, and it appears they were very much expecting the Second Coming of Christ at that time. It also shows that the prophecy analysts of the Crusader era saw the fulfilment of the Messianic age in Revelation chapter 20, as part of the Church Age and not some future period after the second coming.

However war is war. It turns otherwise reasonable people into killers, and looting some of the treasures of the orient was probably a magnet for some minds.

At any rate, the First Crusade became a desperate fight for survival once they arrived in Anatolia. Toward the end of the 200 year Crusader period, higher motives seem even more conspicuous by their absence and the Crusades became an excuse for Western European militias to raid and pillage openly. Even cities such as Constantinople became fair game for looters, as a West versus East schism further developed in Christendom. It becomes apparent that by this time faith in the prophecies and their interpretation had faded, and baser desires had triumphed.

Despite this sorry end to the Crusades and a long hiatus in between; a thousand years on at the turn of the 19th and 20th centuries, many scholars were again looking closely at the prophecies – perhaps for the same reasons - the end of a millennium was not far off! What was going to happen?

The world had never been free of war, but the tensions in Europe and the prospect of the first truly industrial scale conflict (using new and monstrous self-propelled contraptions on land, sea and air) were both mind boggling and horrifying. Perhaps more significantly, Zionist activism took advantage of the oversight of Palestine and Trans Jordan by Great Britain; a responsibility given under the auspices of the League of Nations.

An agreement was finalised in 1922-23 and followed the significant events of 1917 when Jerusalem was liberated from the Ottoman Turks; a truly historic and prophetic accomplishment under the command of General Allenby and his British Empire forces during the Great War.

Ultimately, after many trials, skirmishes and outright war - together with the transitioning of the League of Nations into the United Nations after World War 2 - there emerged the unlikely conditions for the declaration of the State of Israel in 1948.

This prime example of the 'budding of the fig tree', prophesied so long ago, was a sign that the people were back and a promise was being fulfilled!

Given the obstacles over centuries, it was a modern miracle which again had believers scrambling for their Bibles.

This unlikely occurrence is in contrast to the stance the UN is currently taking, and the minimalist support and mixed messages from long time national friends. They are heavily outnumbered in today's UN makeup anyway, but it all shows that despite the will of men, the Almighty has a plan and works his magic accordingly. Summed up; it can only be said that 'God moves in mysterious ways his wonders to perform'.

Now a question of timing arises, because the return is linked to the 'seven times', but also to the concept of an 'age'.

A thousand year millennium was 'anything but' to the biblical prophets - and if that sounds like tautology, it is meant to be. Nor did the ancient astronomers measure the 'ages' that way. In fact the title of this section could be better expressed as 'ancient and prophetic ages revisited'.

There are at least two ways of understanding long periods of time in the Bible; one is prophetically 'times' based, but both are astronomical in origin. The second is also woven into prophecy as an overarching age structure, complete with meaning laden messages for the faithful.

The loss of this knowledge was one of the first departures by western Europeans, from a true understanding of time and prophecy. We see the problem beginning before the Crusades, but by the Reformation years of the 16th century, there is vastly more distance from reality because of Catholic – Protestant mudslinging. Finally, the corruption of ancient understanding by 99% of Protestant prophecy expositors in the 20th and 21st centuries, is simply mind-blowing and totally dismaying.

All appreciation of the prophetic rules and milestones save very basic 'fig tree' signs are misinterpreted.

It also appears that the biblical analysts of the Crusader period were struggling with a number of interpretation issues given their circumstances. So examining their situation is likely to be instructive. The discussion and debate that they had over this topic would have resulted in some enlightenment, but also a good deal of controversy and confusion.

As alluded to previously, they seem to have been expecting the imminent return of the Lord, and the first sign was probably that 1,000 years had passed since Jesus' first advent. The Revelation 20 reference would have prompted their actions together with the presence of Islam in the Holy Land in recent centuries.

Whether they saw the 1,000 years as a literal period as so many do today, or alternatively as the first half of the Pisces Age is another question. The literalist approach does not lock into solar system parameters in a way that could be considered prophetic. For example, there are 365.24 rotations of the earth (one year or prophetic 'time' of 365.24 years) in one revolution of the sun, but what is significant about 1,000 revolutions? How does 1,000 link to another independent solar system parameter?

On the other hand they may have been referencing the first half of Israel's tribulation (the first 3.5 times) as the 1,000 years where Satan was restrained and seen the subsequent occupation by Islam as the relatively 'little' while of satanically inspired activity of Gog and Magog; this, as described in Revelation 20:7-10. A reference to *'the thousand years'* could simply have been shorthand for the first *half of the 'seven times' timeline.*

This seems quite plausible under the circumstances.

Now earlier in the chapter called 'Ice and Flood', I mentioned a book by Graham Hancock and Robert Bauval. It proposed something that is believed to be an erroneous finding in regard to the age of the Sphinx at Giza - that the Sphinx (and body) as we see it today, is 10,000 years old.

Despite this, there is a positive spin-off from the research of these authors and their compatriots. They analysed astronomical data which included the heliacal rising in the east of the various signs of the zodiac. They noted that ancient astronomers and writers mentioned this heliacal rising of the zodiacs as part of astronomical time keeping. This view of the zodiacs (as seen just before sunrise and specifically measured at the equinox) is used as a marker for the passage of the ages. It is commonly called the precession of the equinoxes.

I also described how the precessional cycle is calculated to be 25,771.5 years in duration and is due to the earth's axis (currently tilting about 23 degrees) slowly turning with respect to the sun and the background star constellations.

The modern understanding of the zodiacal cycle as defined since 1930 by the International Astronomical Union (IAU) is also an issue, and not the same as was used by the ancient Mesopotamians. In fact the IAU uses a modified datum and 13 signs, including an extra one called Ophiuchus, the Serpent Bearer. This sign slots in between Sagittarius and Scorpio. So if we divide the zodiac cycle up equally using 13 signs, each sign reigns for a little less than 2,000 years, rather than 2,148 for twelve signs. In fact using 13 signs gives us a 1,982.4 year zodiacal age (or millennium). Of course neither the thirteen, or the twelve sign cycle gives us an exact 2,000 year zodiacal age, and so for now, I have included both zodiacal sequences and a nominal 2,000 year one in the biblical master timeline. Figure 4.

However there is a compelling case to use the 12 sign system, when using the 60 based counting system that the Mesopotamians invented. The 12 zodiacal signs are also obviously attractive when dealing with Abraham and the 12 tribes of Israel. A 12 sign zodiacal system is therefore a candidate for prophetic purposes, because there is every indication that the stars and their groupings were used to demonstrate prophetic and messianic truths. This also goes for the seasons, the solstices and equinox crossing points, which are in some ways similar to each earth day and the rising and setting of the sun.

Having stated the above, it is likely that a nominal 2,000 year period has special meaning. This is because the sons of Joseph - Ephraim and Manasseh - became prominent in ancient Israel, and therefore made the number of tribes up to 13. This then provides a potential basis for a compromise zodiacal period for prophetic purposes.

Additionally, the complications in determining which zodiac sign reigns at any particular time are fairly arbitrary. They include such issues as; how far key stars must be above the horizon for transition, the latitude and altitude at the observational location and how to measure the equinox accurately enough for repeatability. Also, whether to take in the variable size of each zodiacal grouping of constellations, or simply use standard 30 degree segments for all the signs.

You can bet your boots, that the ziggurat jockeys in Mesopotamia and the pyramid boffins of Egypt understood that all these parameters needed to be assessed. Now the variability of possible start dates used by the ancients amount to a century or two as far as I am aware, and so for our purposes here I am comfortable with lining up the zodiacs with the life of Christ.

It also appears there are messages in the zodiacal names which have some revelatory aspect to them; this in relation to an overall Divine plan. The meaning ascribed to the various signs and their sequence suggests this. In other words the zodiacal signs are prophetic, despite there being only a brief mention in scripture - the Psalms and Job for instance. It may have been a prior revelation, prominent during the time of the patriarchs and prehistory - and back in the days when writing was not sufficiently developed to record such inspiration. In fact back in the days of pictures and symbols, wall paintings and carving.

This would be similar to other key aspects of the Messianic story, which were revealed prophetically to the ancients and incorporated in seasonal festivals. The stories of virgin births, death and resurrection and atonement are carried in many stories of 'the gods', and they probably derive from common sources in Mesopotamian or Egyptian history. In any event, we know the precessional cycles were described by the Greeks of classical antiquity in the mid 5th century BC, and this probably resulted from much earlier knowledge. It means that the precessional cycle is the shortest of what is called the Milankovitch cycles, but is also the critical one for our study of the ages.

Continuing with our 'precession of the equinoxes' theme, consider the following: I mentioned the priest-king Melchizedek of Salem earlier, with reference to Abraham and how Jesus was compared to him in the New Testament. His name in Hebrew means 'my king is righteousness' and although Hebrew is an old language, we have established it is based on even older languages drawn, it seems, from Ebla, Mari, Assyria and Mesopotamia. This, in the same way all other languages are formed.

I suspect therefore, that the term 'righteousness' may have come from the idea of 'most high', where the attention of the priest-kings were focussed; this in terms of their observation of the heavens and other ideals of elitism, spiritual awareness, exemplary living and closeness to God. I also sense that the 'zedek' part of his name originally meant something more practical than righteousness, and the clue is the similarity in pronunciation and spelling to 'zodiac'. Almost identical in fact, given we know vowels were not used in the earliest writing and that the original word has apparently been transliterated verbatim.

Melchizedek is described as a priest-like figure in the New Testament and he may have come from a line of priests dating back to prehistoric times, or something even more amazing.

So this is how the ancients understood the world and read the ages before writing. Here we have a link to revelatory messages and the timing of the ages written in the sky; a more ancient and inherently pristine understanding, that was taught in the days when the priests and magi knew the definition of a real millennium. This is therefore our next lesson:

Lesson 5: **an 'age' in prophecy is never a literal 1,000 years, but based on either: (a) a zodiacal age which is close to 2,000 years in length or (b) is simply shorthand for a** *'time, times and half a time',* **or 1,278 years.**

It means that a **whole age, or one half of a zodiacal age** are potential candidates for the Revelation 20 example, as is **half of the 'seven times' of tribulation.** In addition, in any particular instance both periods may line up in some fashion.

Now the precession of the equinoxes and the zodiacal cycle over nearly 26,000 years is an overarching cycle of the ages. But it is mimicked in an annual way by the 12 lunar months of the year. The astrological cycle we are familiar with in regard to our birth signs, roughly follows this scheme and is derived from the sequence of constellations we see in the sky at night.

The ancient Egyptians and Macedonian Greeks followed this more closely, and in fact at one time used a ten day week known as a decan or decameron. It was very close to one third of a lunar month and an alternative to dividing it up into four weeks of seven days, as the Hebrews did.

If we consider that one lunar month is a whole zodiacal age in microcosm, it stands to reason that dividing a zodiacal age of 2,000 (nominal) years into three (3) or thirty (30) is a concept the ancients might have used.

Interestingly however, there is an example in scripture of another method for dividing up a zodiacal age. It is one where the age is not divided into ten, but into twelve and then goes on to use a half of a twelfth, as the smallest fraction. This alternative scheme follows the divisions of the overall zodiacal cycle itself and again suggests a Mesopotamian origin. The take home message is; back in the day, a zodiacal age was divided into finer increments for more precise chronological measurements.

Now I have gone to some length here, to use a likely misconception from the Crusader period as a lead in to further analysis of the zodiacal ages. I have done this in part, because of the resistance to the idea that the zodiacal signs could have anything whatsoever to do with the Bible; experience teaching that many see it simply as astrology.

Earlier when I introduced Lesson 2, I suggested that there may be some usage of the term 'day' which potentially refers not to a calendar year, but to a 'time' where a time is an extended 365.24 years as outlined previously. Context is everything of course and dual interpretations may apply. There are also one or two cases where a fraction of the 'seven times' or other prophetic period, may be expressed as a fraction of a 24 hour day.

Furthermore, as explored much earlier, the usage of 'day' in Genesis 1 is consistent with a prophetic vision. There a 'day' is not meant to be a literal day, or a calendar year, or even a 'time'. Nevertheless it is a defined period of time, where each of the six ages have a beginning and an ending and its own theme. The narrative then moves on to the next age or epoch.

This brings me to the solar system reality of 'days', 'weeks', 'lunar months', 'years', 'times' and 'ages' which I touched on earlier. A day is the time taken for one rotation of the earth on its axis with respect to the sun. A seven day week is about a quarter of a lunar month and lunar months are determined by the waxing and waning of the moon. This phenomenon is controlled by the period it takes the moon to revolve around the earth, while the earth in turn, revolves around the sun in 365.24 rotations of its own axis. As its axis wobbles, the millennial (zodiacal) ages are produced to determine the ages.

Thus we have cycles within cycles that define our earth and solar system. It means the prophetic timing involved is unique to us, and if we lived on another planet (and perhaps other beings do) then the Lord would use the parameters that apply to those solar systems to reveal his timing strategy. When you meditate upon this, it becomes rather inspiring and humbling.

It is also an overarching lesson for biblical prophetic timing; one that has emerged as a consequence of this study.

There is another rule which is not particularly intuitive, but very important. We need a geographic reference point for examining the rise and fall of nations which are part of our Jewish prophecies. It is very easy to go online, or to the history books today and find when this empire, or that nation first rose to prominence in their own land. However if we are guided solely by that, we can easily be blindsided.

As important as geography is to identification in a general sense, that is not the primary way the prophets of Judah identified the **sequence** of nations in their prophetic visions.

What is required, is to view the rise and fall of the nations and empires in relation to how they intersected with Israel history. That means the timing of their entry into Judea; and more often than not, by their attack on, or siege of Jerusalem itself.

So Lesson 6 is: **the start and finish of the influence of foreign powers and therefore their identity, is determined by their entry and exit from Jerusalem.**

The prophecies are Jerusalem-centric and not governed by the place of origin, or the history of the nations in their own land. That is not to say that the prophecies are restricted to Israel at all, as we will soon see. But rather, that the Gentile interaction with Jerusalem is critically important in identifying the empires of prophecy and the sequence in which they appear.

The importance of this rule will only become apparent, as we see the difficulties of identifying various aggressors by their prophetic image. *It might seem trivial at first, but it is a real eye opener when applied consistently.*

Having stated the above, there is one case where two end times apply to an empire. In fact, they are more correctly described as a number of parallel kingdoms and are usually described as the Ten Kingdoms in prophecy. One end-time applies to their earlier time in Jerusalem, the second end-time to their interaction with Israel at the end of the age. In each case they interact with the Holy Land at the end of a period, and that period is at the end of a 'time, times and half a time' of prophecy. So in reality it is not an exception, but a twofold fulfilment.

There are other considerations which may be required to I.D. a national transition under some circumstances. They could include the movement of the capital, or a change of language. The latter appears to be an important key to identity.

The next formal lesson relates to the start of the 'seven times' timeline:

Lesson 7: the **'Seven Times' commences with the Babylonian exi**le.

This will be discussed in greater detail in the next chapter, where we will look at the key milestones from the exile onward.

Now these lessons on prophecy will appear rather complex to those who have not encountered them before, or are new to prophetic matters. But understanding how historic events are related in time, not only allows us to get the kingdom sequence correct and identify the national players, it also allows us the possibility of reconstructing the original intent of the prophetic word - even where the text is ambiguous or poorly preserved.

Sadly, the lessons will also be difficult to swallow for those who carry much intellectual baggage from other interpretations.

CHAPTER 10

Age of the Gentiles

*There is a time for everything,
and a season for every activity under heaven*
ECCLESIASTES

Good time, bad times

We have now established that the idea of a 'day for a year' in prophecy comes from the prophets of Judah, especially Ezekiel and Daniel. Daniel is also the prophet of the 'timelines' and John in the Revelation provides backup and much extra detail for the end of the age.

I have also suggested a 'time' is not 360 years, but 365.24 years to a close approximation and that Israel was disciplined 'seven times' beginning at the Babylonian exile – our Lesson 7.

An obvious question arises: if there were seven times of national trauma, was there a 'time' of greatness and prosperity?

There certainly was, although it may not be spelt out specifically in Leviticus or anywhere else. However it does not take much imagination to appreciate the time of greatness began with the Kingdom of David and Solomon. This was in the early to mid-900s BC and included the dedication of the First Temple when all Israel remained united. Of course, it was soon followed by the days of the Divided Kingdoms, where Judah and Benjamin ruled in Jerusalem and the Ten Tribes in Samaria.

Nevertheless, the two Kingdoms remained autonomous and self-rule prevailed, until a sequence of threats and foreign incursions resulted in the capture of the northern tribes by the Assyrians by 701 BC and the Babylonian siege of Jerusalem in 589-588 BC. This ended Israel's golden age and their days of glory.

The 'seven times' of trouble therefore date from the time of the siege. However there is always a lead-up to these events, so I believe it is acceptable to use 590 BC as the first year of tribulation. To use 590 BC instead of the best estimate for the beginning of the siege is a minor adjustment, but the reader may feel that it needs some explanation.

Essentially the siege was only part of the overall Babylonian military campaign against Jerusalem. Shortly it will be explained, how the remainder of the major milestones on the timeline are in fact start dates, for events that take 3.5 literal years before they reach their climax.

So in the case of Nebuchadnezzar, we can legitimately suppose that he would have sent out many warnings of impending Babylonian military action. In fact we know that this was the case and that the Jews were particularly warned in regard to alliances with Babylon's great rival in the south; Egypt.

The Egyptians had made a recent incursion into northern Israel - one of their favourite rallying points from previous centuries - and this put the Babylonians extremely on edge. Around 597 BC when Jehoiakim was king in Jerusalem, the Babylonians took him captive (along with Ezekiel it seems) and for a short while his son Jehoiachin ruled. In fact some of the Jewish elite were removed to Babylon prior to this, including Daniel.

As it transpired, Nebuchadnezzar was not satisfied with Jehoiachin's efforts and had another king, Zedekiah, placed on the throne as a compliant (puppet) monarch. It becomes obvious that Babylon now had overwhelming control, but was still concerned about the threat from Egypt. From the Babylonian's point of view, Jerusalem was theirs and they now expected absolute obedience from the Judah ruling class. Any sign of deceit or conspiracy on Judah's part was met with swift action.

Despite the overt interference of this earlier period, it becomes apparent on inspection, that the prophetic 'seven times' does not begin until the next phase. This is the time when Zedekiah's emissaries resurrect talks with the Egyptian Pharaoh Necho II of the Egyptian 26th dynasty.

As is the way with such things, the Babylonians got a whiff of the negotiations and reacted predictably and with intent. We understand this, because Jeremiah in chapter 42 of his book, warns the Jewish hierarchy to avoid any dialogue with the Egyptians that would antagonise Nebuchadnezzar. Ezekiel also gives out the same warning in his chapter 17 prophecy of the 'two eagles' (Babylon and Egypt) – although he seems to be addressing his fellow exiles at this time.

We also learn that many from Judea subsequently fled south to Egypt and Arabia and that Jeremiah, Baruch and some of the royal entourage were in that cohort. The question is, what happened to make Jeremiah take this step?

The obvious explanation is that Nebuchadnezzar's second (third or fourth) warning would have been before or around 590 BC, whereas Jeremiah's later actions and pronouncement were after this date. By this time it was obvious, that the Babylonians were mobilising their army and were indeed heading again to Jerusalem. It was time to get out of there; to vamoose.

So the event which kicks off the 'seven times' of trouble is Nebuchadnezzar's final attack and siege of Jerusalem, and this period I will commence at a nominal 590 BC and finish after 586 BC and the acknowledged end of the siege.

As other milestones are highlighted, this will become more understandable and the methodology confirmed.

It will also become apparent, that if I used a 365 day Egyptian solar year instead of a 365.24 one, the timeline could be matched more closely with 589-588 BC and the best estimate for the start of the siege. I am somewhat reluctant to use the 365 day approach, because of the issues arising from the use of the Egyptian Sothic year I outlined earlier. I prefer not to round off figures to that extent, for fear of unintended consequences.

Despite my concerns, the 0.24 fraction of a year only changes the overall 'seven times' timeline from 2,556.68 to 2,555 years. This would only be a problem, if we were analysing very specific dates at the end of the gentile period, i.e. in our time (the 20th and 21st centuries).

For general discussion therefore, I am rounding down to 2,556 years for the whole 'seven times' as outlined earlier.

The 0.68 of a year lost this way represents 8 months and if those months *were* to be taken into account, they could be distributed with 4 months before and after the 2,556 years, or indeed, the whole 8 months tacked on whichever way made sense for fine tuning. There are also other factors which would impact a more exact timeline, and they include the use of particular calendars and the beginning of their year; whether in January (Roman), northern spring equinox (Mosaic / Mesopotamian), summer solstice (ancient Egyptian and Athenian) or autumn equinox (Greek and Babylonian) calendars.

Refocussing again on Israel's previous 'golden age': 590 BC minus 365.24 years leads us back to 956-955 BC.

This then becomes the commencement of the 'time' of prosperity, when Israel was at its zenith in the Levant. A great time in fact; when Israel gained its autonomy and threw off the Egyptian yoke completely.

There was a bit of insurance at one point, with a marriage to a pharaoh's daughter. But of course we are told that Solomon had many wives, so political expediency did not exactly cost him too much. Cynicism aside, the days of a Pharaoh's presence in the north at Beit She'an and Kinneret (Galilee) was a distant memory. Israel joined with the Phoenician city states and sailed the world looking for raw materials for their metal working industries, jewellery and other trading ventures. The biblical narrative suggests that at the height of Solomon's kingdom, most of Syria to the Euphrates paid tribute to the king.

Now a common estimate for the inauguration of the First Temple is also circa 956-955 BC based on a King Solomon reign of 970 to 931 BC.

It must be conceded though, that there are a wide range of other dating proposals, including a traditional Judaic dating in the 9th century over 100 years later, or earlier - perhaps more probable dates - in the late 11th century BC.

1 Kings 6 states that the Temple was completed just 13 years into Solomon's reign and 7 years after it was started. Solomon also completed his own palace within an extended 20 year period, i.e. according to the most likely interpretation of this account. The start date around 5 years into Solomon's reign is also significant on two other counts. It was the end of David's reign, meaning there was likely a four year co-regnal period, and it was also the end of the 480 years going back to the Exodus, according to the reference in 1 Kings 6:1.

As discussed earlier, these were the days when we are told the presence of God was felt in the Temple and manifest as the

Shechinah (the dwelling, settling of the Holy Spirit). They were blessed and heady days and they lasted for a 'time' of 365 years despite the national split, warnings of idolatry, external threats and the deteriorating conditions toward the end.

So whether 955 BC was the inauguration of the Temple, or some other significant date early in Israel's history is a moot point. It is mainly significant in terms of determining the time of the Hebrew Exodus. Indeed the proposition could be advanced, that the seven times of judgement do not have to follow immediately after Israel's time of greatness. This would allow the wonderful time of the kingdom period to be detached and moved backward in time, in order to embrace David's early reign - even that of King Saul.

The important thing is that it does not materially impact the discussion of the 'seven times' going forward.

I will now round out our formal lessons with three more. One of these results from our study of Ezekiel 4 in the previous chapter and from Daniel in his book. There will then be Ten Prophetic Lessons in all to guide us; something that neatly parallels the Ten Commandments.

Our **Lesson 8** therefore states: **430 (40 + 390) is a number which represents ALL Israel.** And: **3 times 430, i.e. 1,290** (as used by Daniel in chapter 12 of his book) **is a historic milestone dedicated solely to the Israel - Egypt relationship**.

In Ezekiel we learnt the prophetic principle of a 'day for a year', but were also introduced to the numbers 40 and 390.

The significance of these numbers and their meaning is well hidden and not at all obvious in relation to the prophetic timeline. Daniel just provides the $1,290^{th}$ 'day' as a milestone; he gives us no explanation or description of its significance, or what it might represent.

However the 430 years is also the number of years that Abraham's descendants spent in Canaan and Egypt. That is, after he left Harran and the Northern Levant. Suffice to say, once these numbers are understood, they reveal some remarkable lessons about biblical history from Abraham to the present.

There is another rule which is important to get our heads around. It is where a personal description for a specific person (and in a particular timeframe) is also used in a dual way; to describe a line, or lineage of people as well. The notion can also be extended to cities and regimes, where a city is used to identify a nation, kingdom or empire. It is a common way to refer to nations anyway, but is well worth emphasising.

Lesson 9. A personal description is often used to describe a **whole lineage of people, or an empire** with the same characteristics. **The personal name is often the initiator,** or other prominent individual **who personifies the line in question.**

For example 'Adam' means 'mankind', but also Adam, the father of Cain and Abel. Likewise, the 'antichrist' means a whole lineage of figures opposed to God (and in fact pose as substitutes for the true Messiah) but also satanically lead individuals who are identifiable in the context of history.

Now I made a comment suggesting **that each timeline milestone covers several years rather than a single one.** In fact this proposition will be formally labelled our **Lesson 10.**

As previously indicated, each of the 'day' milestones of Daniel appear to cover a **seven year span.** Furthermore, the first year is often the most significant and pivotal one, although peak activity is often in the centre of the period.

I acknowledge there is limited historical information from which to work and that this rule it is not easily proven. The milestone where we get the strongest indication of this, is the very last one on the prophetic timeline; the only one still future. However Jesus' ministry on earth and all the key milestones along the main timeline, provide strong support as well.

We will deal with all of them in due course, but support also comes from one or two other places in scripture. For example, in the Apocrypha we have the very interesting prophetic work called 2 Esdras, which is a non-Masoretic sourced translation of the second book associated with the prophet Ezra. In chapter 7 (amongst other interesting things) we are told that when the 'day of wrath' comes (or words to that effect) that '... *it will last for about a week of years*'. Now there is the 'about' to qualify the statement, but there is no doubt that this is a special instance of the 'day for a year' prophetic principle and a clear indication that this time will last around seven years.

It therefore follows that all the other milestones are likely to follow the same pattern *of 'about' seven years* and that *prophetic 'days' such as the 1335th day of Daniel,* also refer to a seven year period. The 'about' is also consistent with the idea that a seven day week is 'about' a quarter of a lunar month.

So this lesson is really a special case of Lesson 2, but one I have highlighted and given a separate slot, because of its importance throughout the prophetic timeline.

Summing up: The initial year as the primary marker for each seven year milestone follows naturally for me - if somewhat arbitrarily. It has also become obvious as I have worked through the history, that there are cases where the middle of each seven year period is also very significant, as can be the last year. So this will all be an ongoing consideration as we move along.

There are other contenders for the Ten Lessons which may be encountered in passing, but the above seem the most fundamental for this analysis. Other rules and guidelines could be proposed and further suggestions may emerge during the study.

At this juncture I feel I should insert a general disclaimer in regard to inspiration and vision. Hopefully the reader will have gathered throughout the previous chapter, that I believe I have the honour of sharing some *real* inspiration from the Lord. These specifically relate to the prophetic timeline and associated prophetic principles, but will also include insights into some of the identities in the prophecies to come.

So I will share more inspiration, but I will also interpret things as I understand them, given my overall experience. In these instances it is about conclusions which I feel are right and correct, rather than a direct word of knowledge. This includes all the wisdom I can recall from literary sources, together with that obtained directly from students of scripture, pastors, laypersons and acquaintances.

After all, we are all in debt to the devout men and women of God we sometimes have the privilege to meet and hear.

In regard to fallibility, the Apostle Paul writes in a very open way, sometimes admitting that those he is writing to are well aware of his weaknesses and sometimes 'foolish' ways. In his letters, he is known to distinguish between where he has a word direct from the Lord, or where in fact, it is more along the lines of personal advice, albeit based on sound principles.

This discrimination between the two sources of his teaching will unsettle some folk, who prefer the idea that all scripture is inspired at the same level and is of equal value. But I am also sure that others will recognise that where they have had personal experiences from God; that the Lord speaks on various levels according to their need and the job he has given them.

On rare occasions he reveals his word dramatically and at other times prods gently. Then there are the more frequent times where he seems to say 'you can work it out for yourself, you now have the knowledge and wisdom - get to it'.

In the latter case the scripture is still inspired, because we are being taught sound, spiritual, life based principles. But it is in a practical, easily understood, common sense way.

In this vein, the history sections of this volume are examples where a lot of research has been required to extract relevant facts. Mind you, I have a real sense, that much of this knowledge has been placed in my path to trip over and remember when needed. But the fact remains, while I hope there are no significant errors or omissions, there is always the possibility some corrections may be needed, or some opinion modified.

So back to the last years of the First Temple period, where Israel was fractured into two and subject to two separate captivities / exiles; Assyrian and Babylonian. A time where many were forced to vacate their 'promised land' and Jerusalem.

It is interesting to note, that the Bible does not dwell in detail on this exodus or escape *to* Egypt. The one *out* of Egypt under Moses was a triumph of sorts, at least a successful escape to their own land. This 'flight *to* Egypt' is not covered in glory like the Mosaic Exodus. Apparently the Jewish paparazzi of the day failed to capitalise on Jeremiah's foresight and leadership.

It is not that the Bible says *nothing* about this flight to Egypt; it is simply that the exile of so many Jewish elite to Babylon commands our attention.

It is also probable, that the number of Jews and associated Ten Tribe folk who managed to flee south to Judea from the Assyrian conquest of 722 BC, would have been considerable. This will depend of course on the overall population and how many from the Ten Tribes were taken into captivity. I suggest it also depended on prior migration and escape, using the relationship with the Phoenician city states and their maritime colonising and trade activities. However we can safely assume, that by this time Judea was now occupied by Judah and Benjamin and a mix from the Northern Tribes. The 'two sticks' of Ezekiel were already reuniting as they fled south as the Jews.

One imagines though, that with the final Babylonian assault on Jerusalem looming ca. 590 BC, it was more difficult to reinvent the catastrophe as some sort of Pyrrhic victory. There was simply no escaping that fact, as they headed south to Egypt, Arabia and Africa, any 'which way' they could. No marvellous Mosaic escapade from a foreign land, but in fact an embarrassing and hurried withdrawal to the safety of that self-same country. This was an unmitigated disaster. Nothing to celebrate here!

This love-hate relationship with Egypt needs to be looked at in the following light. At least since the reign of Amenhotep III in the early New Kingdom, the evidence today (DNA analysis and the physiology of Egyptian mummies) shows that Egyptian royalty had some Caucasian features and hair colouring.

That does not imply they came from Europe themselves; but rather their ancestry was in part derived from the old kingdoms in northern Syria, or nearby in Asia Minor and the Caucuses.

The people from these regions were apparently also the ancestors of the Phoenicians and the Minoan Mediterranean civilisation. Like the Mycenaean Greeks, they may also have been related to folk from the Russian Steppes. Similarly, many Europeans are descended from ancestors who lived in Anatolia, the Russian steppes and the Caucuses; the Indo-European people.

It has been suggested that European DNA can be linked with the Black Sea region, with over 80% of British men and 50% of European males showing some link back to these people. This does not mean they are necessarily descendants of Egyptian royalty, but rather that they share common ancestry.

Furthermore, specific northern Levant / Anatolian ethnicity on the Egyptian side was limited in time. Overall, the ethnicity of their kings varied greatly throughout history. It included (as a base) Nubian Africans, but also Libyans (plus the New Kingdom kings described above) followed later by Assyrians, more Nubians, Persians, Greeks, Romans and Muslim Arabs.

So it makes sense, that the Jewish relationship with Egypt, would depend to a considerable extent on who was in power.

Tell-tale signs of the Jewish presence, over the following centuries, include the Coptic writings of Jewish writers found as far south as Aswan and their involvement in all the usual facets of Egyptian society. Highlights, included the translation of the Torah and other scriptures into Greek during the Ptolemaic Period of the third century BC. This translation was necessary as Greek became all pervasive, and expatriate Jews lost their familiarity with Hebrew. In addition, the Gentile Christians of later years also sought those same books in the Greek language.

But where did most go? Well some stayed. I am fairly confident there are still descendants of Jews in Egypt, based on Coptic family names and associations - this despite the closure of the synagogues and the diverse ethnicity within the country.

But we do not hear about the 'wandering' Jew, because he sat around on his backside. This particular time is also likely to be an early episode in the great Sephardic migration to the west, and where Spain may have received its name of Iberia derived from the influx of immigrants. It was through the Spanish Peninsula (particularly in the north in Galicia) that so many eventually emigrated to Western Europe, Britain and the Americas.

Of course, the peak of this migration occurred over a thousand years later, on the back of the Islamic conquest of the region - and the subsequent recovery of the Spanish Peninsula by the armies of western royalty. If we doubt it was chaotic at times, we need only look at recent events where Syrians, Iranians, Afghanis and Africans have been pouring into Europe as refugees escaping persecution, or looking for a better life.

But migration into Europe was not the only migration, and this seems to have been established with a fair degree of certainty by Jewish sources. It has been portrayed by film makers Simcha Jacobovici and Elliott Halpern in their 'Quests for the Lost Tribes'. Migration out of the neo-Assyrian Empire appears to have taken multiple routes to the north, east and west as well as to Egypt and Arabia. There is mention of five places which were starting, or gathering points within the empire.

According to the research the film makers followed, the routes north and east ended as far away as the border of India and Myanmar (Burma) for some groups - and China for others.

These people were more likely from the Assyrian captivity and therefore from the Ten Tribes, rather than Judah. Nevertheless we know that a Jewess, Esther, married into Persian nobility and that a Jewish presence remained in that part of the world well after the Babylonian exile.

Support for this view comes from the long held knowledge, that by the time of the Roman Empire, the lands directly to the east, such as Mesopotamia and Persia were included in the Parthian Empire. This empire apparently included Israelite elites complete with 'wise men' (magi) and it seems three of these gentlemen made their way to Bethlehem at the time of Jesus' birth. In my view, they were almost certainly a Jewish or Ephraimite remnant of the Assyrian and Babylonian exiles. Apparently some of the Parthian descendants still live in Afghanistan today, practising a combination of Islamic and primitive Mosaic Law - with all that entails in regard to the penalties in the Torah.

However the majority of the Ten Tribes seem to have headed west and north into Europe. This was the wilderness in those days; sparsely populated in comparison to the centres of civilisation in the Middle East. It was remote as well. Homo sapiens and Neanderthal habitation and art, Mesolithic artefacts and Neolithic monuments all attest to a very ancient European prehistory. Despite this, it seems it was a largely forested and undeveloped continent; not exactly terra nullius, but with a heck of a lot more room to move than in the Near East.

Diaspora and the Brit thing

In relation to the Assyrian exile and the Ten Tribes of Israel, I was introduced to the concept of 'British Israel' teaching when I was quite young. It seemed a plausible idea, with a bit of novelty about it as well and some mystery attached to boot. It was a particularly popular theory in previous generations, when the British Empire was at its height and controlled a quarter of the globe.

Essentially the question posed was: what happened to the Ten Tribes of Israel and where did they go? The answer provided was 'to the British Isles of course!'

I have not followed recent discussion by British Israel (B.I.) followers, so the following comments are basic and may be a little dated. It is not so much a summary, but rather examples of various arguments used by proponents to validate the theory.

Now I have always had some misgivings with the concept in its simplistic form, but have also been convinced that there is some truth to be had, if broadened out and rightly understood. One of the reasons for this, lies with the fairly obvious close relationship with Britain - and more recently America - with Jewish people and their interests. The liberation of Jerusalem and Palestine from the Turks in 1917 and the ongoing association, both militarily and strategically is part of this.

Of course, it could be argued that many European nations have had similar ties and histories with their colonial empires, and also a significant Jewish presence as part of their heritage. Nevertheless in the last couple of centuries, with various pogroms in eastern Europe and the rise of Nazi Germany, it has been without doubt the English speaking countries (aided by the Latino countries of the New World) that have offered the greatest sanctuary and provided the most positive support.

In fact the tragic consequences of World War 2, have reinforced this ethnic rebalancing out of Europe in a very significant way. The choice has been either to emigrate to Israel, or to the many New World countries around the globe.

Given my interest has been on the historic issues outlined earlier in this manuscript and latterly by my preoccupation with the prophetic word, I could have approached this segment without mentioning B.I theory at all. It is after all a controversial subject. But it is also a base from which to explore the overall Israelite Diaspora going forward.

The discussion will be necessarily brief, but hopefully add to the understanding of the greater picture in later centuries.

There is a term '*all* Israel' used from time to time in biblical references, to cover the wide cohort of the Israel people. There is also one ancient account of the Ten Tribes of Israel and their dispersion, that would be a travesty to ignore.

As true as that is, the B.I. argument revolves around prophecies pertaining to Jacob (the father of the tribes) which they believe have been fulfilled by people, or nations, outside the strictly Jewish tradition; this, as we popularly understand it. Their argument is that many Tanakh prophecies (particularly those found in Genesis) do not apply exclusively to Jews and modern Israel, and cannot possibly do so, given that the prophecies *must already be fulfilled.* This quite logically follows, given the prophetic era is coming to a close.

The prophecies cited include specific promises to Joseph, Jacob's favourite son and his sons, Ephraim and Manasseh.

The B.I. understanding is that the prophecies were only fulfilled in the last few centuries as Ephraim (the British Empire and Commonwealth of Nations) and the rising U.S.A (recognised as Manasseh by some) have risen to prominence and willingly, or unwillingly been thrust into the limelight.

In fact I think those that advocate this scenario, are looking deliberately or subconsciously, to identify the 'stone kingdom' of Daniel 2 (or its precursor) simply because of the expectation that the 'stone kingdom' will be involved in 'end of the age' peacekeeping - and the role of 'policemen of the world'.

The above is not the only interpretation proposed in broader discussions, but one of many, depending on interests and biases.

However, much of this depends on whether the key prophecies to Joseph's offspring, were fulfilled when the Kingdom of Israel (the Ten Tribes) gained their own Kingdom and capital in Samaria. After all, is it necessary that the prophecy be fulfilled with Joseph's progeny only, or rather, that Ephraim and Manasseh were simply the lead tribes in the Northern Kingdom, i.e.

with Ephraim at the head and Manasseh prolific (with the largest group)?

In overview; the migration (or escape) of members of the Ten Tribes from the neo-Assyrian Empire is broadly in line with other refugee movements out of the highly controlled and populated empires of the Middle East and Egypt. Some were escaping slavery and forced labour, others no doubt simply had aspirations of greater freedom, self-determination and opportunity. Their common aim was to seek greener pastures.

In that regard, the Israelite captivity and subsequent bid for freedom would have been typical enough, and the drama kicks off with the 'Lost' Ten Tribes of Israel being captured by the Assyrians ca. 722 BC and then absorbed into their empire. As the Babylonians rose to the ascendency (and subsequently the Persians) these people in their tens of thousands - and perhaps far more - did a runner and escaped to the west, to finally migrate into Europe and beyond. As power shifted south-east from Nineveh to Babylon and then further east to Persepolis in Persia, the westward route would have been the most attractive.

There are dozens of clues suggested to support the migration, although it is fair to say, that advocates of British Israel theory have thrown a whole grab bag of clues and associations in as evidence. From memory, some of this is not particularly convincing and lacks the rigour expected in a strong, historically based argument. Having said that, there is certainly enough material to make one wonder, and enough smoke to suspect fire.

An example of teasing, but suspect material is the following: Denmark (Danmark) is taken as a name originating from the tribe of Dan. Jutland in the same North Sea locality is seen as a corruption of Jude, or Jew and the Anglo-Saxons are considered 'sons of Isaac'.

I do not know if there is any real evidence for the identification of the first two examples, beyond some artefacts which are probably the result of the Christian conversion of leaders of the various tribes. For example it is known that many kings, including Scandinavians, developed their own genealogies going back to Noah and Adam and converted specifically to be part of the 'in-crowd', when trading with the Holy Roman Empire. It virtually guaranteed them the prospect of wealth.

The Scandinavians were also coastal people and seafarers and therefore interacted with others relatively easily.

The Saxons on the other hand, are linked by B.I. advocates back to the Scythians of the Russian Steppes and Kazakhstan region today, and one particular Scythian tribe called the Sacae have been fingered. They were one of several plains tribes, which also included the Sarmatians, Massagetae and Sogdana. 'Sacae' may derive from a word meaning archer (Skudat).

In any event, the Sacae are referred to by ancient historians such as Herodotus (5th cent. BC) and Flavius Josephus and the latter tells us, that in his time 'there are but two tribes in Asia and Europe subject to the Romans (Judah and Benjamin) while the Ten Tribes are beyond Euphrates until now...' This is seemingly derived from his contemporary historical knowledge, but supporting the Ezra source we will examine shortly. It means we cannot write off the obvious; that some tribes-people beyond the Roman influence *were* in fact of Israelite descent.

The Scythian migration also has a biblical connection, because the Greeks renamed Beit She'an in northern Israel, as Scythopolis. It was part of an incursion from the Scythian heartland of the Russian Steppes, and it may have included some returning Israelites. This migration (or invasion) of the Scythians into Persia and the Levant is described by Herodotus.

According to Herodotus; the Scythian Massagetae pushed the Sacae westward. They in turn displaced a group called the Cimmerians living north of the Black Sea, who then headed south. The details are disputed by other scholars, who suggest that the Cimmerians were driven into Persia by the Scythian Sacae, via the route east of the Caucuses (as with Scythopolis) and that they ended up in Anatolia that way.

Regardless of the minutiae, one wonders whether one or more of these groups, were originally refugees from the Assyrian captivity, including Israelites. In that scenario, they were being forced back into Persia and subsequently fled to Anatolia.

Despite the problem of tribal identities, statements like these indicate that the Ten Tribes now live in some form beyond the

Mediterranean empires. This suggests an association with what might be described as, Iranic tribes in the east, or Celtic / Gaelic tribes in Europe; the latter penetrating as far as the Baltic Sea.

It makes sense as well. For a brief period the Israel tribes had their own Kingdom, but at heart they were pastoralists, explorers and artisans with wandering feet. In addition, people of Hebrew ancestry may well have had common ethnic roots with the plains people, going way back to pre Abrahamic times.

Now Celtic (from the Latin) and Gaelic (from the Greek) are names referring to people on the move. The alternate name to Lake Kinneret is Galilee, showing the migrant (Galatae) influence there. The inference is; these folk were also returning Jews or Samaritans from the north, albeit from a later period. The remains of the 4th century synagogue at Capharnaum, with its Greek inscriptions lends general support to the idea.

The various Celtic groups are considered indigenous to Europe. However with an overall east to west migration, due to the population pressure from the old empires, it is my view the Eurocentric definition is likely to be a simplistic observation and not truly representative of *all* their ancient history or ethnicity.

It surely does not preclude the absorption of wave after wave of east-to-west migrants, from the overpopulated ancient kingdoms, as the major source of new settlers into the Celtic lands - something which would have occurred from very early mining days onward.

This meshes with the DNA evidence cited in the previous section, linking the Black Sea peoples with Western Europe.

We know trading across Europe took place in both directions, and that copper and tin mining as far afield as Britain and Ireland were a source for manufacturing in the Mediterranean and Levant back as far as 1300 BC. The history of tin mining and metal trading with the Cassiterides (Britain) and the ancient copper mines in Wales (such as Parys Mountain on the Isle of Anglesey) is fascinating for geologists and historians alike. These sources were exploited after earlier copper mines in Cyprus and the tin mines of Afghanistan had been exhausted.

In fact trade and migration would have occurred at least as early as the old Assyrian Kingdom ca. 2000 BC, with waves of refugees following during the Eastern Mediterranean cataclysm which was the Thera eruption. A 'dark age' in the Eastern Mediterranean followed, dating from ca. 1200 to 900 BC and this spawned the Sea People. It is ventured, that this was a result of earthquake (very likely in Anatolia) probably volcanic eruptions and years of drought resulting in a severe disruption to society.

The difficulties of that extended period, close with the captivity of the Israelite people by the neo-Assyrian and Babylonian Empires from 722 BC to 586 BC.

It means that over a period lasting 1,500 years minimum, there are identifiable circumstances encouraging refugees, explorers and traders to exit the Near East and move to Europe.

The other intriguing matter is how they moved. A considerable number may well have arrived by ship, via the Mediterranean and joined Celtic tribes directly in Western Europe. This would have injected large numbers of people in a relatively short time and parallel the more commonplace occurrences in our contemporary world. Others used the Danube and Eastern European river systems for emigration, travel and trading.

Documented counter migrations, include that of the Galatians into Asia Minor (Anatolia); the same people the Apostle Paul preached to in the first century. They comprised three tribal groups who migrated to that area from the Balkans through Thrace, in the Great Celtic Migration of 279 BC. One of the tribes were a Belgae group and therefore presumably from Western Europe. However it is believed these people were related to other tribes living in the southern Russian sector originally, and I understand this is not just B.I. propaganda.

Now if the Belgae were related to other tribes in the Russian sector, then the Belgae may have originated in the Black Sea region themselves centuries before - and were in fact returning to Galatia on a back migration from Europe. It is another indication that the west to east migration was not the dominant one.

Earlier, I mentioned the head gear of the Celtic Anglo-Saxons and how it represents sun and moon worship. Again that is not proof positive they came from the Eastern Mediterranean, but it is a possibility that cannot be arbitrarily dismissed. It is known that an earlier Sea People in the Mediterranean, also practiced this custom. So whether it was through a direct link, or at arm's length, the likelihood is that some of the tribe may have emigrated from the east through the Mediterranean region, with a possible link back to the Scythian Sacae.

Indicators put forward by B.I. proponents in Western Europe are the name, Britain (BRT in Hebrew) meaning covenant; although apparently derived through the Latin and Greek.

Also Ireland; a combination of 'R' or mouth in the Egyptian hieroglyphs with 'land' from Scandinavia, suggesting the 'Ire' were rather proud of their oratory talents and the Vikings either recognised their eloquence, the relevance of the Christian message they had to offer, or were simply bored by their incessant chattering. Whatever the case, the Irish are forever associated with the Blarney Stone!

Further indicators presented are the primacy of the British (and European) Royal Houses, together with the Coronation Stone, aka the Stone of Destiny, now housed more or less permanently in Edinburgh to pacify the Scots. At times of royal baton changing, the Stone is retrieved temporarily and placed in the Coronation Chair in Westminster Abbey, London.

The Coronation Stone is also said to come from Scone in Scotland, or perhaps originally from Ireland. But some say it was of Middle Eastern origin, and the self-same legendary stone that Isaac laid his head on when he received the vision recorded in Genesis 28. A possible link to the 'stone kingdom' perhaps?

There are also the various stories from the Irish Chronicles, that suggest Jeremiah and the Jewish Princesses arrived in Ireland from Egypt and married amongst the Tuatha De Danann. Scotia, a queen from which Scotland gets its name, was also from Ireland, and has been identified by some with the Egyptian Princess Meritaten of the Akhenaton royal family. Others have speculated that she was from Israelite royalty and there are various anecdotal accounts cited in support of that. A possible Black Sea region, ancestral connection has also been proposed.

Because the kings of Scotland later became kings of England, we potentially have a Sephardic connection. On the other hand the relationship with the Germanic and other European royals throughout history, means that the British Royal House could also have Ashkenazi blood as well. The history of Victorian Britain, of course, is fascinating in this respect.

A group of people who are claimed to have migrated to Ireland, are the famous Milesians; a society who produced noted philosophers back in Anatolia. They are said to have travelled the earth and arrived in Ireland, where they fought with the Tuatha De Danann. The implication here is that they came from Miletus on the Aegean coast in the 6th century BC.

Another, almost mythical character connected to Britain, goes by the name of Brutus. Said to be originally from Rome, he apparently engineered an escape of enslaved Trojan relatives from the Greeks, who then fled to Britain by sea.

These groups and others from Western Anatolia, are therefore prime candidates for once Hebrew and related people, who escaped from the Assyrian Empire. This, based on possible ethnic linkages, timing and geographical location.

Other miscellaneous connections include; the tales connecting Glastonbury in western England with both mining, the Celtic Church and the famous Joseph of Arimathea. Also cited are the similarities between Hebrew and some Celtic languages such as Welsh.

So the number of strands of evidence presented are enormous, but of course open to other interpretations. How much substance there is to ancient Celtic characters and their exploits is anyone's guess, but suffice to say, the Greeks do not have a monopoly on heroic ancestor figures.

The underlying difficulty of course, is establishing the history beyond a shadow of a doubt. Where errors have been made, they are most likely in regard to the assumption that whole peoples, such as the Scythian or Cimmerian tribes were Israelites, when in fact, only subsets such as families were integrated. This will also apply to the Celtic tribes.

To be clear, the above is my simplistic understanding of the propositions and arguments involved - and attempt to list them in condensed form. I hope that at the very least, this quick appraisal has established the certainty of early migrations out of Mesopotamia, Egypt and the Eastern Mediterranean.

I emphasise, it is not the purpose here to go through the propositions and analyse what may be acceptable and what has to be discarded. But for those interested in the subject, some further material is included in the Bibliography.

Broadly then, journeys from the eastern Mediterranean to Britain, France or Africa were nothing to Phoenician, Greek and other sailors of those days. It is also likely that there were people from Europe living in the middle of the Atlantic in the Azores (Acores) dating from Carthaginian days, and perhaps much earlier. We know that the Neolithic history of the Orkneys and New Hebrides extends back to the same time. In fact controversial new research is attempting to show that America was visited by Europeans, even at the height of the Ice Age.

Given the range of known exploration and very ancient possibilities, a voyage of a couple of thousand kilometres from the Greek Islands to the British Isles in comparatively recent times, would have been a cinch. It is all about incentive; how much you value your skin, or how empty your stomach and wallet happens to be. It is also about opportunity; and the Ten Tribes lived adjacent to the Phoenician city states and their awesome maritime capability with linkages to Carthage, Tarshish and the West. Later, their neighbours were the Greeks of western Anatolia who traded the Mediterranean and beyond.

Speculation aside, the strongest biblical evidence referring to the Ten Tribes exclusively and their migration (or escape) from the Assyrian Empire is recorded in 2 Esdras 13 in the Apocrypha. This account is also the key driver for the proposed direction, i.e. westward.

It is a vision given to Ezra with an interpretation. We read from verse 39:

*39 And as for your seeing him gather to himself another multitude that was peaceable, 40 **these are the ten tribes which were led away from their own land into captivity in the days of King Hoshe'a,** whom Shalmane'ser the king of the Assyrians led captive; he took them across the river, and they were taken into another land.*

*41 But they formed this plan for themselves, that they would **leave the multitude of the nations and go to a more distant region,** where mankind had never lived, 42 that there at least they might keep their* *statutes which they had not kept in their own land.* **⁴³** *And they went in by the narrow passages of the Euphra'tes river.* **⁴⁴** *For at that time the Most High performed signs for them, and stopped the channels of the river until they had passed over.* **⁴⁵ Through that region there was a long way to go, a journey of a year and a half; and that country is called Arzareth.*[j]

⁴⁶ *"Then they dwelt there until the last times; and now, when they are about to come again,* **⁴⁷** *the Most High will stop*[k] *the channels of the river again, so that they may be able to pass over.* **Therefore you saw the multitude gathered together in peace. ⁴⁸ But those who are left of your people, who are found within my holy borders, shall be saved.**[l] **⁴⁹** *Therefore when he destroys the multitude of the nations that are gathered together, he will defend the people who remain.* **⁵⁰** *And then he will show them very many wonders."* (RSV)

This passage suggests that the Ten Tribes, although separated, were not disjointed, but maintained communication within Assyria and planned an escape to a distant and relatively unoccupied land, once the opportunity presented itself.

As noted above, it may have included tribes who had escaped north, but were subsequently pushed back into Persia and the Levant and then headed to Anatolia.

The term 'Arzareth' is commonly interpreted as 'Another Land' and therefore of no known geographical location. However, after historical research relating to Asia Minor, Ephesus and the Seven Churches of Revelation, I believe I can now pinpoint the centre of the region quite precisely. Arzareth can be broken into Arzar-eth which is similar to Nazar-eth, B-eth-el or Be-it She'an, where the 'eth' or 'et' is a suffix, or particle which denotes a sense of entity or physicality. The rest is usually a description of the location, people or other identity.

Interestingly, Ephesus is located in a region of Western Anatolia which was once called Arzawa. It was a political region dating from the 15th century BC and centred on the Kestros River (Kucuk Menderes). Its capital, Apasa, later to become Ephesus. The Hittites conquered Arzawa and divided the region up into three; Mira to the south on the Meander River (which later became Caria) a northern district called Seha River Land along the Gediz River and an eastern province called Hapalla.

In later years, the Mycenaean Greeks and others in the Arzawa region gained a modicum of independence and lived uneasily beside their militarily stronger Hittite neighbours to the east.

The Kucuk Menderes River has long since silted up and left the old port site several kilometres from the sea. However the archaeological site which is Ephesus today (with its government buildings situated on the hill) is only a couple of kilometres removed from the location of old Apasa. It is therefore to this region, that the Ten Tribes eventually escaped and the indication is, that like Ezra's fellow Jews, 'those who are left of your people' are returning, or will return in 'the last time'.

In brief; their first opportunity for a massive breakout from Assyria, surely came when the neo-Assyrian Empire transitioned into the Babylonian Empire. If we accept that some tribal people went east into Asia and that some people in Afghanistan, India and even China have a tradition to this effect, then this ancient written account describes a very coordinated move westward into Anatolia and eventually to the Aegean coast. In fact it suggests it was a large proportion of the refugees.

This then ties into the history of the Seven Churches of Revelation (initially synagogues) of the first century. The ruins of the towns with which they were associated can be found in western Turkey, with Ephesus the most famous.

It becomes apparent why these assemblies were of prime interest to Paul and the other Apostles. The worshippers were either expat Jewish kin, or more distant Ten Tribes folk (even if now ethnically mixed) and the synagogues notable, because they were established by a composite diaspora. In addition, the central positioning of this area in Anatolia, in relation to western Asia and Europe (together with the Ten Tribes aspect to this significant episode) will almost certainly play a part in the familiar Jewish Ashkenazi and Sephardic stories. It was a great hopping off point if you did not mind a bit of travel.

Now there is considerable dispute about whether 2 Esdras is an authentic Ezra document, and given its considerable prophecies, some have sought to dismiss it as having been written in the Christian era. However this and other prophecies and descriptions within its pages, suggest to me that most of it, if not all, dates to the post exile period.

The usual compilation and editing issues of later times will confuse the issue of provenance, but given that proviso, I believe we have a genuine pre Christian document.

A quotation late in the book gives a rough date toward the end of Esdras' (Ezra's) life. Here Ezra has another vision from the Lord while sitting under an oak tree. 2 Esdras 14:7-12

7"And now I say to you: Lay up in your heart the signs that I have shown you, the dreams you have seen, and the interpretations that you have heard; for you shall be taken up from among men, and henceforth you shall live with my Son and with those who are like you, until the times are ended.

*10For the age has lost its youth, and the ties begin to grow old. **For the age is divided into twelve parts, and nine of its parts have already passed, as well as half of the tenth part;** so two of its parts remain, beside half of the tenth part."* (RSV)

Verses 7 to 9 are beautiful in their own right. These are words from the Lord, to a weary old man who has been extremely faithful and completed so much. He is now about to go to be with his fathers and the Son.

Who exactly wrote the words and sentiment of the Lord down is not explained. They are written in the character of the time, and one can sense they do justice to the faithful efforts of Ezra over many decades. That is what is important.

In my view it is very much worth getting a copy of this book and reading it for oneself. In fact how it escaped the main cannon of most Protestant translations is beyond me. Importantly for the discussion here, we learn from verse 10 to 12 when this prophecy was given. It says that 9 out of the 12 parts of the age *plus* half of the 10th have been completed.

Since this work is about timelines, this particular milestone was guaranteed to arouse my interest. What age are we talking about? The 2,556 years of the Age of the Gentiles?

But 9.5 of 12 parts would place the time way too late in the Christian period; in fact around AD 1433. Then the thought occurred, what about the zodiacal ages? If indeed it was written after the time of Jesus as some suggest, then that would be the 2,000 years of Pisces. But again 9.5/12 of 2,000 would be even later; in the vicinity of AD 1583 and the Elizabethan era.

Then of course the penny dropped and I checked out the obvious era; the previous 2,000 year age of Aries (the Lamb) - the time I have called the Age of Sacrifice in Figure 4.

Working back BC, we have (12-9.5)/12 x 2000 which is 417 (418 BC). That date will shift, depending on how we define the length and end points of the age and here I have just used the simplest definition. For example if we used AD 30 as our nominal zodiacal milestone instead of AD 1, then the date would become 388 BC. This seems too late for Ezra. On the other hand if we use AD 30 and a zodiac age duration of 2,148 years (12 zodiacal ages) the equation becomes:

(12-9.5)/12 x 2148 – 30 which is also 417 years (418 BC).

There are obviously other combinations that could be tried here, but the first and third options illustrate, that a time during Jesus likely ministry period could fit the criteria as the turning point from the age of Aries to Pisces. In any case we have at least two options that give us a date that could work.

So the range 418 - 388 BC looks a little late for Ezra, but then again, nine and a half parts of twelve is always going to yield a ball park figure only. The end of the Babylonian exile would have been about 520 – 516 BC and in the next chapter we address the difficulties of dating the various edicts and decrees of the Persian kings. Those kings begin with Cyrus, who was the Medo-Persian monarch who gave the initial permission for the Jews to return to Jerusalem and rebuild the Temple.

There is another quote in 2 Esdras which shows its internal consistency: 2 Esdras 7:28-29 (ca. 418 - 388 BC):

'For **my son the Messiah shall be revealed** with those who are with him, **and those who remain shall rejoice four hundred years.** And after these years my son the Messiah shall die...' (RSV)

Now the structure of the sentence is rather rough, but the intent seems to be that those who remain (after Ezra) will be waiting for the coming of the Messiah and will rejoice when he arrives. This would be based on the Daniel 9 prophecy. The (approximate) zodiacal date of writing and the 400 years are obviously reasonably compatible and mutually supportive.

For our purposes here, it is sufficient to state that Ezra arrived in Jerusalem with his band of exiles in 458 – 457 BC.

So perhaps this ties in rather well and that he *did* live a further 40 years or more, dedicated to the priestly and prophetic leadership needed at the outset of the Second Temple period.

This reference to an 'age being divided' is also an Old Testament example, showing (absolutely) how the ancients measured the ages. Yes, the prophets used a 'time' for measuring specific prophetic periods, but most never considered an age to be anything but a zodiacal age!

Now I cannot prove that this Ezra narrative has never been tampered with, or added to at a later date. There could have been an attempt to fake it as a 5th century BC document. But it seems genuine on the face of it, and given the Torah is agreed to have been translated into Greek at Alexandria in the 3rd century BC from an original Hebrew source, I suspect there is every reason to conclude it is primarily a pre Christian document.

Turning once more to the Ten Tribes escape back over the Euphrates and in a westerly direction toward Arzareth (Arzawa) in Asia Minor, it says they lived there to the 'last time' and now 'begin again' to return to Israel.

This is rather vague in terms of timing, but given the original Assyrian captivity occurred ca. 722 BC, it indicates that at least some of the people were returning (or were planning to return) to Israel in Ezra's time in the late 400s BC. This would make sense given the news would eventually reach them, that their Jewish brethren had been released by the Medo-Persians and at least some in Asia Minor would have opted to reunite.

It would be more fulfilment of the 'two sticks becoming one' prophecy of Ezekiel 37. However the 'last time' also suggests, that some were not so anxious to return to eretz Israel and may have ventured far further afield and for far longer; this as the prophetic vision of Revelation 12 demands.

It could well be that the more spiritually attuned folk, had some premonition, vision, or 'word of knowledge' that suggested the trials of the nation were far from over.

British Israel (B.I.) theory aside, other western European nations have been discussed by writers of the genre, as various candidates for the tribes and have rated special mention accordingly. Regardless of tribal and national identities, it makes a lot of sense to see Jews and Ten Tribes people in Europe as part of the leadership of at least some western nations. This would mimic their apparent role in the Parthian Empire of Roman times. However attempting to show that all the citizens of the various countries are related ethnically to a particular tribe of Israel is unlikely in the extreme, even if great societal cohesiveness is demonstrated and a prophetic identity hinted at.

The identification of the USA is a difficult one with the immigrant population coming from all the European countries (and right around the world). In fact one is tempted to see the Americas (not only the US) as an ethnic smorgasbord, and if anything a composite of people very much like the British Commonwealth of Nations; a grouping pre-empting the 'stone kingdom' of Daniel.

But then, that pretty much describes Israel in the early days. Israel was always multi-ethnic, ever since the days of Jacob and his two wives and their two Egyptian handmaids.

That diversity grew during the Exodus saga without doubt, and the glue that held them together was their common history, beliefs and desires. Those beliefs were in theory a national spiritual identity, despite that proposition always being strained, with only the devout few, faithful to their calling.

If anything identifies a people, it is how they operate and we know that while the Jews were faithful to tradition, the Ten Tribes were idolatrous in the extreme for much of the time.

Despite this, Jesus had an affinity with the Samaritans, who would have included Ten Tribes folk that opted to return to the land from Ezra's time onward. Jesus' attraction was not because they were particularly observant, but because he noted they took care of each other and showed compassion to strangers.

On the other hand, we get a sense in the New Testament, that the Jews were nitpicking traditionalists who focussed on observance and sometimes forgot what life was all about.

So Ezra specifically identifies the Ten Tribes and describes their escape to Anatolia and settling there, but before their further migration to Europe. At this point there are apparently few from the tribes of Judah and Benjamin with them.

However we are also aware of the general Jewish migrations into Egypt and on to Spain on one hand, and to Northern Europe on the other. This is the Sephardic and Ashkenazi groups mentioned above. This occurred commencing with the Babylonian emergency and continued through Greek and Roman times into the medieval Islamic period.

At the end of this prolonged process of migration into Europe, the mix seems to have morphed **from a distinct Ten Tribes cohort** and **one or two predominantly Jewish migrant groups** out of the Near East, into **just two identities** at the western end of their journey: the **Sephardic Jews** of Western Europe and the **Ashkenazi Jews** of Northern Europe.

Where these people practise the old Israelite feast celebrations they are identified as the 'Jews' today. Where they have been absorbed into the Gentile world, but nevertheless retain their faith, they are Christian. If we contemplate this scenario, I believe we are getting close to the reality. It includes an integration of the original Israelite tribes to a considerable degree, but also a gradual assimilation with the Celtic and Germanic world.

Therefore to limit our concept of 'Jewishness' to the Judaic tradition and those who grace the doors of synagogues only (as is almost universally the practice) may be useful from a religious point of view, but is to underestimate the number of people involved in the overall Diaspora of Israel related people.

This is quite evident watching ancestry television programs, where so many well-known people find they have obvious Jewish ancestry a few centuries ago. Even more to the point, a casual investigation of Sephardic name lists from Galicia in Spain reinforces this proposition to the 'nth' degree.

Finally, the Jewish leaders of the Christian sect of the first century, recognised that Gentiles of non-Hebrew origin were being selected by God through the Shechinah / Holy Spirit anointing – 'whosoever will, may come'. It was a revival of the First Temple faith. Therefore in a perfect world devoid of the unbelief and immorality plaguing western society today, the worst we would have in the Judeo-Christian story are different roles within the ranks, but a great dialogue nonetheless between Jew and Christian, with everyone batting for the same team.

In the case of the European migration, the influences that the Hebrew (Jewish / Israelite) people have had on the west, I suggest, have been enormous and pivotal to that broad society. It is not so much that the various tribes have totally lost their identity, but rather that they are now living and working as one.

Back in the Land of Israel however, our 'Jews' are surely becoming a model of the original United Kingdom of David. They are the ones that the prophets address in the many Jerusalem centred prophecies of the end of the age.

More than this, the devout are very much expecting the arrival of their Messiah as described in Zechariah chapter 12. The surprise for them will be his identity.

Tribulation

So the starting point for the 'time of the gentiles' began ca. 590 BC as Nebuchadnezzar organised his new attack on Jerusalem and Judah. It is an especially significant point in Jewish history; the start of a 'tribulation' or trial of a whole people and the end of their right to absolute self-determination.

Although some young Jewish men, such as Daniel and Ezekiel were exiled in earlier Babylonian actions against Judah (the prophet Daniel in 605 BC and Ezekiel a few years later) the Jewish elite who were left and did not flee to Egypt, were taken to Babylon and 70 years of exile after the final siege of the city.

As expressed in Lesson 10, all the key milestones seem to follow a similar pattern and mark tumultuous events during the seven times of trial. They also appear to be 'about' seven years in length with the result that, at the mid-point of three and a half years, there is a central and sometimes critical point. In the case of the Babylonian siege of Jerusalem, this was 587 BC when the city was overcome and finally fell the following year.

Subsequently, Babylon itself was overcome by the Medo-Persians as recorded by Daniel, and later we read about the return of the following generations of Jewish leadership to Jerusalem. They were freed under Cyrus, Darius and later Persian kings and a Jewess, Esther, was at the heart (literally) of this ongoing miracle release at the end of the exile. Her role is one of the more amazing in the history of the Jews.

The accounts of the return are found in Ezra, 1 Esdras and Nehemiah, where we read about the numbers of those who returned and the problems they had with the 'trans Euphrates' troublemakers. These were people bitterly opposed to the rebuilding of the Temple and the city of Jerusalem. And there is little doubt, that the 'trans Euphrates' people, were the descendants of those who had formerly paid tribute to the Judah-Israel kings, centuries earlier. That memory would have been particularly galling to them and it becomes obvious their resentment continued for many years.

Of course by this time, everyone was under the thumb of the Persian kings who were firmly in control of Babylon, together with vast swathes of Asia and Africa from India to Egypt. They also occupied the Black Sea region and part of Eastern Europe under Darius. But Daniel passed away in exile and interestingly, there are at least five grave or memorial sites to prove that. There are a couple of places in Iraq, another two in Iran and one further removed to the north-east in Kyrgyzstan, Central Asia.

And perhaps it was just as well that he was never able to return; the times were dangerous when Judah returned to the land. It was never free, there was always an overlord, continual strife and rebellion became the norm and the sense of spirituality of the early Temple times never quite returned. If the Jews had known their seven times of tribulation were going to last as long as it did, they would have despaired; seventy years only of exile, but two and a half thousand years of pain.

Returning our attention to the timeline using a 'time' of 365.24 years, it follows that a 'time, times and half a time' (3.5 times or 1,278 years) must now be used beginning in 590 BC.

Before we do though, there is another consideration revolving around the BC to AD transition we must take into account.

Both astronomers and historians agree, that when counting backward from 1 BC, 590 BC (for example) is only 589 years before Christ. This is true for all dating purposes, not only for astronomical observations. It is due to the fact that no allowance was made for counting backwards and forward between the BC and AD chronological systems, i.e. no 0 year was inserted.

So if we subtract 589 from 1,278 we get a date of AD 689.

At the outset, I have to emphasise there is something extraordinary about this timeline. Every significant milestone is centred on Jerusalem, or something associated with that city.

This milestone is as amazing as all the others, and is mentioned time and again in both Testaments. In fact it is pivotal. Many of us know the significance of AD 70 and the destruction of much of the city under Titus. We may even be aware of the rebellion lead by Simon bar Kokhba in AD 135, when the city and Judea were once again crushed. But what about AD 689?

Islam rose in the days of Muhammad, who was born in Mecca around AD 570 and died in Medina in June, AD 632. He claimed to have many visions from the Archangel Gabriel, and it is also claimed by his followers, that the whole of the Qur'an (Islamic Holy Book) was from that source.

Caliph Umar (Omar) bin Al-Khattab of Damascus was a senior advisor to Muhammad. He succeeded Abu Bakr in the Rashidun Caliphate and occupied Palestine (the name given to Israel by the Romans) and Jerusalem in AD 637. In popular Islamic accounts, he was a responsible man who opposed Muhammad until his conversion to Islam, and treated the Christian and Jewish inhabitants of Jerusalem with some respect. Umar bin Al-Khattab was also instrumental in transforming the Arabian Islamic state into a world power; conquering much of the Middle East including the Sassanid Persian Empire and Egypt.

But was he such a benign figure? Later we will see the biblical prophets were shown a vastly different view of this period, and regardless of the description of Umar's personal traits by Islamic writers, suggests that deceit and lies were the order of the day.

The history of successive attacks on Jerusalem by Sassanid and Islamic armies in the early 7th century, show that in the space of a few years, tens of thousands of lives were lost.

Umar is revered by Sunnis more so than Shia Muslims, because of the latter's concerns on how he came to power. It seems though, that the first Caliph, Abu Bakr, did in fact select him as his successor. In any case, Caliph Umar was assassinated in Medina in AD 644 and so it was (most likely) up to one of his successors, Abd al-Malik ibn Marwan (the 5th Umayyad Caliph) to work his will in Jerusalem and on the Temple Mount.

Abd al-Malik cleared the area of unwanted structures by AD 689; structures which may have included remnants of the Herod Temple or a temporary synagogue. This clearing and levelling of the Temple precinct was in preparation for the building of the Muslim edifice known in the West, as the Dome of the Rock.

The construction commenced ca. AD 689 and seems to have been completed around 692. With the associated AD 695 to 705 construction of the al-Aqsa Mosque on an existing raised platform - and using material from the 6th century Justinian Byzantine Church of St. Mary - we can say; 'on a wing of the Temple he will set up an abomination'. The mosque also incorporated a prayer house built by Umar, or an earlier Umayyad Caliph.

If this is so, it suggests that 'sacrifice and offering' as per the Prophet Daniel's prophecies, were not 'taken away' permanently in AD 70 or 135, but much later in AD 637 to 689. It means that it was not the Romans that caused the 'abomination of desolation' of Daniel, but the Islamic Caliphates. After all, building the abomination is what caused the spiritual desolation.

In terms of world history and the interaction between Jewish, Christian and Islamic faiths, this is the most significant event, smack bang as it is, at the very centre of the seven 'times of the Gentiles', or 'time of Jacob's trouble'. For the first time since the Joshua lead return, it was impossible for Jews to worship at the site. No wonder it is mentioned several times by Daniel and recounted by Jesus in his Olivet discourse.

Daniel 9 verse 27 also directly describes this point in history as an overarching prophecy. The 'seven times' are also evocatively described as the 'tribulation' elsewhere in scripture, and this desolation event on the Temple Mount is the devastating beginning of the second 'time, times and half a time'.

If we now complete the 'seven times' epoch with another 'time, times and half a time' of 1,278.34 years:

We have AD 689 + 1278 years which comes to **1967 and the Six Day War** when the city of Jerusalem was reunited.

This extraordinary result and almost unbelievable event, marks the final recovery and reunification of the city by the Israelis. The War ran from the 5th June until the 10th June 1967 and seemed like a miracle to most of us at the time.

It is only under grace that Muslims gather at the Temple Mount today, because it is most definitely under Israeli government and military control and has been since 1967.

Remember, our initial premise is that 590 BC (the beginning of the tribulation) was only the start of operations against Jerusalem in Nebuchadnezzar's final campaign. It continued to 586 BC and no doubt beyond. In the same way, the Temple Mount (Mount Moriah) was cleared around 689 (the central point of the times of judgement) but the actual construction continued over the next few years. Similarly in 1967 the Six Day War is just the start, I believe, of the end of the tribulation. The whole episode appears to end in 1973 and the Yom Kippur War.

I will suggest something further about the years 1967-1973 later. It has implications about the totality of the people, Israel.

Now to reconcile all this with scripture, requires a new understanding on the part of most of us. It impacts on the prophetic interpretation of Daniel's prophecies, but also on what Jesus meant when he talked about the destruction of the Temple. In the later verses of Matthew 23 and early verses of chapter 24, Jesus is apparently talking about the destruction of AD 70 under Titus. From verse 15 in the same chapter, he is definitely referring to Daniel's 'abomination of desolation'; illustrating the chronological unfolding of the 'seven times' at its mid-point.

The Muslim occupation of the Holy Land under Caliph Umar seems relatively benign on the surface, when compared to the Roman attack in AD 70. Apparent reassurances were given to the Byzantine rulers of the day of an orderly withdrawal on their part. This is emphasised by Muslim correspondence and Muslim historians today. But again, is this the true story?

An analysis of Daniel 11 suggests this was one of the biggest deceptions in history, and perpetrated before the Mediterranean peoples (now living under at least a nominal Christian society) understood the cruelty that had already occurred in Arabia and the craftiness in general of the Arabian Muslims.

In the Qur'an, we get a sense that Muhammad's personal journey was one that at first showed some affinity with Jewish and Christian sensibilities and beliefs. But this became strained toward the end, until outright hostility becomes evident. To the extent in fact, that in dealing with infidels, an 'anything is permissible' mentality becomes the order of the day; where heavy taxation, slavery, sexual predation, torture and beheadings are condoned and even encouraged.

Of course, life for non-Muslims living within the Caliphates at any particular period, depended on personalities and circumstances, with the occasional Caliph showing more consideration to his subjects than others. However generally speaking, there were two codes of conduct; one between Muslims (and sufficiently based on Jewish and Christian precepts to hold their society together) and another, very much darker code for the treatment of the infidel / outsider.

The sacking of Jerusalem by the Sassanid Persians in 614, shocked the Jews and Byzantines and therefore the oppression which was to come under Islam and its double standards had yet to be experienced by them and probably seemed unlikely. The idea that a third, apparently monotheistic people, who said they followed Abraham and called Jesus a prophet would be anything but another example of the same brand, seemed distant.

In the end this proved wishful thinking at best. Ultimately the naivety and spiritual neglect within the ranks, resulted in unbelievably swift destruction for Jew and Christian alike.

I have no doubt, that the believers at the time who *were* concerned about the ideology of this new gang on the block, were derided as 'Islamophobic', or words to that effect. Once their brethren had lost their spiritual compass and were inclined to accept the new system and ideology as an equal, then understanding and discernment disappeared and judgement arrived.

But that is the beauty of this God given timeline. It not only clarifies statements in scripture that are otherwise ambiguous and open to several interpretations, it also sheds a spotlight on key moments of history.

We see for example, that the 2,556 years of judgement are interspersed with times of absolute horror.

The first, of course, was the sacking of Jerusalem and the 70 years of Babylonian captivity right at the beginning. The occupation by the Seleucid Greek regime and the 'desecration' of the Temple saga of Antiochus IV was another. The 40 years of terror leading up to the AD 70 destruction of Jerusalem by the Romans and the bar Kokhba AD 135 uprising, were other times of grave stress, danger and destruction.

As the presence of Jews in the land decreased, the judgement and persecution followed them. From Jerusalem to Europe, it lead to the greatest horror of all; the Holocaust of World War II.

All this; absolutely everything is included in the 'tribulation' timeline!

Once this 'time of Jacob's trouble' is properly understood, it comes as no great surprise, that the greatest tragedy of all comes just before the end of the 2,556 years of tribulation.

Indeed, just as things are looking up and a return to the land and nationhood for the Jews is looking more certain, a hideous last judgement occurs. Not just for the 6 million in the concentration camps from all over Europe, but the 70 or more million people in total throughout World War II and many more millions in the precursor, the 'War to end all Wars' – World War I.

No wonder Jesus described the last 2,000 years as 'great tribulation' and 'if those days had not been cut short, no one would survive'.

Again you may look at Daniel's visions, or those in Revelation chapters 11, 12 and 13 and say, 'well I can understand what you are trying to say, but I do not really see how you have gained this interpretation from the prophecies alone, or indeed have anything concrete to validate the start and stop times'.

I will provide further clues as we go along, but I have to admit, that the circumstances surrounding my personal enlightenment on the matter and the logical nature of what followed, just seemed to be guidance from the Lord and given in rather difficult personal circumstances.

Nevertheless it *is* all there in scripture, although not easy to tease out. The understanding is for our secular dominated times at the end of the age. It is to ensure that we know that there is a plan of God; but that time is short and that the age 'has lost its youth'. In that sense these words are prophetic, and have not been revealed by study alone.

So the key to understanding is; that there is not one 'time, times and half a time', but two – and they are consecutive. This critical piece of the puzzle is not at all clear, after only a casual reading of the prophecies where the references are found.

But once we go back to the Torah and read Leviticus 26 - as we have done - it starts to make sense. We see the consistency of God's word right from the earliest times. As the disparate and confusing pieces of the prophetic jigsaw are gathered and positioned according to the timeline, things gradually take shape. It is actually quite remarkable.

Now keeping in mind the references to 'time, times and half a time' in both Daniel and Revelation; the ones that were presented in the previous chapter, we will now move along.

The key passages are Daniel chapters 7, 8, 9 and 12 and Revelation 11, 12 and 13. Suffice to say, I have been shown some patterns in the timeline and these are interlinked with specific prophecies of considerable interest.

Summing up: the various visions of Daniel provide a variety of perspectives which help to confirm the identity of the major players; the empires and kingdoms. They also give the impression, that he is desperately trying to recall all that he has seen in his night visions and write them down as accurately as possible.

Nothing must be omitted, or incorrectly described. He fears that no one else may ever be entrusted with the task. What a responsibility! What a burden!

The goat and the prophet

The first of the linking prophecies that have come to my attention, are found in Daniel chapter 8. Here he describes another long period timeline, starting with the 'goat', the Macedonian Greek army under Alexander the Great. This is the prophecy of the 2,300 'days'; literally 'evenings and mornings' patterned on the Jewish day.

> *Unbelievably, **I again woke up startled one night (bolt upright) with the sound of a voice, seemingly from nowhere.*** It was some weeks after the first occasion in late 2011, and at a time I was still beset with personal troubles. It meant I was at a low ebb on one hand, but in a surreal place on the other.
>
> I remember thinking that I could not remember dreaming like one usually does. It was uncanny and lasted for weeks, possibly months. *Even when awake, or half awake, it was like there was utter blankness in my mind - as if it was a clean slate, or an open canvas. The back of my eyelids was the only picture I saw.* My own home seemed to have become a 'thin place' of Celtic tradition, minus the inspiring background surrounds of nature.
>
> In fact what I heard was a steady, but urgent voice saying; **'David, it is a separate vision. Use a date from the beginning of the chapter!'**
>
> *It was the same clear voice I'd heard weeks earlier.*

The instruction that came to mind was obvious. Use a significant date from Alexander's own time and start from there; something early from Alexander's attack on the Medo-Persians.

You see I was aware from 35 years ago that many different interpretations had been posited for this prophecy, but none seemed satisfactory. Most commentators were looking at start dates from Daniel's era, others proposed that 2,300 evenings and mornings meant half that many days (1,150) and therefore that many years. In fact it seemed every imaginable possibility had been proposed, including the usual literal day interpretations which could potentially (with a lot of manipulation) be linked with the Greeks such as Antiochus Epiphanies.

It was one of those prophetic stonewalls which seemed to have stumped everyone, including the historicist interpreters and myself as a young man.

So as the sleep evaporated from my mind and I walked out to my living area, I decided to consult Mr. Google once more.

I found that Alexander the Great and his army crossed over the Bosporus, and attacked the Persian Empire in Asia Minor (today's Turkey) in 334 BC. Amazingly it was my first attempt to obtain a response, and it included the year as well.

Now remember, we are required to subtract one year, because the astronomers and historical bean counters say there is no 0 BC as we count backwards. It means that 334 BC reduces to 333 actual years before the start of the Christian era.

We then have 2,300 – 333 years = 1967.

I still marvel at the result to this day. It brings a tear to the eye. It catches my breath. It just seems too simple, too much! All glory to the One who reveals His secrets in His own time.

This prophetic result was the more powerful to me, because in 2010 I stayed at the Mosaic Hotel in Istanbul, and enjoyed a few days sightseeing around the magic of the Bosporus, the Blue Mosque, Hagia Sophia, Tuthmose III's Obelisk and the Markets. I then caught up once again with family I had met earlier on Santorini. They were lovely days, broken only by a hurried return aeroplane flight to Ephesus and the beautiful southwest coast of Turkey.

I was still multitasking and following Saint Paul's steps, wondering what happened to Mary as well, and contemplating my recent experiences of Old Testament places.

OK, calling it multitasking is overdoing it I know, but then for a male at least, it was an attempt - and I rest my case.

Less than two years later in 332 BC, Alexander was in Gaza on his way to Egypt and the rest as they say, is history. This prophecy is uncanny confirmation pointing to the liberation of the Mount and return (militarily) to the Jews!

It also illustrates how prophetic alignments are only confirmed in hindsight and revealed when required. Grattan Guinness may have made the connection if he had lived fifty years later, however in his time he was still working with a 360 day year and the 1917 liberation of Jerusalem in World War I.

Since my own enlightenment on the matter, I have found that one or two others have also proposed this Alexander the Great initiated period, ending in 1967. I cannot say 'great minds think alike' because in my case I was alerted to this fulfilment in a way I could not avoid, rather than by intellectual reasoning.

The pertinent scripture is Daniel chapter 8, starting from verse 9:

9 Out of one of them came another horn, which started small but grew in power to the south and to the east and toward the Beautiful Land. 10 It grew until it reached the host of the heavens, and it threw some of the starry host down to the earth and trampled on them. 11 It set itself up to be as great as the commander of the army of the LORD; it took away the daily sacrifice from the LORD, and his sanctuary was thrown down. 12 Because of rebellion, the LORD's people[a] and the daily sacrifice were given over to it. It prospered in everything it did, and truth was thrown to the ground.

13 Then I heard a holy one speaking, and another holy one said to him, "How long will it take for the vision to be fulfilled—the vision concerning the daily sacrifice, the rebellion that causes desolation, the surrender of the sanctuary and the trampling underfoot of the LORD's people?"

*14 He said to me, "It will **take 2,300 evenings and mornings**; then the sanctuary will be reconsecrated."* (NIV)

This is an interesting answer to the question posed in verse 13, because the span of years goes well beyond the 1,278 years of a 'time, times and half a time'.

In fact it includes some of the first 3.5 times and *all* the second. The prophecy not only covers the time up to the desolation of the Temple, but the whole time it is under the control of the 'horn' or king 'who started small, but grew in power to the south and east and toward the Beautiful Land'.

My interpretation is that this 'horn' or king, grew in-situ in the general area west-south-west of the Holy Land and the Arabah, and expanded in every direction. That is; south into Africa, east into Mesopotamia, Persia and beyond and westward into the Beautiful Land.

In other words the central point is Arabia and the timing very much later than most dare think.

However this description has been interpreted by some analysts from a Greek perspective, which is from the north-west and the question is; could they be right?

When we look at the history of the Caliphates in detail, it becomes apparent that there is room for both views - the reason being that the capitals of the Caliphates moved around considerably. Importantly, toward the end of the Islamic period the capital was indeed centred in the Greek world!

There is much more to explore here, but returning to Alexander the Great, there is a story from Flavius Josephus that Alexander visited Jerusalem and spoke to some of the priests there. They showed him the prophecies of Daniel; prophecies in which he featured. I think it safe to say that Alexander would have been both flattered and motivated. His visit to the Oracle at Delphi early on, showed his interest in knowing the will of the gods in regard to his campaign. His time in Jerusalem no doubt spurred him on to see other oracle sites as well.

In fact his visit to Egypt was short; but one out-of-the-way place he went to, was the Siwa Oasis in the Egyptian (Libyan) western desert. This was an uncharacteristic detour by Alexander and fraught with danger, but the Oracle of Amun was well known long before his time.

He and his army then returned to Asia to continue his attack on the Medo-Persian king, Darius III, via the Sinai Caravan Way and King's Highway to Mesopotamia. This account is another clue indicating a possible tie up with Amun, Egypt, the Jerusalem priesthood and the worship of the unseen God.

But there is still another prophecy which came to my attention; this time on-line. It is not a biblical one, but because of its seeming accuracy and alignment with the timeline outlined, I think worth sharing. A German Jew and rabbi, Judah ben Shmuel from Regensburg, lived at the turn of the 12th - 13th centuries from AD 1150 to 1217. He was part of a mystic group called the Chassidei Ashkenaz and not associated with the more common Kabbalistic teaching of the time. He made a forecast it is said, using the Gematria of the Hebrew scripture (the study of numbers associated with letters, words and phrases) as well as astronomical observations. At least this is the consensus, although I suspect there might have been more to it.

His prediction was that a Caliphate in Turkey (now known to us as the Ottoman Empire) would rise in the Islamic world, and that in 1517 it would take control of the Holy Land and occupy Jerusalem for 8 jubilees (400 years), i.e. until 1917.

Now, throughout this book I am also using a jubilee of 50 years – seven times seven with an extra jubilee year to finish. I believe the writer of the Torah intended this. Shmuel stated that this period would be followed by 1 jubilee of 'no man's land' i.e. divided rule in the Holy Land **which brings us again to 1967** and the end of the gentile age in Jerusalem.

Unfortunately, Shmuel's book on the subject has only been passed down to us through quotations from other writers and then only in fragments. Importantly we have his summary of findings, but we do not know how he actually arrived at those through his studies, or more direct inspiration. A pity, I am sure it would be absolutely fascinating and enlightening reading.

There is just one more timeline that finishes in 1967. It is the fourth timeline that I have come across. Indeed the last one to appear as I have been typing the words for this book and seeking completeness and confirmation of the 'seven times'.

As far as I am aware it is not something that has been identified before; the nature of the material being rather esoteric to say the least. But it seems completely at one with the 'abomination of desolation' picture and the instigator, the 'false prophet' of Revelation. I will outline it shortly, so that we might be without doubt that this is indeed the Lord's timetable.

In addition to four separate timelines, there is the astronomical aspect of the end of the zodiacal age. The end of the age appears to coincide closely with the end of the 2,556 year, 'seven times' and tribulation age. Previously we discussed the technical aspects of measuring the zodiacs, and decided that it is the transition from the Age of Pisces to the Age of Aquarius which marks the end of our contemporary times.

A poignant and rather interesting aspect to all this, is that around 1967 (the end of our Gentile age) a hit song arrived on the music charts. It was called 'Age of Aquarius' and came from the musical 'Hair' which itself was a great success. It is quite a haunting tune and the lyrics suggest that 'This is the dawning of the Age of Aquarius'.

Now the alignments of the planets Jupiter and Mars proposed in the song as significant, seem more astrological than biblical, and as far as I am aware have no special significance in the context here. It is only a song after all, and there are other considerations which would suggest that the Age of Aquarius begins considerably later. But the fact that the song came out at the very end of the 'time of Jacob's trouble' is beyond coincidence... surely? Is this another sign of the end of the Gentile Age *and* the end of the Age of Pisces, the fish, and therefore of prophetic significance?

I have now reached the stage where nothing would surprise. Perhaps with so much variation in determining the potential dawning of a new zodiacal age, we can at least accept the introduction of this song as a sign that Aquarius is dawning, while at the same time concede the age may still be some way off.

The fact that the zodiacal sign of Pisces has presided above us since our Messiah arrived, is surely no accident of history either. It brings new significance to Jesus' words to his disciples 'I will make you fishers of men' and the feeding the 5,000 with loaves and fish comes to mind. The fish was also used by the early Christians as one of their secret signs during the pagan Roman persecution - an acronym in Greek for 'Jesus Christ, God's Son, Saviour'. It is 'the sign of Jonah' and has been used by Christians ever since; adorning the windows of many vehicles and causing a stampede to the markets every Good Friday.

There are other signs of the advent of Aquarius as well. It was in the 1960s that decimalisation began in many English speaking countries. It started with currency and has continued from that time into other areas at a steady rate; this, as we move away from the old 60 based system of Babylon and its associated traditions.

I am therefore now confident that the timeline as outlined is the skeleton, if you like, of Jewish / Israelite history since the sacking of Jerusalem by Nebuchadnezzar. I believe it also underpins a correct understanding of the last two and a half thousand years on a world scale. Not from the aspect of a mighty empire, but from the point of view of a relatively small nation caught in the middle; a wayward and disappointing people in many ways according to the judgement of their own prophets and scripture. No doubt their punishment was all the more severe compared to other nations and peoples, simply because so much was expected of them as God's chosen.

Nevertheless they *have* survived and had a large impact through many nations and empires. They were judged, found wanting and went through this incredible time of trial and refinement; all in order to eventually be useful in the plan and work of the Almighty. I am sure that this is the story.

And it is the same with our personal journeys too – the third baptism – the baptism by fire. The baptism of fire is not the baptism we wish to dwell on. But it is the one that burns off the dross. It is painful, but needed. It prepares us for the future, while making our lives worthwhile. It is about learning to live up to our job description.

This is not just some theoretical discussion on biblical prophecy, or a religious numbers game. It was revealed to the prophet Daniel by the Archangel Gabriel, who in turn was directed by a Divine figure, the 'Son of Man'. It was later confirmed through another Jewish prophet, the Apostle John; the disciple that was Jesus' closest friend in all the world.

Those visions to John unveil so much more in regard to the end times, and it follows that everything was revealed through the same 'Son of Man' and is therefore the Lord's word and meant to be taken seriously. 'I go before you' he says in Isaiah 45:2-3 where we find wonderful inspiring words first meant for Cyrus the Great, but ones we can take to heart ourselves.

Is it all too difficult to believe? Does it contradict the notions of too many knowledgeable experts and commentators?

It certainly is not contemporary secularist thought and it is not popular futurist teaching either. In the end the proof is in the pudding, and I leave it to the reader to contemplate carefully and judge accordingly.

CHAPTER 11

Animal Kingdoms

For my Son the Messiah shall be revealed with those who are with him, and those who remain shall rejoice four hundred years. And after these years my son the Messiah shall die... ESDRAS / EZRA (ca. 418 BC)

The first 'three and half times'

In the previous chapter 'Age of the Gentiles', we surveyed the broad span of history that was the 'seven times' of tribulation of Israel. This began with the sacking of Jerusalem by Nebuchadnezzar in the early 6th century BC and ended with the reunification of the city in 1967. In the next two chapters, the prophecies will be matched up against history in more detail.

The chapter titles will follow chronologically and here we will focus as much as possible, on the first 1,278 years up until AD 689. In that timeframe, there were several 'beast' (or animal) kingdoms as described by the prophet Daniel. Again the table at Figure 4 summarises the timeline.

However it will not be the first 'three and a half times' followed by the second 'three and a half times', in the strictly consecutive manner that my applied science mind would prefer. This is because I think it advisable, to deal with each vision (and its prophetic implications) as fully as possible, before moving on to the next one. More often than not, there is more than one 3.5 times of prophetic events in any specific prophecy and this makes a strictly consecutive approach unworkable. It means there will be a degree of repetition in this and the following chapter, and I hope this will work to the advantage of most readers, given there is much to absorb and meditate upon.

Before investigating Daniel's prophecies of the 'animal' kingdoms, we will look at an earlier vision he had in very testing circumstances. This was actually a vision of Nebuchadnezzar of Babylon and comes to us in Aramaic (as does Daniel 7) rather than the Hebrew of later chapters.

Faithful servants throughout this world who have experienced true stress, know that in those humbling times where their ego is stripped away and they feel crushed beyond words, that the Lord is finally able to get through to them. It is at such times that the Lord appears to use individuals effectively and give insights for the blessing of many. The first few verses of Daniel chapter 2 are an extraordinary example of this. Daniel 2:1-13:

2 In the second year of his reign, Nebuchadnezzar had dreams; his mind was troubled and he could not sleep. ² So the king summoned the magicians, enchanters, sorcerers and astrologers[a] to tell him what he had dreamed. When they came in and stood before the king, ³ he said to them, "I have had a dream that troubles me and I want to know what it means.[b]"

⁴ Then the astrologers answered the king,[c] "May the king live forever! Tell your servants the dream, and we will interpret it."

⁵ The king replied to the astrologers, "This is what I have firmly decided: If you do not tell me what my dream was and interpret it, I will have you cut into pieces and your houses turned into piles of rubble. ⁶ But if you tell me the dream and explain it, you will receive from me gifts and rewards and great honour. So tell me the dream and interpret it for me."

⁷ Once more they replied, "Let the king tell his servants the dream, and we will interpret it."

⁸ Then the king answered, "I am certain that you are trying to gain time, because you realize that this is what I have firmly decided: ⁹ If you do not tell me the dream, there is only one penalty for you. You have conspired to tell me misleading and wicked things, hoping the situation will change. So then, tell me the dream, and I will know that you can interpret it for me."

¹⁰ The astrologers answered the king, "There is no one on earth who can do what the king asks! No king, however great and mighty, has ever asked such a thing of any magician or enchanter or astrologer. ¹¹ What the king asks is too difficult. No one can reveal it to the king except the gods, and they do not live among humans."

¹² This made the king so angry and furious that he ordered the execution of all the wise men of Babylon. ¹³ So the decree was issued to put the wise men to death, and men were sent to look for Daniel and his friends to put them to death. (NIV)

It really is not possible to imagine this situation fully. It is virtually beyond comprehension, and yet the distress, danger and confusion were faced by all the advisors to the king at the royal court. They included Daniel and his elite Jewish friends.

Those in places of authority and responsibility in government, industry and society, will understand a little of what these advisors were going through. There was vastly more at stake than being made to look foolish, or even being sacked from the court. Daniel had already been through impossibly difficult circumstances and had his faith thoroughly tested. So he spoke to the king's officer Arioch, who was about to implement the king's wishes and put everyone to death. He asked for time, so that the dream might be revealed to him in the first place, but also that an interpretation might be forthcoming.

He then explained what was happening to his friends Hananiah, Mishael and Azariah, so that they could also be supportive and pray for mercy. Amazingly, during the night 'the mystery was revealed to Daniel in a vision'. In the morning he asks Arioch to take him to the king, Daniel 2:31-45

31 "Your Majesty looked, and there before you stood a large statue— an enormous, dazzling statue, awesome in appearance. 32 The head of the statue was made of pure gold, its chest and arms of silver, its belly and thighs of bronze, 33 its legs of iron, its feet partly of iron and partly of baked clay. 34 While you were watching, a rock was cut out, but not by human hands. It struck the statue on its feet of iron and clay and smashed them. 35 Then the iron, the clay, the bronze, the silver and the gold were all broken to pieces and became like chaff on a threshing floor in the summer. The wind swept them away without leaving a trace. But the rock that struck the statue became a huge mountain and filled the whole earth.

36 "This was the dream, and now we will interpret it to the king. 37 Your Majesty, you are the king of kings. The God of heaven has given you dominion and power and might and glory; 38 in your hands he has placed all mankind and the beasts of the field and the birds in the sky. Wherever they live, he has made you ruler over them all. You are that head of gold.

39 "After you, another kingdom will arise, inferior to yours. Next, a third kingdom, one of bronze, will rule over the whole earth. 40 Finally, there will be a fourth kingdom, strong as iron—for iron breaks and smashes everything—and as iron breaks things to pieces, so it will crush and break all the others. [41] Just as you saw that the feet and toes were partly of baked clay and partly of iron, so this will be a divided kingdom; yet it will have some of the strength of iron in it, even as you saw iron mixed with clay. [42] As the toes were partly iron and partly clay, so this kingdom will be partly strong and partly brittle. [43] And just as you saw the iron mixed with baked clay, so the people will be a mixture and will not remain united, any more than iron mixes with clay.

[44] *"In the time of those kings, the God of heaven will set up a kingdom that will never be destroyed, nor will it be left to another people. It will crush all those kingdoms and bring them to an end, but it will itself endure forever. [45] This is the meaning of the vision of the rock cut out of a mountain, but not by human hands—a rock that broke the iron, the bronze, the clay, the silver and the gold to pieces.*

"The great God has shown the king what will take place in the future. The dream is true and its interpretation is trustworthy." (NIV)

There are a number of known interpretations of this and the following visions, and here I will be following the most popular (especially amongst historicists) because it seems to fit beyond reasonable doubt and accords with Daniel's interpretation.

We see the vision is of a statue. The head is made of gold and represents Nebuchadnezzar's Babylonian Kingdom and Empire. The next kingdom is represented by a chest and arms of silver, and history has revealed it as the Medo-Persian Empire, which overran Babylon a few short years later. One arm is interpreted as the Median kingdom, the other the Persian

The third kingdom is of bronze and is represented by the belly and thighs of the statue.

It follows that this was the Macedonian Greek Empire; the one which held sway over so much of the known world, after Alexander the Great's invasion of Asia and Egypt.

Of course this interpretation would have meant little except in hindsight, and this is a characteristic of biblical revelation. We are told what we need to know for our time and nothing more. However we are not left hanging, without further confirmation and so this picture is backed up by other visions as well.

We then come to the legs of iron which are a great image of the Roman Empire rampaging over all of the Mediterranean, Europe, North Africa and the Middle East. The two legs signify the Western Empire based in Europe and the Eastern Empire based in Asia and Africa; this, since the time of Julius Caesar.

The ten toes follow and become the Ten Kingdoms of the original empire into which Rome is broken in later days. The toes are a mixture of iron and clay, indicating a variability and fragility which ultimately leads to trouble and weakness. Given we have five toes on each leg it suggests that five kingdoms are associated with Rome and Western Europe and five with the Eastern Empire and Constantinople. As I mentioned earlier, it also suggests that some of the eastern kingdoms are most likely in Asia and Africa (Egypt). For the most part these kingdoms were also former provinces of Greece and Medo-Persia.

Then Daniel's interpretation goes directly to the final result, which is the setting up of a Kingdom which will never be destroyed and will last forever.

It is important that we understand the historical scope of this prophecy. The statue arguably only depicts the first half of Gentile occupation, but the following narrative takes us through to the very end. It is not a regional, or local one that happened in the 6th, 5th or 4th centuries BC. It spans thousands of years and is consistent with our 'seven times' of the Gentile occupation (tribulation of the Jews) period of 2,556 years - but does not spell out the length of the period. When we look at the Daniel 7 and 8 visions, we find that they also span the total 'seven times' period or a significant portion of it, but *do* give us numbers.

The exception is chapter 12, which outlines the complete seven times in a uniform overarching manner. However chapter 12 is a very brief description, somewhat cryptic in translation and other numbers are introduced toward the end of the chapter, to make it less than comprehensible at first glance.

Recapping; the neo-Babylonian kingdom was active from the mid 7th century, but for our 'seven times' purpose and Israel's final demise; dates from 590 - 586 BC in Jerusalem.

Babylon was then overrun by the Persians about 50 years later. The Persians occupied the land until Alexander and the

Greeks invaded in 334 – 332 BC, who in turn retained control until the Romans arrived around 62-49 BC.

From there the control of Rome was paramount as the pagan Roman Empire and continued until AD 476, when Rome fell to the Germanic tribes. This occurred despite the capital being moved to Constantinople in the early 4th century.

At some point after AD 476 (perhaps around AD 538 to AD 565) the 'Ten Kingdoms' now represented by the Byzantine Empire, held sway in Jerusalem until the early 7th century (637) and the Islamic conquest. A few years later the Muslims finally cleared the Temple Mount and removed all vestiges of Judaic or Christian worship in AD 689.

So the imagery of the statue is fairly self-explanatory, with the head and arms symbolising Nebuchadnezzar's Babylon and the Medo-Persian Empire respectively. The bronze gut and thighs do not give much away, except that Greece is elsewhere shown to split into four, and arguably there are four sides to the human torso. A bit of a stretch, but not inconsistent, given we are constrained by the analogy of the human form.

The two legs are very apt in describing the Roman Empire; with a Western European and Eastern Asian empire, and its demonstrated ability to stomp on everybody in and around the Mediterranean world. The vision then ends with the ten toes of iron and clay. They represent the Ten Kingdoms which follow the two legged 'iron' kingdom toward the end.

If there is any overlap between the first 3.5 times and the second, it will be here with those toes. However it is important to remember that the ten toes (kingdoms) persist to the end, i.e. 'until the God of Heaven sets up a Kingdom that will never be destroyed, nor left to another people'. They persist to this day.

What is 'key' is that they do not rule in Jerusalem after the year 689, and in fact we know the surrender was signed in 637. Except for the Crusades, we only hear of them again at the end of the age, where they are involved with the defeat of the final beast kingdom in Europe, the Middle East and Holy Land.

The following visions of Daniel also go further than the first 3.5 times. But the reader has to be very alert to realise just how much further. In Daniel 7 we have the first of the 'animal' visions of the empires, with just a hint of what is to follow. Said another way; with the four animal (beast) kingdoms and ten horns of Daniel, we are looking at the first 'three and a half times' in detail, but with a lot more subtly following in the interpretation section. Also keep in mind, this is a Jerusalem centric timeline and therefore not the way someone from Rome, would have understood history. If we remember this, the prophecies and associated history will become clear and we will not fall into the trap that the Reformers did, with their Rome-centric historicist interpretation from the 7 times midpoint onward.

It was perhaps natural for them to see Rome that way given the Daniel 2 statue vision. Unfortunately they only saw the Ten Kings (as the toes) finalising the whole 'seven times', instead of realising that the Kings also featured in Jerusalem 'for a short while' at the end of the first 3.5 times. They missed something crucial; their hatred for Rome blinded their understanding.

A primary question to be answered is: are the timelines presented by each vision identical, or are we talking about different peoples and times altogether? This compares to the way that the 'time of the Gentiles, 'time of Jacob's trouble' and the 'tribulation' could be mistaken for different times or events, whereas they are actually different expressions for the same period.

I and most historicists, believe Chapter 7 simply repeats the description of the four kingdoms or empires of Daniel 2; kingdoms which dominate consecutively throughout the period. This time they are depicted as animals, but refer to the same history. The visions are repeated to reinforce the message - partly because of human frailty in memory, comprehension or scriptural transmission - but also to give more information. Certainly the greater detail and different perspectives give us the equivalent, as it were, of a 3D view.

I mentioned earlier, that Daniel would have been under pressure and was desperate to record things faithfully. The individuality of the visions, but also their common theme would have ensured he did not forget too much.

One of the alternate interpretations that has been proposed in the past for these visions is worthy of mention. It illustrates the difficulties of identification and the multiple possible solutions.

Instead of the kingdoms being successively Babylon, Medo-Persia, Greece and Rome, it has been proposed by some that the succession was Babylon, Media, Persia and Greece.

I offer this simply as background, and there have been many more iterations and permutations suggested as well, including bundling the whole picture into the future.

Let us now look at Daniel 7 in its near entirety:

"In the first year of Belshazzar king of Babylon, Daniel had a dream, and visions passed through his mind as he was lying in bed. He wrote down the substance of his dream.

² Daniel said: "In my vision at night I looked, and there before me were the four winds of heaven churning up the great sea. ³ Four great beasts, each different from the others, came up out of the sea.

⁴ "The first was like a lion, and it had the wings of an eagle. I watched until its wings were torn off and it was lifted from the ground so that it stood on two feet like a human being, and the mind of a human was given to it.

⁵ "And there before me was a second beast, which looked like a bear. It was raised up on one of its sides, and it had three ribs in its mouth between its teeth. It was told, 'Get up and eat your fill of flesh!'

⁶ "After that, I looked, and there before me was another beast, one that looked like a leopard. And on its back it had four wings like those of a bird. This beast had four heads, and it was given authority to rule.

⁷ "After that, in my vision at night I looked, and there before me was a fourth beast—terrifying and frightening and very powerful. It had large iron teeth; it crushed and devoured its victims and trampled underfoot whatever was left. It was different from all the former beasts, and it had ten horns.

⁸ "While I was thinking about the horns, there before me was another horn, a little one, which came up among them; and three of the first horns were uprooted before it. This horn had eyes like the eyes of a human being and a mouth that spoke boastfully.

And:

¹¹ "Then I continued to watch because of the boastful words the horn was speaking. I kept looking until the beast was slain and its body destroyed and thrown into the blazing fire. ¹² (The other beasts had been stripped of their authority, but were allowed to live for a period of time.)

¹³ *"In my vision at night I looked, and there before me was one like a son of man,[a] coming with the clouds of heaven. He approached the Ancient of Days and was led into his presence.* ¹⁴ *He was given authority, glory and sovereign power; all nations and peoples of every language worshiped him. His dominion is an everlasting dominion that will not pass away, and his kingdom is one that will never be destroyed.*

The Interpretation of the Dream:

¹⁵ *"I, Daniel, was troubled in spirit, and the visions that passed through my mind disturbed me.* ¹⁶ *I approached one of those standing there and asked him the meaning of all this.*

"So he told me and gave me the interpretation of these things: ¹⁷ *'The four great beasts are four kings that will rise from the earth.* ¹⁸ *But the holy people of the Most High will receive the kingdom and will possess it forever—yes, forever and ever.'*

¹⁹ *"Then I wanted to know the meaning of the fourth beast, which was different from all the others and most terrifying, with its iron teeth and bronze claws—the beast that crushed and devoured its victims and trampled underfoot whatever was left.* ²⁰ *I also wanted to know about the ten horns on its head and about the other horn that came up, before which three of them fell—the horn that looked more imposing than the others and that had eyes and a mouth that spoke boastfully.* ²¹ *As I watched, this horn was waging war against the holy people and defeating them,* ²² *until the Ancient of Days came and pronounced judgment in favour of the holy people of the Most High, and the time came when they possessed the kingdom.*

²³ *"He gave me this explanation: 'The fourth beast is a fourth kingdom that will appear on earth. It will be different from all the other kingdoms and will devour the whole earth, trampling it down and crushing it.* ²⁴ *The ten horns are ten kings who will come from this kingdom. After them another king will arise, different from the earlier ones; he will subdue three kings.* ²⁵ *He will speak against the Most High and oppress his holy people and try to change the set times and the laws. The holy people will be delivered into his hands* **for a time, times and half a time**.*[b]*

²⁶ "'But the court will sit, and his power will be taken away and completely destroyed forever. ²⁷ Then the sovereignty, power and greatness of all the kingdoms under heaven will be handed over to the holy people of the Most High. His kingdom will be an everlasting kingdom, and all rulers will worship and obey him.'

²⁸ "This is the end of the matter. I, Daniel, was deeply troubled by my thoughts, and my face turned pale, but I kept the matter to myself."
(NIV)

The first animal (beast) is a lion with wings. Now the lion was an emblem of the neo-Assyrian Empire and the Babylonian Kingdom / Empire was a product of that previous empire. The wings are indicative of an eagle and in Ezekiel 17, where we see a 'two eagles' prophecy warning to the exiles, we find that one eagle represents Babylon and the other Egypt. In the latter kingdom the eagle is more familiar as the Horus bird or vulture, and is represented prolifically in their hieroglyphs.

We then find the wings are torn off, the animal is stood up and a human mind is given to it.

This is where I struggle. But we know Nebuchadnezzar had a statue made, which he expected everyone to bow down to and worship. The wings being removed is a sign of the downfall of the Babylonian kingdom, but there are also other things going on here, in regard to the Babylonian 'beast' - and in fact all the successive 'animal' empires. This observation particularly includes the final empire as we will see.

What they all have in common is that whether King, Emperor, Caesar or Caliph they now all demand to be worshiped as demigods - or as God. It is therefore emphasised, that at the core of these empires, there will now be a human being. A person who to all intents and purposes, demands worship and complete obedience, just as Nebuchadnezzar did in his kingdom.

This is the 'human mind' component to the vision and we will see shortly, that it ties in seamlessly with the vision of John in Revelation 13 describing the last beast kingdom.

Moving on, we find the next animal kingdom is represented as a bear with three ribs in its mouth - probably representing kingdoms it has devoured. Two of those are possibly Elam and Babylon, although a wider interpretation could encompass anything from India to Egypt. An apt enough image of the Medo-Persian Empire, if not as definitive as elsewhere.

The next kingdom is a leopard which has four wings and four heads and to those who have looked at these things before, we recognise Macedonian Greece and Alexander's blitzkrieg invasion of Asia. We also recognise the four main kingdoms into which it was subsequently divided. Alexander died quite young and he handed the empire over to his four generals; Lysimachus, Cassander, Ptolemy and Seleucus.

Lysimachus took Thrace (roughly Bulgaria today) and western Asia Minor, Cassander was happy with Macedonia and Greece; Ptolemy seized Egypt, Palestine, Cilicia, Petra and Cyprus, while Seleucus make do with the rest of Asia (Syria, Mesopotamia, Persia and India). Plainly, no one went without!

In the third century BC, the Macedonian Greek Empire abutted the Qin province in today's Western China. It is speculated that the Greek influence resulted in the hundreds of sculptures found at Xi'an of the 'Terracotta Warriors'; the replica armies of Qin Shi Huang, the first Emperor of all China.

In addition, the first Buddha sculptures in India came from Greek craftsmen, and were likenesses of young European men in the beginning. The influence of the empire was therefore truly remarkable and unfortunately, not always for good.

Returning to verse 7 we come to the fourth beast, more terrifying than the rest, with teeth of iron that crushed and devoured its victims. It had ten horns or kingdoms that emerge from it at the end, and we can see it corresponds with the earlier statue vision of chapter 2 and its ten toes. In short it is difficult not to conclude that the fourth beast is Rome and ties in exactly with the Daniel 2 prophecy. However there is another horn, a little one, which rises amongst the ten horns and subdues three of the previous kingdoms, i.e. takes control of them.

We are told the horn has the eyes of a man and a mouth that speaks boastfully. This little horn has more often than not been interpreted as the Holy Roman Empire, Vatican and the Pope by many analysts; and particularly so by Protestant commentators ever since the Reformation. I have to admit that knowing the sort of garb we associate with the Pope and the Cardinals, it is easy to conjure up a picture of a stout horn with eyes painted on it, dressed in crimson and purple.

This view has been defended by looking at the power blocks and peoples which might be identified as the ten 'toes / horns' which resulted from the breakup of the Roman Empire. Also how they link to the prophecies of Daniel - but particularly Revelation and the relationship to Babylon the Great.

In this regard, the identity of the three kingdoms which are subdued by the little and boastful horn in the prophecy, have been of particular interest. It is suggested that these were kingdoms in the European heartland adjacent to Rome itself; kingdoms who were opposed to the establishment of the Vatican and its increasing power. The kingdoms usually singled out are the Ostrogoths, Vandals and Heruli; all tribes which copped the wrath of the Vatican and subsequently faded from history.

If we look at the prophecy from a European viewpoint, then interpreting the little horn as the Vatican makes some sense, except that none of the three kingdoms sprang out of the Roman Empire. The Ostrogoths and Heruli came from far Eastern Europe and north of the Black Sea respectively, and the Vandals from the Baltic. So none of the three qualify, with a military sortie by Darius the Great to the Danube the only connection.

But if the prophetic timeline is Jerusalem-centric, then all this is speculation and can be disregarded. Once we do that, we soon realise that there are other nations / regions which are candidates for the three uprooted kingdoms. Moreover, there are several reasons why the Vatican cannot be seen as a legitimate kingdom, with the viable army and significant power that the other beast kingdoms of Daniel possessed.

Other historicist interpretations have started from AD 476, AD 538 or similar significant dates around Roman history.

Unfortunately, they do not tie sufficiently into the historical facts surrounding Israel. In AD 476 the last (supposed) Roman Emperor, Romulus Augustus was deposed by the Germanic King Odoacer and this is normally agreed to be the end of the 500 years of domination of the original Roman Empire.

Several years later, Roman Church authority became supreme in AD 538, after a letter of Justinian I (the Justinian Decree) acknowledged the Bishop of Rome as the head of all the churches and bestowed political, civil and ecclesiastical power to Pope Vigilius. Justinian reigned from 527 to 565 in Constantinople and is sometimes referred to as 'the last Roman', because he was the last ruler to use Latin officially.

Importantly, he regained much of the territory of the Western Empire before his passing, while thereafter the Empire decayed and lost its former glory. This suggests that Justinian has a better claim to be the last emperor, than Romulus Augustus.

The significance of these dates is that they occur a few years before AD 689 and the Dome of the Rock, and accord with the prophetic concept of the Ten Kingdoms coming out of the Roman Empire *before* the 'boastful horn' becomes prominent.

It is also Jerusalem and not Rome which is being addressed here, and this means the Ten Kingdoms become recognisable immediately prior to the Muhammadan takeover of the Holy Land and much of the Middle East. It ties in neatly with Daniel's prophecy and those in Revelation. The 'ten horns' then exist in parallel with the following super power, symbolised by the 'little horn with eyes and a mouth'.

However there are a couple of potential problems here. The Eastern Roman Empire was considered to last until the fall of Constantinople in 1453 by its citizens back in the day, while modern historians use the name Byzantine for the whole period from Constantine onward. This issue will be discussed later.

The prophetic question is one we have already addressed and is about the identity of 'the little horn'. Is it a continuation of the Roman Empire in a later form and the Papacy as Protestant historicist expositors currently suggest? Or is it the Islamic Empire as strongly indicated in the previous chapter?

When we look at Daniel 7 from verse 24 onward and the interpretation given to Daniel, we see that the 'time, times and half a time' mentioned here, does not refer to the four kingdoms ending in the beast who will devour the whole earth by crushing it (Rome) but instead, to the king who will subdue three of the ten post-pagan Roman kingdoms.

This is critical to correct interpretation, although not definitive on its own. The 3.5 times spoken of here, can only be the second half of the 7 'times' of tribulation, because the first half is the period of the first four empires (plus briefly the Ten Kingdoms) that occupy Jerusalem leading up to AD 637-689.

The Daniel 7 description appears to focus on the first 3.5 times and does so for the most part. However the chapter actually covers the whole of the seven times of tribulation. Furthermore, the upstart of a king with eyes and a boastful mouth, represents a kingdom that rules for the same period of time, as *all* the early empires combined. In fact the entire second 1.278 years up to 1967 – but note well - **in Jerusalem** - not Rome.

If the little horn was the Vatican, why would it be effectively said to start its dominance in the year 689, when that event is significant as the demise of Roman / Byzantine / Ten Kingdom Christian and Jewish worship in Jerusalem?

From a Jerusalem perspective it makes no sense at all. This is precisely when the post Roman Byzantines were beguiled / compelled to exit the Holy Land!

We are therefore forced to acknowledge, that Islam took over from Rome at the critical time of the 'abomination of desolation' and therefore is the end-time Beast Empire.

And while it is true that the second 3.5 'times' above is much the same for the Ten Kingdoms as it is for the Islamic Empire (except the Ten Kingdoms get a century or so start) in the great scheme of things, they are contemporaries *outside* eretz Israel and occupy a parallel universe. Yes, they are collectively the 5th animal kingdom (the horns in Daniel 7) and the second phase of the Roman beast for a 'little while' and 'one hour' as Revelation 17 describes it - and the Church is also tied to them. But they are *not* the final and Eighth Beast Kingdom.

So the relationship is extremely fractious and hate-filled and the Ten Kingdoms play second fiddle to the Islamic Empire where it counts – in the Holy Land. They only rate the briefest of mention by the prophets throughout the second 3.5 times.

For similar reasons, the little horn cannot be a king from the Seleucid, or any of the four subsequent Greek kingdoms, because they did not last anywhere near as long as the required 3.5 times / 1,278 years of prophetic history. In any case the Greek Seleucid period is well and truly part of the first 1,278 years and *not* the second. This is where it is imperative to focus on the events in Jerusalem and not some fuzzy timeline elsewhere in Rome, Cairo, London, Paris or New York.

We are also told in the interpretation that 'he will speak against the Most High and oppress his saints and try to change the set times and laws'. Islam is the only regime in the region, which has had a completely lunar calendar with no tie to the solar year. Their months, including Ramadan, revolve through the seasons and move backward eleven and a quarter days each year. Even a luni-solar calendar is designed to align closely with the seasons - albeit only exactly every 19 years.

And to be completely clear, Islam has totally different feasts to Judaism or Christianity. Notably Passover / Easter and Shavuot / Pentecost are nowhere to be seen.

At the risk of repetition; all I remember hearing in my youth was the notion, that the little horn of Daniel was the Vatican presiding over the Holy Roman Empire. It has been a mantra which has lead Protestant historicist interpreters astray since the Reformation. Although it has not been easy to digest the fact that the 'little horn' is in fact Islam, I have finally managed to redress this perversion using the lessons the Lord has provided.

I will now move on to Daniel 8 and review in detail the vision that Daniel recounts there. He sees himself beside the Ulai Canal, near the Citadel of Susa in Elam. This was part of the Persian Achaemenid Empire in his day. Daniel 8:

> *8 In the third year of King Belshazzar's reign, I, Daniel, had a vision, after the one that had already appeared to me. ² In my vision I saw myself in the citadel of Susa in the province of Elam; in the vision I was beside the Ulai Canal. ³ I looked up, and there before me was a ram with two horns, standing beside the canal, and the horns were long. One of the horns was longer than the other but grew up later. ⁴ I watched the ram as it charged toward the west and the north and the south. No animal could stand against it, and none could rescue from its power. It did as it pleased and became great.*
>
> *⁵ As I was thinking about this, suddenly a goat with a prominent horn between its eyes came from the west, crossing the whole earth without touching the ground. ⁶ It came toward the two-horned ram I had seen standing beside the canal and charged at it in great rage. ⁷ I saw*

it attack the ram furiously, striking the ram and shattering its two horns. The ram was powerless to stand against it; the goat knocked it to the ground and trampled on it, and none could rescue the ram from its power. ⁸ The goat became very great, but at the height of its power the large horn was broken off, and in its place four prominent horns grew up toward the four winds of heaven.

⁹ Out of one of them came another horn, **which started small but grew in power to the south and to the east and toward the Beautiful Land***. ¹⁰ It grew until it reached the host of the heavens, and it threw some of the starry host down to the earth and trampled on them. ¹¹ It set itself up to be as great as the commander of the army of the L*ORD*;* **it took away the daily sacrifice from the L**ORD**, and his sanctuary was thrown down.** *¹² Because of rebellion, the L*ORD*'s people[a] and the daily sacrifice were given over to it. It prospered in everything it did, and truth was thrown to the ground.*

¹³ Then I heard a holy one speaking, and another holy one said to him, **"How long will it take for the vision to be fulfilled—the vision concerning the daily sacrifice, the rebellion that causes desolation, the surrender of the sanctuary and the trampling underfoot of the L**ORD**'s people?"**

¹⁴ He said to me, "It will take **2,300 evenings and mornings***; then the sanctuary will be reconsecrated."*

The Interpretation of the Vision:

¹⁵ While I, Daniel, was watching the vision and trying to understand it, there before me stood one who looked like a man. ¹⁶ And I heard a man's voice from the Ulai calling, "Gabriel, tell this man the meaning of the vision." ¹⁷ As he came near the place where I was standing, I was terrified and fell prostrate. "Son of man,"[b] he said to me, "understand that the vision concerns the time of the end."

¹⁸ While he was speaking to me, I was in a deep sleep, with my face to the ground. Then he touched me and raised me to my feet.
¹⁹ He said: "I am going to tell you what will happen later in the time of wrath, because the vision concerns the appointed time of the end.[c]
²⁰ The two-horned ram that you saw represents the kings of Media and Persia. ²¹ The shaggy goat is the king of Greece, and the large horn between its eyes is the first king. ²² The four horns that replaced the one that was broken off represent four kingdoms that will emerge from his nation but will not have the same power.

²³ "In the latter part of their reign, when rebels have become completely wicked, a fierce-looking king, a master of intrigue, will arise. ²⁴ He will become very strong, but not by his own power. He will cause astounding devastation and will succeed in whatever he does. He will destroy those who are mighty, the holy people. ²⁵ He will cause deceit to prosper, and he will consider himself superior. When they feel secure, he will destroy many and take his stand against the Prince of princes. Yet he will be destroyed, but not by human power.

²⁶ "The vision of the evenings and mornings that has been given you is true, but seal up the vision, for it concerns the distant future."
²⁷ I, Daniel, was worn out. I lay exhausted for several days. Then I got up and went about the king's business. I was appalled by the vision; it was beyond understanding. (NIV)

In my study of Daniel, I cannot get over his personal experiences and self-sacrifice and consequent trauma and anguish. The last verse sums it up. 'I lay exhausted for several days.... I was appalled by the vision; it was beyond understanding'.

Not only had his life been in danger so many times, but in being selected as the Lord's primary vehicle to tell the world what the future held, he went through such excruciating mental torment and agony. It makes a laughing stock of the psychological support it seems many people require today, from an array of shrinks, medicos, social workers and pharmacists - and that just to lead rather ordinary lives.

In the New Testament there are one or two scriptures which in effect say 'God loves you and has made you for a purpose'.

There is no better example of purpose than Daniel. He had all the attributes that the Lord could use to bring his prophecies and timeline to the chosen ones; faithfulness, prayerfulness, sensitivity to the Spirit, courage in adversity, determination and patience, meticulous attention to detail and a gift for numbers and writing - to name a few.

We are all unique in our own way, but Daniel was especially born, prepared and selected from amongst his peers.

So the vision opens with a ram with two horns, one being Media and the other representing Persia. The ram is charged by a goat (no less) from the west. It has a prominent horn and we know from history that the goat and horn represent the Greek Macedonian Empire and Alexander the Great.

I have already discussed how Alexander's Empire was split into four and we learnt how miraculously the 2,300 'evenings and mornings' interpreted as years, covers the precise interval, from the time Alexander the Great crossed the Bosporus, until 1967 and the Six Day War.

We also know, that although the Temple has not been rebuilt, both Jerusalem and the Mount area *are* now in the ultimate control of a very patient Israeli people. That is as it should be, until the 'Most High' provides the right environment for the sanctuary to rise again - if that is indeed, His will.

If we focus on verses 9 to 12 we learn that there is 'another horn' which starts small, but grows in power to the south and east and toward the Beautiful Land. It throws down the starry host that were the saints, and sets itself up to be as great as the Prince of the hosts. It takes away the daily sacrifice from the Prince (and the saints) and brings down the sanctuary.

Now if the mystery of the little horn is a little clearer in our minds after Daniel 7, then these verses should absolutely confirm the identity of this ruler.

The horn here, seems to have characteristics remarkably like the 'little horn' of the previous chapter.

His empire is not brought down until the exact same year, 1967, as is the case with the little horn of chapter seven. He exalts himself to the level of the Most High and above the Prince of the saints. The timeline of 2,300 years here in chapter 8, therefore makes it clear that this horn (kingdom / empire) is dominant throughout the second 3.5 'times'.

Predictably, I have now ventured deep into the second 3.5 times in this chapter, contrary to my best intentions. But it is important to emphasise the critical interpretation issues surrounding the overall timeline.

Now, perhaps the only possible argument against the interpretation of the successive kingdoms that I am presenting here, is that in this vision we do not see anything that represents Rome in chapter 8. How could this be if the 'horn' is not Rome?

When I was travelling in Jordan and visiting places along the old King's Highway, it was quite apparent there were Roman influences. We know that apart from enhancing the existing Greek cities with Roman niceties; like freestanding amphitheatre, public baths and government buildings, that road building was also a priority throughout the area and the Sinai.

However the region was already well established by the Greeks from the start, with Alexander travelling this way from Egypt to Mesopotamia. He established Jerash as a holiday / retirement town for his troops and in essence, the Greeks controlled all the Middle East thereafter under the four kingdoms.

For the Romans though, this was boundary country. There were boundaries with the Parthians and the Arabians, where no boundary existed under the Greeks.

Even as a tourist, it becomes evident that despite Roman enhancements to Greek infrastructure, the next obvious influence after Alexander and the early Greeks is Byzantine; with Constantine era churches and art including many mosaics.

During the first century of the Christian era, it is said that believers vacated Jerusalem and fled across the Jordan to escape Roman persecution. It would suggest that the Romans had an arrangement with their eastern neighbours, of using the trans-Jordan and adjacent areas as a sort of demilitarised zone.

So my sense is: that between the days of the Greek Seleucid era and the Roman (but predominantly Greek speaking) Byzantine Empire, the locals in Jordan mainly experienced the Greek influences in everyday life and language.

The Roman (Latin) presence while feared, was more muted, distant and militaristic.

In any event, our timeline makes it clear, it has to include Rome. There is no alternative, neither in Daniel or Revelation.

The last scripture we will study in this section are the early verses of Revelation 13. In fact 13:1-10 which describes a 'beast out of the sea':

'The dragon[a] stood on the shore of the sea. And I saw a beast coming out of the sea. It had ten horns and seven heads, with ten crowns on its horns, and on each head a blasphemous name. ² The beast I saw resembled a leopard, but had feet like those of a bear and a mouth

like that of a lion. The dragon gave the beast his power and his throne and great authority. ³ One of the heads of the beast seemed to have had a fatal wound, but the fatal wound had been healed. The whole world was filled with wonder and followed the beast. ⁴ People worshiped the dragon because he had given authority to the beast, and they also worshiped the beast and asked, "Who is like the beast? Who can wage war against it?"

*⁵ The beast was given a mouth to utter proud words and blasphemies and to exercise its authority **for forty-two months**. ⁶ It opened its mouth to blaspheme God, and to slander his name and his dwelling place and those who live in heaven. ⁷ It was given power to wage war against God's holy people and to conquer them. And it was given authority over every tribe, people, language and nation. ⁸ All inhabitants of the earth will worship the beast—all whose names have not been written in the Lamb's book of life, the Lamb who was slain from the creation of the world.[b]*

⁹ Whoever has ears, let them hear.
¹⁰ "If anyone is to go into captivity,
   into captivity they will go.
If anyone is to be killed[c] with the sword,
   with the sword they will be killed."[d]
This calls for patient endurance and faithfulness on the part of God's people. (NIV)

Here we come across another vision of an animal. The animal is a composite beast which has the body of a leopard, but with feet like a bear and the mouth of a lion. However it has seven heads and ten horns and one of the heads recovers from a fatal wound. In Daniel we see a beast with ten horns, but never with seven heads. It is imagery only seen in Revelation.

Despite this, there are distinct comparisons to be made with Daniel's visions. From Daniel 7, we know the lion represents the neo-Assyrian Empire which morphed into the Babylonian kingdom, while the bear is the Medo-Persian Empire. The leopard (the body) is Alexander's Macedonian Greek Empire.

The only thing to note with this new animal, is that the various features (and therefore empires) are noted in reverse order; starting with the horns (Ten Kingdoms), followed by the Greek leopard, then the bear feet of Medo-Persia and ending in the earliest kingdom; the one with the lion's teeth - Babylon.

Again, I am reviewing this fairly closely, because after verse 10 of Revelation 13 we are introduced to another 'beast', which follows this composite one above. It comes out of the earth.

The second beast along with the 'beast out of the sea', is interpreted as a later iteration of Rome, by many non-Catholic commentators. That view has now been around for a very long time and is extremely entrenched in historicist thinking.

Now there is much about this first beast, which suggests it represents more than the triumvirate Babylon – Persia - Greece.

Firstly it comes out of the sea, and we are told in the interpretation that this means the kingdoms come from highly populated parts of the earth.

We know this was the case with Babylon in the middle of Mesopotamia, the Medo-Persian Empire which covered half the known world and the Greek with its original centre in Macedonian Greece in eastern Europe. But it is also true of the following kingdom, the most terrifying of all of the four beasts of Daniel 7; the one representing the Roman Empire.

Secondly and critically, we are told this beast has ten horns, which are exclusively used to describe the kingdoms which are an extension of the 4th 'beast' of Daniel; Rome.

So the first beast that comes out of the 'sea' of humanity in Revelation 13:1-10 *has* to include Rome, making it the 4th kingdom embodied in the beast.

And just quietly, John had good reason not to be too descriptive of this beast; it was in his interest to be as cryptic as possible. It was the beast kingdom he was living under with Caesars such as Tiberius, Caligula and Nero. They were unpredictable tyrants, not to be messed with, and John spent a lot of time in prison and fear of his life because of the Roman state.

Furthermore in verse 5, we are told that this composite beast which comes out of the sea '....was given a mouth to utter proud words, and blasphemies and to exercise his authority for **forty two months!**'

Wow, there is our wonderful timeline to the rescue, Forty two (42) months is another way of expressing the 3.5 times or

1,278 years. It means that this beast occupies the whole of the first 'time, times and half a time' – exactly as the beasts of Daniel 7 do. Therefore it *must* include the Roman Empire, for this composite animal to occupy the entire age until AD 689.

The image in Revelation 13:1-10 has 'seven heads' as well as the ten horns of Daniel 7, and we also find the 'seven heads' description in Revelation 12. In chapter 12, there is also a pictorial description of Israel being harassed by a dragon (the terrible beast). So all three chapters portray the 'ten horns' and two of them, the 'seven heads'.

Now I remember trying to work out how the 'seven heads' related to the 'ten horns'. In fact, I must say, initially without much success. Some amusing combinations came to mind in an attempt to squeeze some meaning out of the numbers; combinations such as 4 unicorns and 3 bullocks rolled around in my head. I mean what is a bloke meant to think?

Fortunately once we realise that the seven 'heads', aka empires, or kingdoms are also referred to in Revelation 17 (where we are told the final 8th beast kingdom is a resurrected form of one of the earlier 7 kingdoms) a solution presents itself.

That solution is of **sequential kingdoms / empires**.

Until now, we have only seen a sequence starting with Nebuchadnezzar's Babylon and ending with the Ten Kingdoms in Daniel prophecies. When we count these through, we discover there are five sequential empires / kingdoms, providing we view the Ten Kingdoms as a single parallel, contemporaneous group on the timeline, or recognise the empire that represents them.

So if the Ten Kingdoms can be considered a collective 5th empire, the question becomes: who are the other two 'heads', where did they spring from and where do they fit into the picture before the final 8th beast kingdom arrives on the scene?

The answer to this problem will be provided shortly and put a final stamp of authority on this interpretation.

Together, all the clues form a strong foundation for our identifications in the next chapter, where we discuss the second half of the seven times of tribulation in detail, i.e. the second 1,278 years.

Messianic prophecy

So far, we have covered the overall dating of the age from 590 BC to AD 689 and touched on the Babylon the Great period running up to 1967. The pivotal year of 689 has been covered in some detail, because it is precisely in the middle of the 2,556 years; dividing the first '3.5 times' from the second. It would not surprise if there are many other milestones along the way however, and it so happens that there are two which I have identified.

They are both significant and the first one is particularly important, because it ties into the Daniel 9 prophecy of the 70 'weeks', which points to our Jewish Messiah.

Reading the Gospels and the Apostle's letters closely, it becomes apparent that many Jews and Samaritans of 2,000 years ago, understood that Daniel's prophecy in chapter 9 was predicting the imminent arrival of their Messiah. Their interest was heightened because they were experiencing the severe yoke of the Romans and were looking forward to liberation. We read that some had a real expectation, that the subjugation and tyranny they were experiencing could well be over.

What many would have struggled with; was the concept that their Messiah was no ordinary oil-anointed king, but a Holy Spirit (Shechinah) anointed miracle working Priest-King. Someone who was bent on changing attitudes, from war and hate, to peace and salvation. This was not going to be accomplished as a political movement, or through military might, but by fulfilling the Abrahamic covenant and building a grass-roots, Kingdom of Heaven on earth.

It was about bringing an end to the age of Aries and animal sacrifice, and completing the Passover promise of the Lamb.

Some suggest that the Jews were looking for two messiahs; one a king, the other a priest, in much the same way that their society was set up during the Davidic era. We learn through the Dead Sea scrolls, that the Qumran community still thought this way nearly 1,000 years later leading up to the time of Jesus.

I suspect the first century BC reality was probably a shadow of the First Temple period, where the Shechinah experience and implicit (God given) knowledge of the Messiah was known. It may only have been a liturgical skeleton of former times, but had potential to blossom once more.

As something of a priest-king himself, this was an experience that even King David was aware. Sure, he had Levite priests and Samuel was prominent, but every indication in the scripture of his overall demeanour - and the worshipful nature of his psalms and songs - suggest to me, he had a full-gospel Holy Spirit experience.

Unfortunately he also committed some serious wrongdoing, and his response to that again confirms his anointing in a heartfelt way. 'Oh Lord do not take away your Spirit from me' he desperately cries. It demonstrates the difference between a traditional and liturgical worship system, where people go through the motions, compared to a real relationship with God.

So the first milestone is centred on a date, which I found at first glance to be meaningless – AD 50, the exact centre of the first 'time, times and half a time' or 1,278 years.

Was it a guess to try this? The line between inspiration and an idea worth investigating, has become blurred for me in this prophetic exercise. The Lord, it seems, speaks through all things providing we have an ear to hear. But I knew that AD 70 was the year that the Roman General Titus attacked Jerusalem and carried off the Temple treasures to Rome. And I also knew that somewhere around AD 30 or 31, was the likely crucifixion / resurrection year of Jesus. In fact the year AD 30 is the generally accepted year by historicist interpreters of prophecy.

There is something more; the significance of 40 years to the Hebrews as a period of trial, 40 years in the wilderness during the Exodus, 40 days of rain back during the Flood and the twelve times 40 (480 years) from Solomon's reign back to the Exodus. These all point to something special about the number forty. It was always 40 days or 40 years, when exceptional trials and testing was involved.

Now some of these examples of 40 days or 40 years may be as much figurative as anything, but here is one that is more than that. Whether an exact 40 years is another question. It appears that the Jews would still counted 40 years, even if the first and last years were only a fraction of the calendar year.

So the Year of our Lord (AD) 50 is again in the middle of a period of time (1,278 years) but it is also in the middle of a 40 year period of trial, firmly anchored by the destruction of Jerusalem by the Romans in AD 70.

The thought that came to mind was simply this: could my new knowledge of the timeline, help to confirm the year of Jesus' death and resurrection?

This question is something which has been the subject of much speculation over the years and I think the answer is 'yes'.

The start date of Daniel's 70 week prophecy has always been controversial and I am sad to say, used in Orthodox Jewish circles for two millennia to confuse people in regard to the prophecy and its intended target subject. Even worse, academics in the west seem largely convinced that prophetic visions and inspiration are impossible, such that any and every scheme imaginable is concocted to bypass acknowledgement of the fact.

Postdating the obvious era of writing is a favourite trick, but substituting unlikely start dates and improbable royal figures as the object of prophecies, are all considered fair strategies.

Of course, if one can present prophecies which unambiguously link the history of nations, events and kings covering an age from Nebuchadnezzar to the near future as a complete integrated master plan, you might expect more respect for the prophets. One can only hope that a few will be so persuaded.

At any rate, after the Babylonian exile, the only kings who were in a position to authorise a rebuild of the Jewish Temple, or a reconstruction of Jerusalem, were the kings of Medo-Persia - the ones who titled themselves the King of Kings. They were the ones who ruled the empire that Judea was a minuscule, but important part. And so it is from this time and lineage, we must find our kings and decrees which tie into the Daniel 9 prophecy.

Now the 'day for a year' principle is a non-negotiable starting point, and the prophecy itself indicates that only 69 weeks (483 years) are used to point to the Messiah. The last week (or 7 days) represents something more complex: an overarching seven times tribulation prophetic period, as previously outlined. But it also describes the 3.5 year period of Jesus' ministry before he was 'cut off'; this, to a high degree of certainty.

Many commentators and writers today, take the Nehemiah account to rebuild the walls of Jerusalem (and the city) as the guide to the start date of the prophecy. This appears reasonable enough, given it is the only account of the rebuilding of Jerusalem to be addressed at any length in the Bible.

But we know that the earlier, and first command of King Cyrus to rebuild the Temple, had to be repeated because of opposition; particularly from the 'trans Euphrates' people to the north in today's Syria, Lebanon and Samaria. Because of this, a second decree had to be issued; apparently by King Darius.

Of the later decrees suggested as options for rebuilding the city (as opposed to the Temple) did they also have to be issued more than once? Were those decrees opposed by rivals as well and needed to be restated and supported before work began?

If so, the Nehemiah description relates to a second attempt to continue, or recommence the work.

There is another problem I must address. The scholars who believe that the Nehemiah story and dating is the correct place to start the count (444-445 BC) are then required to adjust the 70 week prophecy timing to point to Jesus.

Unfortunately, the method they use conflicts with the timing interpretation used here. The adjustment that is used by followers of the Nehemiah starting date, require the prophetic years to be treated as Babylonian years of 360 days each. This is consistent with previous historicist understanding, and in this instance, is partly justified on the grounds that portions of Ezra are in Aramaic from the Babylonian exilic period.

Thus adherents to the 360 day Babylon year, then reduce the 483 years (69x7) by the factor 360/365.24. When this conversion is completed, a date of AD 32-33 for Jesus' crucifixion / resurrection is the result. If this conversion is not performed and 365.24 actual years are used, the date blows out to AD 37 and is not realistically supportable.

Now despite my concerns, the argument justifying the use of a 360 day prophetic year based on a reckoning using Babylonian timing (while inconsistent in the context I am using) might be said to point to the life of Jesus as the 'Anointed One' to an acceptable degree of accuracy.

The problem here is; from the outset one of my 'rules of engagement' has always been to use a 365.24 day year and to turf the fictitious 360 day Babylonian concept in the bin.

Frankly, the 360 to 365.24 day conversion is a fake arithmetic manipulation, because the Jews and everyone else employed the Metonic convention since the 8th century BC, and this is based on relating lunar months to 365.24 day, solar years. This is confirmed by later historian-mathematicians, who have also used Gregorian 365.24 day years proleptically to reconcile the kingly reigns of the past. In fact this erroneous 360 day sleight-of-hand is not required, if the 365.24 day year interpretation can be tied in exactly with the AD 30-31 to AD 70 proposition.

It turns out that it can, but it requires two assumptions:

- That the starting point is a Persian decree directing Ezra to begin rebuilding the city in 458-457 BC
- That the end of the 69 'weeks' of the prophecy points to the beginning of Jesus ministry period and baptism, and not his crucifixion and resurrection

Unfortunately, the circumstances surrounding the 457 BC starting point are not as well documented as that of Nehemiah's later efforts to complete the rebuilding of Jerusalem. Nevertheless, it is recognised that (457-458 BC) is fixed in time and is the year when Ezra and his colleagues arrived in Jerusalem.

The second assumption is perfectly in harmony with the overall interpretation of milestones already presented. It begins a seven year period, in the same way all our other milestones are proposed to work. The only difference here, is that Jesus' life was cut off mid-way through the period.

It would be great if there was absolute clarity around the nature of the Ezra decree. Certainly some things we know with confidence, but specific instructions to rebuild the city are not spelt out as clearly as in the Nehemiah story.

Given this, any further evidence that confirms the circumstances and timing is welcome. But if we can tie 457 BC firmly to the beginning of the 69 weeks pointing to Jesus' ministry, it will also give us some confidence that we have those years fixed in time as well.

It sounds simple, but it turns out there are many issues to resolve. Firstly, questions have been asked, as to whether the kings of Persia have been referred to by their correct names in the book of Ezra. Ezra is pretty obviously a compilation of various sources and letters, some of which are in Aramaic, with other portions in Hebrew. If the names are correct, then the chronology of some events seems to be provided out of sequence as flash backs.

Some descriptions of the return events are also uncannily similar, although with different participating leaders. This is a bit unsettling and much the same is seen in 1 Esdras, the first book of that name in the Apocrypha. It is a version of Ezra that is derived from a non-Masoretic Hebrew source and probably originally from the Greek. It also has slightly different content.

Fortunately we also have some detail of the Persian Achaemenid Empire and their royal dynasty from archaeology, Persian records, and Greek and Egyptian contemporary sources.

As noted earlier, the Empire stretched at one time from India and central Asia, to Lower Egypt. There were also incursions into the Russian steppes and Europe, including notable battles in Greece and Romania. (To be fair, some of this Achaemenid documentation was not available in the early 20th century).

In brief, the four decrees which have been identified for the rebuilding of the Temple and subsequently the reconstruction of Jerusalem are:

Decrees to rebuild the Temple:

Ezra 1:   537 BC   (ca. 1st year of Cyrus as king of Babylon)
Ezra 6:   520 BC   (around the 2nd year of the reign of Darius)

The Ezra 6 date may in fact be as late as 516 BC. It would also be close to the end of the 70 years of exile when 42,360 Jews return to Jerusalem to rebuild the Temple.

Decrees to rebuild Jerusalem:

Ezra 7:   458-457 BC (reign of Artaxerxes)
Nehemiah 2: 445-444 BC (reign of Artaxerxes)

All these dates are accurate to within 12 months, as far as I can determine, with the exception of the 520 BC decree which is more uncertain.

In regard to the express reconstruction of the city, Ezra arrives in Jerusalem in the 5th month of the seventh year of the reign of Artaxerxes (458-457 BC). He does so with hundreds of returning Jews, and after receiving permission, support and extraordinary authority from the king.

Ezra 7:12-13

'Artaxerxes, king of kings. To Ezra the priest, teacher of the Law of the God of Heaven: Greetings,

Now I decree that any of the Israelites in my kingdom, including priests and Levites, who volunteer to go to Jerusalem with you, may go.'
(NIV)

Letters in Aramaic, to kings Xerxes and then Artaxerxes from the opposing Trans-Euphrates elite (led by a commanding officer by the name of Rehum) continue to frustrate progress on the city. They have no desire to see the Jews re-establishing themselves in Jerusalem at all; neither rebuilding the Temple or the city. At one point this causes Artaxerxes to stop the Jews and Benjamites from attempting any reconstruction.

One letter emphasises that Jerusalem is a rebellious city, troublesome to kings and provinces and that is why it was destroyed. It includes the words: Ezra 4:16

'We inform the king that if the city is built and its walls are restored, you will be left with nothing in Trans-Euphrates'. (NIV)

And Artaxerxes response, Ezra 4:21

*Artaxerxes: 'Now issue an order to these men to stop work, **so that the city will not be rebuilt** until I so order'* (NIV)

To my mind, this implies that the Jews were already rebuilding the city. Certainly Rehum, the commander and his Trans-Euphrates associates, implied that the Jews were doing so in their letter to the king. They were probably exaggerating their case, but with over 42,000 returning originally, some rebuilding would have been necessary from the start. However the question here is; whether they were also repairing the walls.

The Ezra return would have required another construction program, to house the newcomers who came with him. Other folk would have been returning from elsewhere, as well and 2 Esdras specifically identifies returnees from Asia Minor.

This extra housing effort, therefore, could have delayed work repairing the walls and gates of the city. It explains Nehemiah's later disappointment, when he records that the walls were still in a state of disrepair. He then emphasises his eagerness to rectify that, leading up to the 445 BC proclamation.

But does that make the 445 BC decree from Artaxerxes to Nehemiah, the one we should use as a start date for Daniel's 70 week prophecy? In my view, and that of others, the obvious immense opposition from neighbouring states, should not disqualify an earlier edict on the grounds of non-completion.

It is obvious that the Trans-Euphrates opposition of Rehum and Shimshai, and their bold appeal to king Artaxerxes to halt proceedings in Jerusalem, was initially successful. In fact as outlined in Appendix 3 where there is added detail, we find some indication in 1 Esdras, that these concerns about rebuilding the city may date back to Darius' reign as well.

So the Ezra record is not as clear as we would like, in regard to the two or three returns from Babylon that appear to be outlined there. This would be in part, because a subsequent compiler is not quite sure how to knit the various sources together at his or her disposal. The reassuring thing is; that the picture is becoming clearer after reviewing all the data, and the timeline I have the privilege to outline here provides further guidance.

Indeed, based on all the correspondence, there is a compelling case to suggest that with the considerable initial backing of Artaxerxes (and mirrored by the concerns of Rehum and friends) Ezra had the king's blessing to rebuild the city. That Artaxerxes subsequently changed his mind for a while and put everything on hold, is of little consequence.

I am also reminded of a previous comment which I made concerning Ezra. It was to the effect that, 'I cannot understand why 2 Esdras was not included in the main cannon of the Protestant translations'.

One can understand that there was a decision to be made regarding the inclusion of either the book of Ezra, or alternatively 1 Esdras from a non-Masoretic source. They both cover much the same material and history; albeit 1 Esdras has a slightly fantastical anecdote, where Jewish young men compete for King

Darius' approval. They attempt to do that by spinning the best yarn about wisdom, and this takes up a fair bit of the chronicle.

Perhaps on that basis and the fact it is from another translation, was enough for most Protestant Bible compilers to decide for the more straightforward Ezra version. As a result, the book of 2 Esdras was dropped from most translations as well.

Ezra never impacted my consciousness as a great prophet and leader of Judah, until I read and reread the exile story and realised he was not only a leader of men after Zerubbabel before him, and Nehemiah thereafter, but also a priest and prophet after the great prophets of the exile period. Indeed back to Isaiah, Jeremiah and Ezekiel who shouldered the burden of warning Jerusalem of its fate.

This was reinforced when I absorbed the prophecies of 2 Esdras and their impressive content. In fact I was awe struck.

Especially as it seems to me, they were in fact written during his lifetime and not by some scribe in the Greek, or early Christian period, as some suppose. To be perfectly frank, I now recognise him as a great prophet of God, in the manner of all the leading prophets of the exiles.

In fact the impact has been such, that I am now convinced that the Ezra lead return to Jerusalem under the Artaxerxes decree of 457 BC *is* the one marking the start of the Daniel 9 prophecy of the 70 'weeks', and the coming of the Anointed One. Everything points to Ezra.

Daniel looked back to Leviticus for the seven times of judgement, to Jeremiah for the 70 years of exile and to Ezekiel for his support of the 'day for a year' rule. Ezra looked back to all of them; but especially Daniel.

In 2 Esdras 11 and 12 we see a rather extended prophecy of a three-headed eagle. When it was explained to Ezra that this eagle was the fourth 'beast' of Daniel's chapter 7 prophecy (the one we now recognise as the Roman Empire and its associates) it becomes apparent that Ezra and his prophecies are the next link in God's chain of revelation; a chain which continues into the first century and the Jewish-Gentile Christian period.

And so with that preamble, let us now look at some sample scripture from Daniel 9, which directly points to the coming of the Messiah. Daniel 9:1-3 and 20-27

> *"In the first year of Darius son of Xerxes (Ahasuerus)[j] (a Mede by descent), who was made ruler over the Babylonian[b] kingdom— ² in the first year of his reign, I, Daniel, understood from the Scriptures, according to the word of the* LORD *given to Jeremiah the prophet, that the desolation of Jerusalem would last seventy years. ³ So I turned to the Lord God and pleaded with him in prayer and petition, in fasting, and in sackcloth and ashes."* (NIV)

Historians believe there are three Persian kings referred to as Ahasuerus and that one was Xerxes I and another Artaxerxes. If we compare other known lineages of the Medo-Persian Achaemenid kings with this passage, we note that it is nigh on impossible that Darius is the son of Xerxes, during our period of interest. The line goes Darius I, Xerxes I and Artaxerxes.

It could be true for later kings of the same name, but we are focussed on the period from Cyrus to Artaxerxes and so we have a probable example of Levite, or Jewish scribal mix-up at a later period. It further emphasises the confusion of later compilers / historians for the period after the exile.

Nevertheless, we are told that Daniel confesses his sin and that of the whole nation of Israel. He recognises that the disaster that has come upon them (the exile to Babylon and enslavement) is a result of wrongdoing and unfaithfulness. It is while he is in this prayerful frame of mind, that he has another encounter (by way of vision) with Gabriel.

> *²⁰ "While I was speaking and praying, confessing my sin and the sin of my people Israel and making my request to the* LORD *my God for his holy hill— ²¹ while I was still in prayer, Gabriel, the man I had seen in the earlier vision, came to me in swift flight about the time of the evening sacrifice.*
>
> *²² He instructed me and said to me, "Daniel, I have now come to give you insight and understanding. ²³ As soon as you began to pray, a word went out, which I have come to tell you, for you are highly esteemed. Therefore, consider the word and understand the vision:*

²⁴ *"Seventy 'sevens'[c] are decreed for your people and your holy city to finish[d] transgression, to put an end to sin, to atone for wickedness, to bring in everlasting righteousness, to seal up vision and prophecy and to anoint the Most Holy Place.[e]*

²⁵ *"Know and understand this:* **From the time the word goes out to restore and rebuild Jerusalem until the Anointed One,[f] the ruler, comes, there will be seven 'sevens,' and sixty-two 'sevens.' It will be rebuilt with streets and a trench, but in times of trouble.**

²⁶ **After the sixty-two 'sevens,' the Anointed One will be put to death and will have nothing.** *The people of the ruler who will come will destroy the city and the sanctuary. The end will come like a flood: War will continue until the end, and desolations have been decreed.*

²⁷ *He will confirm a covenant with many for one 'seven.'[h] In the middle of the 'seven'[i] he will put an end to sacrifice and offering. And at the temple he will set up an abomination that causes desolation, until the end that is decreed is poured out on him."* (NIV)

Because punctuation can make quite a difference in meaning (especially in translation) I will present another version of verses 24-27 from the Revised Standard Version.

²⁴ *"Seventy weeks of years are decreed concerning your people and your holy city, to finish the transgression, to put an end to sin, and to atone for iniquity, to bring in everlasting righteousness, to seal both vision and prophet, and to anoint a most holy place.[c]* ²⁵ *Know therefore and understand that* **from the going forth of the word to restore and build Jerusalem to the coming of an anointed one, a prince, there shall be seven weeks. Then for sixty-two weeks it shall be built again with squares and moat, but in a troubled time.** ²⁶ **And after the sixty-two weeks, an anointed one shall be cut off, and shall have nothing;** *and the people of the prince who is to come shall destroy the city and the sanctuary. Its[d] end shall come with a flood, and to the end there shall be war; desolations are decreed.* ²⁷ *And he shall make a strong covenant with many for one week; and for half of the week he shall cause sacrifice and offering to cease; and upon the wing of abominations shall come one who makes desolate, until the decreed end is poured out on the desolator."* (RSV)

If we examine a subset of the more conservative Hebrew to English translations; the NIV, Young's Literal, RSV, ESV and

JPS (Jewish Publishing Society) Tanakh, we find the five translations can be grouped into two classes.

The NIV and Young's are similar in outline with the 7 weeks and 62 weeks grouped together for the overall period and pointing to the same 'anointed one'. The Greek Septuagint translations that I have, also follow this approach.

On the other hand the RSV, ESV and JPS Tanakh seem to support a different view, with an 'anointed' coming after seven weeks and then presumably another 'anointed' coming after a further 62 weeks. So it all seems to hinge on punctuation and trying to determine the writer's intent. It shows the translators are struggling with the ambiguity of the Hebrew.

If the RSV, ESV and JPS are correct, then we should be looking for a king, or priest-king 49 years from the initiation of the decree going out to rebuild the city. It would be great to know if we could identify such a person, because this might also be a useful clue to lock in the start of the 70 week timeline.

Unfortunately, there is no Persian king who appears to perfectly fit the bill for a new 'anointed one', appearing 49 years after any of these edicts. And Judea was under Persian control until the 334 BC Greek conquest by Alexander the Great.

The most likely suspect, Darius II, reigned from 424 BC to 404 BC as the Persian monarch, and in this period Darius was also the last pharaoh of the Twenty Seventh Egyptian Dynasty.

This points to the likelihood that the 7 'weeks' and '62 'weeks' should simply be added together.

Another possible explanation, might be a visit to Jerusalem by the reigning king for a dedication ceremony toward the end of the completed work. The king would have been Darius II and a visit would have coincided with the seven weeks (49-50 years) on the basis of the first jubilee period after the Ezra return. It would mean a visit 49 or 50 years after 457 BC (408-407 BC); or 10 years after Ezra's ministry concluded, according to our earlier analysis of 2 Esdras 14:10-12.

In John 2:20 we find a mention of some Jews who were arguing with Jesus. As part of that 'discussion', they state that it took 46 years to rebuild the Temple standing before them. If that is the case, perhaps 49 years to refurbish Jerusalem and rebuild the walls and gates, would be consistent with the work required and momentous enough for a king's visit.

In passing, most translations say that after the 62 'weeks' the anointed one is 'cut off', whereas in my 1999 copy of the JPS Tanakh it says the anointed 'will disappear and vanish'. Very strange terminology indeed it would seem, but if there is more to being 'cut off' in the Hebrew than seems obvious, perhaps very apt in hindsight. After the resurrection and ascension, there was simply no body to be found anywhere!

Now if we start our timeline from the Artaxerxes decree of 457 BC (actually 458-457 BC) we then need to add (7 x 7) + (62 x 7), i.e. 483 years. However we must make our BC to AD adjustment. Meaning 457 BC represents 456 years only. So we have: 483 − 456 = AD 27 (AD 26-27).

We are told that *after* this year will be the year of Jesus' first appearance as a preacher, prophet and the Messiah, at around 30 years of age. Through the ministry of John the Baptist, Jesus is then baptised in the water of the Jordan River and at the same time becomes the Anointed (Messiah) by the baptism of the Holy Spirit. His ministry commences at this point, but he is also 'cut off' at the end of his ministry a few years later.

Now we do not know exactly how long his preaching, teaching and healing ministry continued, but there are hints from the Feast occasions that are mentioned in the gospels, together with his journeys and whereabouts, that it was likely 2 to 3 years.

I suggest three and a half years given the prophetic significance of that number. The concept that we use the beginning of the period as the milestone, but that 3.5 years later is of central or peak significance, is in perfect accord with our analysis of all the milestones on our prophetic timeline. It means that the 3.5 years of his ministry period, need not be included in the 69 'weeks' of the prophecy, but becomes a part of the 70th 'week'.

Now the last verses of Daniel 9 (verses 26 and 27) are somewhat obscure. We learn in a consecutive manner that:

Verse 26:

o After the 69 weeks the Anointed will be 'cut off'
o The people of a 'ruler' will destroy the city and the sanctuary and that war and strife will continue

Verse 27

o   He will confirm (strengthen / continue) a covenant with many for one 'seven'
o   In the middle of the 'seven' he will put an end to sacrifice and 'offering
o   On a wing of the Temple he will set up an abomination

That the Anointed (Messiah) will be cut off and disappear is quite clear. The 'ruler who will come' in verse 26 is almost certainly General Titus on behalf of his father, Roman Emperor Vespasian. This portion describes the siege and destruction of Jerusalem in AD 70. The alternative view (which I think is not so likely) is that the 'ruler' here could be the same one described in the following verse (verse 27).

However verse 27 is more difficult to interpret and describes a ruler who continues a covenant, then puts an end to sacrifice by setting up the 'abomination of desolation'. So verse 27 does not run on from verse 26 easily, and the two rulers are almost certainly different identities. The ruler (or rulers) *after* the abomination of desolation, i.e. beginning in the middle of the week (the seven times) continue with the oppression of the Jews / Israel in their own land.

This 'desolation' is in no way comparable to the desecration that Antiochus VI (Epiphanies) performed back in the Seleucid period, when a pig was sacrificed in the Temple. Neither is it in any way similar to the desecration of Jerusalem under Pontius Pilate, where ensigns and images of Tiberius Caesar (idols to the Jews) were placed in the precinct to antagonise the locals. The 'abomination of desolation' is a complete removal and levelling of the old, and construction of a new monument and false edifice on a 'wing' of the mount. It causes 'an end to sacrifice and offering'; something that had never occurred before.

So although 'confirming a covenant for one seven' in verse 27 has been interpreted by some commentators as referring to the Anointed exclusively (and the end of animal sacrifice because of the cross) this view is stretching the meaning. The fact is, that verse 27 *has* to refer to the 'antichrist' confirming another covenant; the one between Israel and God. The 'abomination of desolation' demands it. This is the *overarching* 'seven times' of trial and tribulation outlined in Leviticus 26, showing that the primary message of verse 27 is the focus on the rulers who follow and are 'antichrist' in nature.

Nevertheless, God rules over this seven times of disciplining, knowing that through his Messiah's ministry and sacrifice, his chosen people are covered while they remain faithful.

In fact, with this overarching interpretation we have an example where a 'day' is not a year, but a 'time' i.e. 365 years.

This is quite acceptable prophetically, because it is still a defined length of time and a prophetic period at that. Daniel 9:27 in the NIV uses the phrase 'He will confirm a covenant with many for one 'seven'. In the middle of the 'seven' he will put an end to sacrifice and offering.'

Young's Literal translation is similar:

'And he hath strengthened a covenant with many – one week, and in the midst of the week causeth sacrifice and present to cease,'

Thus, this brief verse ties everything together in a rather significant and especially meaningful way.

Despite this *specific* 'antichrist' interpretation for verse 27, the overall implications for the seventieth week and the whole of Daniel 9, is a two-fold revelation; one where there is an 'abomination' midway through in each case. For Jesus' ministry it is the destruction of his 'temple' (his body) after 3.5 years to end sacrifice and complete atonement as the Lamb of God. For the overall 'seven times of tribulation', it is the desolation of the Temple by the (Islamic) ruler, who continues (strengthens / confirms) the time of refinement from the midpoint onward.

Both interpretations cause sacrifice to cease and therefore a dual fulfilment is very likely, especially as the chapter is predominantly about the coming of the Anointed.

The remaining three and a half years that would have continued Jesus' ministry, are believed by some observers to tie into Jesus' second coming. I provide information in Appendix 4 which gives some support for that view.

Others prefer to see it as a block of seven years, running into the early apostolic missionary work after the resurrection.

Regardless of a particular seven year interpretation in terms of the Jesus' earthly ministry, or any further ministry to come, the verse 27 description of the ruler who appears in AD 689 and continues the oppression of Israel, outlines the message of this book in microcosm. It describes in a few words, that the 'seven times' of tribulation imposed in Leviticus 26 is the basis for the times and trials of Israel, until those years are fulfilled.

To be clear, verse 27 shadows the deceitfully instigated agreement drafted by Umar Al-Khattab's men and signed by the Byzantines in AD 637. It is therefore not future as many teach. Ultimately though, it was God's plan, and the antichrist ruler allowed to continue the harsh agreement between God and Israel for another 'three and a half times'. The antichrist was a mere implement and someone who will be dealt with in due course.

Wrapping up: the language in Daniel 9:27 is virtually *identical* to Daniel's description of the ultimate desolation of the Temple in AD 689, which is hinted at in Daniel 7:25, described clearly in Daniel 8:11-12 and 11:31-32 and adamantly confirmed in Daniel 12:7-13 in one sweeping statement.

In terms of prophetic fulfillments, the idea of multiple fulfillment is not contradictory, neither is it a prophetic copout.

Prophecies can also have a national fulfillment, but also a Messianic one. They are also likely to have fulfillments following the concept of body, mind and spirit. For example, the spiritual could follow the expression of the faith, the mind the political application, and the body, a literal physical fulfillment.

Given the interpretation presented above, there is a distinct likelihood that Jesus' adult, Holy Spirit inspired ministry lasted 'about' three and a half years, and that his death and resurrection occurred as early as AD 30 or 31 at the Passover Feast.

Note: *If the reader is aware of the arguments for a later crucifixion and resurrection year and have accepted that view, I have provided a discussion on the topic from various pieces of historic and geological evidence in Appendix 4. The aim is to show that the evidence for an earlier date for the Passover and Crucifixion is at least as good as, that for an AD 32 or 33 date.*

Now Daniel describes the origins of his night visions in considerable detail. He saw the 'Ancient of Days', a metaphor for

Almighty God, but he also saw someone who was like a 'Son of Man', who he describes in chapters 7 and 10 most vividly and awesomely. Using a popular contemporary term borrowed from Indian tradition, we could say Jesus was an Avatar of the Almighty. Indeed *the* Avatar of God.

Interestingly, Jesus of Nazareth (Yeshua) nearly always called himself the Son of Man. So the Son of Man is not, and never was a temporal human being, but indeed a Divine Saviour, the First Born of all Creation, and this is why Jesus used that description. It is the same designation as Son of God.

I suspect most Christians would assume he used the term to emphasise his humanity. I certainly did and there might be a hint of this. But the Messiah was expected in Judea because of Daniel's Seventy Week prophecy and therefore many were expecting him to save them from the Romans.

In conclusion; for Spirit filled believers, narratives like the following are full of meaning.

The events played out as Jesus approached Jerusalem prior to that fateful Passover. He predicts the fate of Jerusalem, but at the same time knows that as the Lamb of God, he is to fulfil the prophets by putting an end to sacrifice and thereby facilitate our redemption. Luke 19:37-44

37 When he came near the place where the road goes down the Mount of Olives, the whole crowd of disciples began joyfully to praise God in loud voices for all the miracles they had seen: **38 "Blessed is the king who comes in the name of the Lord!"**[b]***"Peace in heaven and glory in the highest!"***

39 Some of the Pharisees in the crowd said to Jesus, "Teacher, rebuke your disciples!"40 **"I tell you,"** *he replied,* **"if they keep quiet, the stones will cry out."**

41 As he approached Jerusalem and saw the city, he wept over it 42 and said, "If you, even you, had only known on this day what would bring you peace—but now it is hidden from your eyes. 43 The days will come upon you when your enemies will build an embankment against you and encircle you and hem you in on every side. 44 They will dash you to the ground, you and the children within your walls. They will not leave one stone on another, because **you did not recognize the time of God's coming** *to you."* (NIV)

Forty years of fear

From the 'Vanishing of the Lord', there was to be 40 years of particularly gruelling and fear filled days leading up to the AD 70 Roman assault on Jerusalem. The 40 years of drama is initially told in the Gospels and book of Acts and includes the events around the crucifixion, resurrection, ascension and the days leading up to Pentecost.

The continuing drama surrounding the Temple is found in sources such as the Jerusalem Talmud and Babylonian Talmud (the latter a compilation of later centuries).

Matthew's Gospel tells us that as Jesus cried out and died, the curtain of the Temple was torn in two from top to bottom as the earth shook.

From studies of annual (varved) sediments at Ein Gedi on the western coast of the Dead Sea, a historically recorded earthquake in 31 BC has been identified. Using this as a marker, seismic activity over an estimated period from AD 26 to 36 is also evident in the strata. If this was felt in Jerusalem about 40 kilometres away (and of course it most certainly would have been) it covers the most likely period of interest for researchers attempting to identify the crucifixion and resurrection year.

I have already indicated that Matthew is the only Gospel to recount another dramatic event at this time. This is where he states that tombs were broken open by the earthquake and the bodies of many people who had died were raised to life and later showed themselves to the inhabitants of Jerusalem.

Given this episode is only referred to in Matthew and nowhere else in the Gospels, there is a legitimate concern about its veracity. Nevertheless it adds to both Jesus' own resurrection as being the first resurrection of prophecy, and further cements the reality of an earthquake at this extraordinary time. It should also be considered in the light of references to the future resurrection event at Jesus' second coming.

There seems to be just two resurrections mentioned by the prophets in the last 2,000 years, and if this was the first event (or the beginning of a resurrection age) then it would indicate that there is only one more to come; the one associated with, and occurring immediately before the 'rapture' of believers still living in the 'last days'.

Now the Jerusalem Talmud and Babylonian Talmud record there were four anomalous signs which appeared during the forty year period to AD 70.

These were the crimson strap, the western lamp and the lot (or ballot) - white or black - which came up in either the right hand or the left hand of the priest. The fourth sign was the Temple doors opening by themselves, despite the best efforts of the priests to stop it happening over the 40 years.

1. Jerusalem Talmud:

"Forty years before the destruction of the Temple, the western light went out, the crimson thread remained crimson, and the lot for the Lord always came up in the left hand. They would close the gates of the Temple by night and get up in the morning and find them wide open"

2. Babylonian Talmud:

"Our rabbis taught: During the last forty years before the destruction of the Temple the lot 'For the Lord' did not come up in the right hand; nor did the crimson-coloured strap become white; nor did the western most light shine; and the doors of the Hekel (Temple) would open by themselves"

There were seven lamps on the Menorah in the Temple, with the westernmost lamp being the most important. Every day the other six lamps would be lit from the 'western' lamp.

For a long time the western lamp burned continuously, which was seen as a miracle. However, from about AD 31–70 it went out every single day, despite the priests best efforts to keep it alight. From the year Messiah said the physical Temple would be destroyed, the western lamp would not stay alight.

Another Jewish tradition is tied to the Day of Atonement, where a scapegoat 'Azazel', bearing the sins of the people would be released into the wilderness. According to the Babylonian Talmud, the scapegoat would wear a crimson coloured strap, and it would become white once it reached the wilderness, indicating that God had forgiven their sins.

It ties in with the verse 'If your sins be as scarlet, they shall be white as snow' (Isa. 1:18). By a miracle, this crimson-coloured strap turned white, thus showing the people that their sins were forgiven. This tradition was enacted each year at the Temple on behalf of the people, but after the crucifixion the crimson strap no longer turned white and it was clear that this ritual was no longer acceptable to the Almighty.

In addition, another goat was often chosen 'for the Lord', as well as the scapegoat that was sent into the wilderness.

The animals were chosen by lot and the officiating priest would reach with both his hands into an urn, selecting both a white stone and a black stone and placing them on the heads of the two goats. The people always hoped that the white stone (for the Lord) would be in the priest's right hand, and this happened about half the time (as the laws of chance dictate).

During the 40 years however, the white stone never once turned up in the right hand of the priest. It turned up in his left hand every single time. These were impossible odds and therefore taken as a sign.

Both the Jerusalem and Babylonian Talmud also speak of the Temple gates springing open by themselves during the 40 years. According to the Jerusalem Talmud, Jewish leaders knew this was a sign of impending doom.

The end of the 40 years and the siege of Jerusalem beginning in AD 70, where an estimated one million Jews died is well documented. Much of the city was destroyed, but Flavius Josephus who accompanied Titus from Alexandria managed to save many of the books from the Temple Library for posterity.

In bas relief on the honorific Arch of Titus in Rome we can still see today the Temple treasures that Titus took back to the capital. Poignantly a seven branched candle stick (menorah) is amongst the treasure trove depicted.

Sadly though, the failure of the western lamp to light in the 40 years prior to the assault on the once 'Blessed City' was an omen of destruction, but not the last indignation that the people of Jerusalem would suffer.

CHAPTER 12

Babylon the Great

"Don't let anyone deceive you in any way, for the day of the Lord will not come, until the rebellion occurs and the man of lawlessness is revealed, the man doomed to destruction"
Apostles PAUL, SILAS and TIMOTHY

The half time beast

The second 3.5 times of tribulation (and arguably the Roman lead-up to this time) was described by Jesus prophetically as a time of 'great' tribulation. It is a term used to distinguish it from the 'tribulation' which preceded it, and was absolutely true for the Jews of Jerusalem, but particularly so for the Christian sect who suffered intense persecution and martyrdom. It has also been true for hundreds of millions more since those early centuries and extended to the twentieth century globally, during the World Wars and Holocaust.

The 'seven times' kicked off with 70 years of exile for the ruling class, but they were able to return to the land fairly quickly and join those that remained. The Medo-Persian takeover of Babylon and the subsequent generosity of the early kings of the period allowed that.

However on return to Judea, the difficult times intensified again under the Greek Seleucid regime. It peaked when Antiochus IV desecrated the Temple and abused the people.

The Maccabean period was a time of rebellion and unrest, with some victories and respite that in turn rolled into the Roman occupation. From there a usurper king, Herod the Great, along with various Tetrarch rulers took control of Judea, Galilee and elsewhere under Roman rule.

It was a recipe for corruption and extortion. The Jews no longer had the autonomy and freedom they enjoyed during the Davidic Kingdom, and were now subject to consecutive and often brutal empires. But it was once the second 'time, times and half a time' began, that all bets were off and the most extensive persecution emerged. It was grinding and excruciating and lasted a very long time.

Life became exceedingly difficult, self-government and access to the Temple Mount ceased, the Mosaic Feast traditions publicly ended and the relatively benign Roman Byzantine rulers were deceived and kicked out of the country. Anybody with an ounce of initiative, realised it was time to join their brothers and sisters and escape to their choice of far flung lands. It was that or slavery and submission to the Caliphate.

Yeshua (Jesus of Nazareth) when speaking to his followers near the Temple of events that would occur after the 'abomination that causes desolation', did so in this manner. Matthew 24:20-21

> *"Pray that your flight will not take place in winter or on the Sabbath. For then there will be great distress, unequalled from the beginning of the world until now – and never to be equalled again."* (NIV)

In previous chapters, I suggested that most of Daniel's prophecies focussed on the first four empires which adversely impacted Judah i.e. Babylon, Persia, Greece and Rome and then for a short time the Ten Kingdoms. While this is true, it became obvious that all the visions touched on aspects *after* the first 'time, times and half a time' to some degree or other.

The vision of the statue of Daniel 2 is very coy in this respect, but the 'ten toes' of the kingdoms which come out of the last days of the pagan Roman Empire, actually parallel the final beast empire until the 'stone' kingdom appears at the end. In other words, the 'ten toes' belonged to the totality of the second 'time, times and half a time', although significantly, they did not dominate in the Holy Land after the early 7th century.

This overall picture was confirmed in Daniel 7, where we found the 'ten horns' were unsettled by another horn (with eyes like a human and a mouth which speaks boastfully) and actually overcame three of them, very early on.

The Ten Kingdoms were therefore transitional from very late in the first 1,278 years, but continued through the next 1,278 years and beyond. Moreover we are told in the interpretation of the dream, that the kingdom of the 'boastful horn' will subdue the holy people for a full three and half 'times'.

So Daniel's vision in chapter 7 truly does cover the whole Seven Times of tribulation; but the realisation sort of sneaks up on you; much as it did in real life to the Jews in Jerusalem in the seventh century. It also crept up on the Roman Byzantine overlords, rulers and governors who were now showing signs of ineptitude and corruption.

Summarising the various aspects of the length of the reign of the Ten Kingdoms we see:

- From the Jerusalem and Holy Land perspective, their reign is only for a 'little while' and 'one hour' according to Rev.17; i.e. from the demise of the Roman Empire in AD 565 until AD 637 - 689 and Islamic supremacy
- Prophetically, the Byzantine Kingdom becomes more isolated over time as one of the Ten Kingdoms, until AD 1453 when Constantinople falls to the last Caliphate
- Beginning in the 7th century, three Asian / African kingdoms are subdued and become provinces of Islam
- From a global point of view, the Ten Kingdoms play second fiddle to successive Islamic Caliphates for centuries and are at times complicit in persecuting the saints – early on in Jerusalem, but thereafter in Europe
- At the end of the age there is a power shift and the Ten Kingdoms turn on the last Islamic Caliphate and destroy it. They again become involved in the Middle East

To arrive at this picture with any certainty, we need to extract all the information we can from Daniel 2, Daniel 7 and later prophecies. The first and last two points only become clear once we study Revelation 17, so bear with me until we study the relevant passages later.

We also saw in the Daniel 7 vision, that the original Babylonian Empire morphed from a lion with wings into the figure of an upright man, i.e. from Assyria to Babylon and the worship of a mere mortal; Nebuchadnezzar. This symbolism also ties Babylon, the first beast kingdom of Daniel, with the latter Eighth Beast Kingdom; Babylon the Great of Revelation 17.

Both are symbolised by the worship of a man, albeit the first is explicitly represented by a statue. In fact, this last beast kingdom is a regime which would change *all* observance, times and dates and subjugate the 'saints' for over a thousand years.

The change of observance, times and dates start with the first Babylon, a moon controlled calendar and loss of spiritual awareness, but the Babylon the Great regime enforces this absolutely and with a vengeance.

As we observed, the first Babylon had a system where the king by decree in the early days - or by using a 19 year cycle of adjustment of the lunar months later on - kept the solar and lunar calendars in alignment to a workable degree. With the last Babylon, Babylon the Great, no such attempt is made, and a strict 354 day 12 month lunar calendar is implemented. This lunar emphasis is typical of Baal worship and ancient Babylon, but also of Babylon the Great.

Now, Jesus' description in Matthew 24 and that of the Apostles elsewhere, absolutely supports Daniel's description - and why not? The Apostles were all Jews by birth, they were baptised with the Holy Spirit and their inspiration and visions were all from the same source, through the 'Son of Man'. The biblical prophecies which include timelines and milestones, are the product of visions experienced by the prophets of Judah.

This is heady stuff! It is regardless of whether the prophecies come to us through the Tanakh or the New Testament. It is a cohesive whole, breathed out by the Spirit of God.

I will emphasise again, something from the previous chapters that is vitally important for those who have come to believe the final 'beast kingdom' (Babylon the Great) is Rome; either as the Papacy, or the Holy Roman Empire. The timeline makes it relatively simple to explain, because it was the Byzantine (former Roman) Empire / Ten Kingdoms which were displaced from the Holy Land in AD 637 and not the other way around.

The prophecies are not about Rome the city, but about Jerusalem and what occurred there - and when.

It is also instructive to put ourselves in the shoes of those saints who were oppressed, starting in the Middle East and then advancing far beyond, as successive Islamic Caliphates overran the former provinces of Rome. Their lives were to be more than disrupted; certainly with the loss of basic human rights and dignity, no freedom of worship, many enslaved and others martyred. Something; for all the ferocity of the earlier kingdoms, had not occurred on such a chronic and global scale.

There was just no end in sight and importantly for the Jews, the earlier 'beast' kingdoms had allowed worship to continue at the Temple – but the last one did not.

So the little horn is not Rome or the Ten Kingdoms. However the Holy Roman Empire and Papacy were integral to the second 'three and a half times' period after pagan Rome fell - but *not* in the Holy Land and largely *beyond* the reach of Islam.

Unfortunately the Ten Kingdoms were also guilty of the sort of authoritarian and oppressive rule which we associate with a 'beast' kingdom. Three were kingdoms / regions which were part of the Eastern tradition and subsequently overrun by Islam, and there is potentially a case for including the Persian Sassanid kingdom which sacked Jerusalem in AD 614.

Now, the Vatican relied on the standing armies of the Kings for defence, but subsequently used them for much more when they were in a position to do so. Some Popes saw themselves as Christ figures and wrote edicts to that effect. There were other times when the position of Pope was hereditary and controlled by powerful and elite families.

Consequently, when the Papacy had an overarching influence within the Holy Roman Empire, it was often used to oppress groups who threatened their brand of the faith.

One example is Pope Innocent III's pursuit of the Cathars in the early 13th century. The Cathars were a heretical, neo-gnostic population in southern France and Italy, whose beliefs probably derived from similar earlier groups from the eastern Mediterranean and perhaps Armenia. They were slaughtered mercilessly and were rumoured to have treasure from the Holy Land.

The second notable example was the persecution of the Protestants of the Reformation in the early 16th century. At this time, many survivors were forced to flee to Britain and the Americas, while others held fast on continental Europe. Eventually though, the Reformation lead to the demise of the absolute religious power of the Papacy, as well as much of its political influence.

Meanwhile, the Eastern Orthodox Churches had plenty of their own problems; with strong ties to kings and rulers in their own sphere of influence. They also compromised their work as a Christian ministry to the people, but it needs to be stressed; that any authoritarian rule that the Vatican or Constantinople may have wielded, is tied to the **Seventh Head** of the Revelation prophecies. These are the **Ten Kingdoms** and their monarchs – not the **Eighth Kingdom** which is Babylon the Great.

In addition the Vatican was not always a willing partner of royal wrongdoing throughout the Ten Kingdom period. The Roman Church and their saints were also persecuted by the Kingdoms, and this will become apparent when we study some 14th century history.

It therefore follows that the Protestant Reformers' identification of Rome as the 'little horn' and the last and Eighth Beast Kingdom was off target; in fact completely in error. However prophetically, the Ten Kingdoms were complicit in 'beast' activity when it suited them; being an extension of pagan Rome at the end of the Roman Empire and contributing to oppression in Europe and the wider world, during later periods. It is therefore understandable why the Protestant Reformation leaders thought the way they did about the Papacy. They recognised that the Ten Kingdoms still persisted and knew that Rome was both their religious focal point and immediate concern.

Perhaps if they had reflected on the history of the Crusades a little more, they might have recognised the power which really epitomised the resurrection of Babylon. In fact, Marin Luther is reportedly one Reformation luminary who recognised that the 'little boastful horn" of Daniel 7 was in fact Islam. However the general doctrine was that the Papacy was the biblical Antichrist.

So although the Catholic Church still holds to doctrine which is at odds with the Bible from a Protestant understanding, the true believers amongst them do not deny Christ. Certainly devout Catholics are bound to understand that our focus on the Lord is central and like true followers of Christ everywhere, continue shoulder to shoulder to minister to the sick, the underprivileged, the oppressed and the poor in spirit - and call out wrongdoing wherever it is found.

However from an institutional perspective, the sale of indulgences in the medieval period was despicable (essentially a subscription plan for salvation) and there are key Marian doctrines with no, or negligible support from the main cannon of scripture, e.g. Mary's divine motherhood, perpetual virginity, immaculate conception and 'assumption' to heaven.

Mandatory celibacy is also a distortion of the Apostle Paul's teaching and has caused much unnecessary heartache within the ranks. It is no wonder Protestants rebelled. Unfortunately in so doing, they overstepped the mark of prophetic credibility and it haunts Protestant Christendom to this day.

In the section on the Age of the Gentiles, Tribulation, we touched on the beginnings of Islam and how Caliph Umar of Damascus succeeded Abu Bakr and occupied Palestine in AD 637. It was also noted that he has been painted as a relatively responsible man by Islamic historians, and that at least initially he treated the Christian and Jewish inhabitants of Jerusalem with respect during the surrender of Jerusalem.

The overall picture from this period is not so rosy however, and the warnings of Jesus were fulfilled by both the Persian Sassanids and the Islamists. Jerusalem first fell to an invading Sassanid army in AD 614, resulting in the looting of the city and the killing of tens of thousands of the inhabitants. This assault weakened the Byzantine defences and allowed a Muslim called Abu Ubaidah to lay siege to the city in 637. It was then agreed that Caliph Umar would come to town to accept the surrender and this is the way it played out.

It becomes plain even with the scant descriptions recorded, that this period was absolutely horrific for the Jewish and Byzantine Christian inhabitants and associates.

From Caliph Umar onward, Islam then spread rapidly into Egypt, Cappadocia (central Turkey), Persia, Georgia, Uzbekistan, Tajikistan and Kazakhstan - much of this before the building of the Dome of the Rock. As we follow the Islamic story from here and the 'great' tribulation of Israel that resulted from this expansion, are there any hints in Daniel which shed light on how the transition to Islam was to occur?

Daniel chapter 11 recounts a whole sequence of military maneuvering in the Levant over many centuries, starting with the Persian Empire. It is somewhat similar to chapter 8; with a mighty king of Greece (Alexander), breaking up the party, but subsequently having his Empire split into four.

We then have a series of 'he said, she said'; with alternate kings of the north and kings of the south vying for control of the Holy Land. However we are not given a lot to identify these kings, apart from whether they are from the north or south and a few rather sparse family details. Essentially it appears that the struggles early on were between the Seleucid kings of the north and the Ptolemaic kings from Egypt, with some of this argy-bargy being resolved in the age old way through marriage. Later on we have Roman and Byzantine control for several centuries, followed by the final Islamic invasion of the 7th century.

I have no particular inspiration to offer for the following comments, but will proceed in the hope that some of it may be useful. There are several interpretations and many commentaries on this most difficult of chapters and some give a verse by verse, blow by blow exposition.

However the interpretations are so widely divergent and inconsistent, that the following observations are mainly my own thoughts. They are also a summary only.

In Daniel 11 it seems the Ptolemaic king of the south is still in control of his territory (probably including Jerusalem) up until verse 15 or thereabouts, when the king of the north attacks.

[15] Then the king of the North will come and build up siege ramps and will capture a fortified city. The forces of the South will be powerless to resist; even their best troops will not have the strength to stand. [16] The invader will do as he pleases; no one will be able to stand against him. He will establish himself in the Beautiful Land and will have the power to destroy it.

[17] He will determine to come with the might of his entire kingdom and will make an alliance with the king of the South. And he will give him a daughter in marriage in order to overthrow the kingdom, but his plans[c] will not succeed or help him. [18] Then he will turn his attention

to the coastlands and will take many of them, but a commander will put an end to his insolence and will turn his insolence back on him.
19 After this, he will turn back toward the fortresses of his own country but will stumble and fall, to be seen no more.

20 "His successor will send out a tax collector to maintain the royal splendour. In a few years, however, he will be destroyed, yet not in anger or in battle. 21 "He will be succeeded by a contemptible person who has not been given the honour of royalty. He will invade the kingdom when its people feel secure, and he will seize it through intrigue. 22 Then an overwhelming army will be swept away before him; both it and a prince of the covenant will be destroyed. (NIV)

Up to this point we seem to be looking at the Ptolemaic and Seleucid kings at war, or attempting to seek diplomatic solutions. But then the narrative appears to move on from the Greek era, around verse 19 (and the Seleucid reign of Antiochus Epiphanies and the Maccabean revolt) into the early Roman period. Eventually Idumean king, Herod the Great, who married into the Hasmonean royal family, comes on the scene and may be implicated in the text.

This transition from Seleucid to Roman rule has the reference to war and the 'prince of the covenant'. This may be a reference to Jesus himself. If not Jesus of Nazareth, perhaps verse 22 simply refers to the Maccabean period and the Judean Hasmonean line.

From verse 23 to 30 we appear to have the various intrigues, political manoeuvring and battles through the Roman period. There are attempts at power grabbing and sometimes sharing between the candidates vying to be Emperor of Rome.

The issue of Egypt, the lure of rulership there and the attractive title of 'Pharaoh' which enticed some 'kings of the north' south, is also a consideration. In this period the 'king of the north' may have been Octavian Augustus and 'king of the south' Mark Anthony and Cleopatra.

23 After coming to an agreement with him, he will act deceitfully, and with only a few people he will rise to power. 24 When the richest provinces feel secure, he will invade them and will achieve what neither his fathers nor his forefathers did. He will distribute plunder, loot and wealth among his followers. He will plot the overthrow of fortresses—but only for a time.

²⁵ "With a large army he will stir up his strength and courage against the king of the South. The king of the South will wage war with a large and very powerful army, but he will not be able to stand because of the plots devised against him. ²⁶ Those who eat from the king's provisions will try to destroy him; his army will be swept away, and many will fall in battle. ²⁷ The two kings, with their hearts bent on evil, will sit at the same table and lie to each other, but to no avail, because an end will still come at the appointed time. ²⁸ The king of the North will return to his own country with great wealth, but his heart will be set against the holy covenant. He will take action against it and then return to his own country.

²⁹ "At the appointed time he will invade the South again, but this time the outcome will be different from what it was before. ³⁰ Ships of the western coastlands will oppose him, and he will lose heart. Then he will turn back and vent his fury against the holy covenant. He will return and show favour to those who forsake the holy covenant. (NIV)

In the following passage, we have something surer that we can identify and place a date against. The phrase 'set up the abomination that causes desolation' points to the Islamic invasion from Damascus. The earlier attack on Jerusalem by the Persian Sassanid regime could also feature here. We also learn that 'with flattery he will corrupt those who have violated the covenant', suggesting that lies and deceit are used to enable the takeover and that the incumbents (whether Jew or Byzantine) have left themselves open to corruption.

In the wake of this abrogation of duty there is slavery and death, or alternatively, escape leaving everything behind.

Overall there is support for the view that; not only are the Islamic leaders clever and devious, but the Byzantines weak, ideologically conflicted and spiritually bereft.

³¹ "His armed forces will rise up to desecrate the temple fortress and will abolish the daily sacrifice. Then they will set up the abomination that causes desolation. ³² With flattery he will corrupt those who have violated the covenant, but the people who know their God will firmly resist him.

[33] *"Those who are wise will instruct many, though for a time they will fall by the sword or be burned or captured or plundered.* [34] *When they fall, they will receive a little help, and many who are not sincere will join them.* [35] *Some of the wise will stumble, so that they may be refined, purified and made spotless until the time of the end, for it will still come at the appointed time.*

The King Who Exalts Himself:

[36] *"The king will do as he pleases. He will exalt and magnify himself above every god and will say unheard-of things against the God of gods. He will be successful until the time of wrath is completed, for what has been determined must take place.* [37] *He will show no regard for the gods of his ancestors or for the one desired by women, nor will he regard any god, but will exalt himself above them all.* [38] *Instead of them, he will honour a god of fortresses; a god unknown to his ancestors he will honour with gold and silver, with precious stones and costly gifts.* [39] *He will attack the mightiest fortresses with the help of a foreign god and will greatly honour those who acknowledge him. He will make them rulers over many people and will distribute the land at a price.* (NIV)

In verses 31 to 39 we see a description of the early Islamic Caliphates, their rise to power and the removal of worship of many gods. In the eyes of Daniel they use any pretext or military ploy available to achieve their aims.

The god 'desired by women' is likely to be Bes, who was an African god resembling a man with feline features. It became a popular cult amongst women in Egypt and was subsequently adopted by the Philistines. As the prophecy unfolds, it would seem that not even the God of Abraham, Isaac and Jacob is spared contempt, despite pretence of allegiance. We learn that the king, in fact, exalts himself and honours the 'god of fortresses (Mauzzim); meaning to accomplish his aims by force.

Of course this last characteristic and some of the others are not particularly exclusive to Islamic leaders, because we know many kings of those times (and in fact more recent times) see themselves as gods to be worshipped and obeyed. Throughout the earlier beast kingdoms of Daniel, all the empires were characterised by the subjugation of the masses and the exalting of the leadership to deity status.

However the Caliphs made a fine art of it; they wallowed in the glory and the decadent luxury.

The prophecy then continues 'until the time of wrath is completed', meaning until the Caliphate is defeated in the Holy Land (at the very least) and the 'great tribulation' is all but over. This leaves us with verses 40 – 45 and one more episode and one more controversy.

I say that, because it seems many still believe this is future.

40 "At the time of the end the king of the South will engage him in battle, and the king of the North will storm out against him with chariots and cavalry and a great fleet of ships. He will invade many countries and sweep through them like a flood. 41 He will also invade the Beautiful Land. Many countries will fall, but Edom, Moab and the leaders of Ammon will be delivered from his hand. 42 He will extend his power over many countries; Egypt will not escape. 43 He will gain control of the treasures of gold and silver and all the riches of Egypt, with the Libyans and Cushites[e] in submission. 44 But reports from the east and the north will alarm him and he will set out in a great rage to destroy and annihilate many. 45 He will pitch his royal tents between the seas at[f] the beautiful holy mountain. Yet he will come to his end, and no one will help him. (NIV)

My sense in regard to this passage, is that it has already been fulfilled by the Ottoman Caliphate during the years it ruled the Levant and Palestine and invaded Egypt. In the Great War (WW I) the Turkish Ottoman Caliphate - which had occupied Palestine since 1517 per Judah ben Shmuel's prophecy - attempted to defend the conquered land with the help of German military might. This was against British Empire forces advancing from Egypt. They were eventually defeated after a heroic cavalry charge by the Australian Light Horse and New Zealand Imperial Forces at Be'ersheva in late 1917. Up to this point the British Allies suffered heavy casualties, losing 10,000 soldiers attacking along the west – east; Gaza – Be'ersheva line.

In addition, we have already noted the exploits of Lawrence of Arabia, a one-time British archaeological professional in the Middle East. He worked with the Arabs as the war erupted, in order to ensure their support against the Turks and protect the eastern flank of the British Empire troops.

The Ottomans built a railway from Damascus in Syria, through Jordan to Medina in Arabia for the annual pilgrimage (Hajj). This was known as the Hejaz Railway and was privately funded using German railway building expertise and Swiss locomotives. This began as a collaborative Islamic effort with the Arab population, but quickly became contentious as World War I erupted and fear of Ottoman military incursions rose. It was one of the strategic issues that occupied Lawrence's attention.

Although this was right at the end of the Ottoman period in Palestine, I think it epitomises the fact that the Arabs - including those in modern Jordan (Ammon, Moab and Edom) - although Islamic, were relatively independent and not directly subject to Turkish control. This is noted in the prophecy.

It is also interesting to note that there was a final battle for Palestine in the Megiddo Valley, and like so many other battles for the land fought at this location, presents as a Divine marker.

Egypt, on the other hand, still has fortifications and structures in the north, showing the strong presence of the Ottomans in that land and subjugation to Istanbul over many years.

Returning once more to Revelation 13, we find much more information about the final beast kingdom in verses 11-18:

11 Then I saw a second beast, coming out of the earth. It had two horns like a lamb, but it spoke like a dragon. 12 It exercised all the authority of the first beast on its behalf, and made the earth and its inhabitants worship the first beast, whose fatal wound had been healed. 13 And it performed great signs, even causing fire to come down from heaven to the earth in full view of the people.

14 Because of the signs it was given power to perform on behalf of the first beast, it deceived the inhabitants of the earth. It ordered them to set up an image in honour of the beast that was wounded by the sword and yet lived.

15 The second beast was given power to give breath to the image of the first beast, so that the image could speak and cause all who refused to worship the image to be killed. 16 It also forced all people, great and small, rich and poor, free and slave, to receive a mark on their right hands or on their foreheads, 17 so that they could not buy or sell unless they had the mark, which is the name of the beast or the number of its name.

18 This calls for wisdom. Let the person who has insight calculate the number of the beast, for it is the number of a man. That number is 666. (NIV)

Previously we looked at the first verses of Revelation 13 and saw the image of an animal coming out of the sea. The 'sea' represented many people and the composite beast in the vision was a combination of the animals of Daniel 7.

The great empires of this time came from populous places and the centres of civilisation – Babylon, Persia, Greece and Rome, and finally the Ten Kingdoms. They occupied the space of the first 3.5 times from Nebuchadnezzar until the 'desolation' of the Temple. We also saw that the sequence of kingdoms had seven 'heads' and yet only five appear in Daniel's prophecies; those that are named above.

From verse 11 onward another beast appears, but this time it comes out of the earth, which can confidently be assumed to be the opposite of the sea. It means that this animal which looks like a lamb and has two horns, emerges from a region which is only lightly inhabited – a relatively deserted place compared to the great civilisations of yore.

In fact it is a place very much like a land 'growing to the south and east and toward the Beautiful Land'; a description we saw in Daniel 8 of the peripheral Ptolemaic heartland and Arabia of the first Caliphate. Indeed, it also describes the original homeland of the end time Caliphate as well; the steppes of Asia.

The two horns will also represent the two principle variants of Islam - Sunni and Shia.

More pointedly, the symbolism of the lamb represents the object of worship of the first Babylon. The head of the animal represents the sun and the horns the moon. The crescent moon has always been described with two 'horns' and Harran, the city where Abraham lived, was a centre of sun and moon worship.

This was noted earlier and interestingly, the Umayyad Caliphate of AD 661 - 750 was also based at Harran during the desolation period of the Jerusalem Temple site.

This imagery also demonstrates a further direct correspondence between the Old Babylon and the new (Babylon the Great) - but also something further. The sun and moon also represent the male and female and the union thereof, and are therefore emblems of sexuality and fertility. In the male dominated context of the Middle East and Islam, they represent the interests of male leaders and absolute control one way or another – including sexuality, where male excess and privilege reigns.

In summary the key attributes are:

- It has two horns like a lamb
- Exercises all the authority of the first beast
- Makes everybody worship the first beast whose fatal wound had been healed
- Causes fire to come down from heaven
- Deceives the inhabitants of the earth with powerful signs on behalf of the first beast
- An image of the first beast is set up to be worshipped
- The second beast is given power to give breath to the image of the first beast, so that the image can speak and cause all who refuse to worship the image to be killed
- It forces all people to receive a mark on their right hands or on their foreheads. The number of the Beast is 666

After the two horns, we see that the second beast exercises all the authority of the first. It means that the second also has absolute power, and supplants the earlier beast 'out of the sea' in the Holy Land. It also exalts its leader to divinity; this in the same manner as the previous composite beast.

It is common to associate the fatal wound with the re-emergence of European influence emanating from Rome; that is, after the old pagan Empire was defeated by the Germanic tribes. In this previous historicist interpretation, the following Holy Roman Empire as a form of the Ten Kingdom federation (or other association / sequence of monarchs, legislative and religious powers of the period) is what is referred to as the 'fatal wound being healed'.

This is definitely not the case, because the beast out of the sea of Revelation 13 is none of those things. It is a composite of Daniel's beast kingdoms of the first 1,278 years, with special emphasis on the 'lion's mouth' (neo-Assyria / Babylon); the first beast portrayed by Daniel. Now the modus operandi of this lion's mouth is not dissimilar to the 'little horn with eyes' of the same vision. The one which had a 'mouth which spoke boastfully'. It is a further indication of the unbreakable association between ancient Babylon and Babylon the Great.

Nebuchadnezzar's Babylon was cut down in the middle of the night by the Medo-Persians. It was dramatic, decisive and fatal. The city was thought impregnable, but the attackers dammed the Euphrates and entered Babilum (the Gate of God) on the dry river bed. Therefore, the 'head' that recovers from the otherwise fatal wound is Babylon on the Euphrates. It is a template for, and resurrected in the form of Babylon the Great.

In Revelation 13:11-18 we see confirmation of this and the rise of a new empire, with god-like leaders of flesh and blood. It comes from apparent nowhere. It absolutely stuns the previous world empires of the Middle East and Mediterranean and harks back to the head of the first beast that comes out of the sea in Revelation 13; the one that begins with ancient Babylon.

Revelation 17 describes the same beast:

"The beast who once was, and now is not, is an eighth king. He belongs to the seven and is going to his destruction'. (NIV)

Indeed, the Abbasid Caliphate was initially based at Baghdad on the Tigris, not far from the original Babylon; a fact that would not have escaped the early Crusaders, who were contemporary with this time. I believe it likely, that unlike later generations, they knew the identity of the 8th Beast Kingdom.

Now some might suggest that the fatal wound is associated primarily with this last kingdom; Islam itself. And indeed Islam suffered a near fatal wound to the head during the Middle Ages; firstly at the hands of the Crusaders, followed quickly by the Mongol armies of Genghis Khan. It could also be argued that the fall of the Ottoman Caliphate in Turkey in the 1920s as a consequence of the Great War, and its attempt to rise today, is a third near fatal wound being healed. Regardless of the significance or otherwise of these episodes, there is no doubt that the resurrection of old Babylon is the subject of the prophecies.

We then learn that the second beast (the one out of the earth) also makes everybody worship the first beast. Again this is referring to the original beast of Daniel that began it all – Babylon, but nevertheless necessarily to the whole line of empires in Revelation 13:1-10. I believe it is primarily referring to the type or manner of worship; i.e. the same adoration of the 'image' of the leader (the claimed deity of the King, Emperor or Caliph).

In addition, the Jerusalem Temple site together with a plethora of synagogues and churches, were converted into mosques by this last empire. Within these commandeered sanctuaries, it became mandatory to worship in a strictly supervised and subservient manner. This is an example where the new regime forces people to worship '*in*' the sanctuaries of the first beast.

I will examine a specific case later, but this potential second phase of fulfilment, applies specifically to the Caliphates.

We are then told that the second beast uses powerful signs on behalf of the first beast, including bringing fire down from heaven to deceive everybody in regard to its power.

It was in the 13th century that gunpowder arrived from China into the Middle East and Europe. Until that time any 'fire that came from heaven' was likely to come in the form of burning sulphur and bitumen propelled by a catapult. Now the secret of fireworks and the use of gunpowder, finally arrived in the west.

It undoubtedly came in the wake of Genghis Khan and his Mongol hordes and was in all likelihood, used to aid his advance into Europe. It revolutionised warfare and allowed the use of guns and rockets to rain terror from greater distances - and in great quantities. This occurred two centuries before the downfall of Constantinople, but would have been a significant new weapon, once the Muslims embraced the technology.

The second beast then gives breath to the image of the first beast, so that the image could speak. This verse (15) is definitely obscure and difficult to interpret, but I believe it simply follows on from verse 12 and is describing the new incumbents; the Caliphs – or in Jerusalem – the Sultans.

It is imagery simply saying that the first beast, the beast out of the sea (with the statue of old Babylon at its head) is now being superseded in the Holy Land by the last beast, with a leadership of human flesh. It reinforces the idea that the original Babylon where Nebuchadnezzar had a 60 cubits high idol of a man raised up, is now being replaced by a living human 'with breath'. The 'idol' is now a person with the 'ability to speak' in the manner of the Caesars and those that went before. This leads to the installation of Caliphs and Sultans in the same vein, in the final (eighth) beast kingdom, that was / is, the Islamic Empire.

It is also interesting, that the pagan Roman Empire of John's day is implied only, in the image of Revelation 13:1-10. And I mentioned this earlier. We know it has to be a part of the imagery of the first beast, but it is only represented by the ten horns.

It becomes obvious that this revelation is not something to boldly broadcast. The Julian and Flavian Roman rulers at the time martyred so many, and John himself was imprisoned on Patmos Island for years and physically abused. Nevertheless he gives us enough information to work out the missing identity for ourselves, and Revelation 17:10 confirms that Rome is the sixth kingdom or 'head' and is current in John's time.

Last, but not least, the scripture presented above in Revelation 13:11-18 ends with the famous number 666 which is the 'mark of the beast' of Babylon the Great.

It has been suggested that these numbers, which are letters from the Koine Greek, would have been very familiar to John and reminded him of the strange writing he saw in his vision.

This mark of Babylon ends up in Arabic, as a war cry for Islam and is used on bandannas, wristbands and flags by its fighters, and gives the new beast identity. To John, it becomes a mark of the second beast of Revelation 13 and a sign of their worship of Man and the System. The letters were chi, xi and stigma (digamma) which also have the values 600, 60 and 6.

This was noted by Palestinian Walid Shoebat who subsequently converted to Christianity. His familiarity with Arabic lead to this comparison and although the xi sign is oriented on its side (and mirror imaged for closest representation), it shows a real likeness to the cross swords and Bismillah (Allahu Akbar / God is Great) sign used by Islamic terrorists / jihadists and other militant leaders and speakers.

More than this, 'chi', the Islamic crossed swords, remind us of the sin of Cain, our sin and in fact the sin of the whole world. The same sin that Jesus our Saviour was nailed to on a Roman cross, as the Lamb of God for the redemption of the world.

The crossed swords also provide a link to the 'man of sin' of the New Testament and I believe it is highly likely that this is how the Apostle John saw the letters in his vision.

It is not that the Greek is a perfect replica, but rather the writing reminded John of those three Greek letters like nothing else he could think of. Again, how do you describe some writing, that did not even exist in your time?

The exercise required here, is to put ourselves in the place of a person who saw something in a vision, that reminded him of some letters in the Greek alphabet. The Greeks studied the Babylonians carefully and not too much would have escaped them. The number 666 would also have reminded John, of things he knew about the original Babylon and their fascination with astronomy and numbers; the sun, moon and stars. Things which are also very much symbols of the new Babylon as well.

The vision that John saw of the letters, is also strangely reminiscent of the words seen by Belshazzar, a successor to Nebuchadnezzar, and the last king of the original Babylon (Daniel 5).

This is when he saw the 'writing on the wall', 'Mene, Mene, Tekel and Parsin'; God has numbered the days of your kingdom, you have been weighed in the balances and found wanting and your kingdom is divided. A dire warning to the last Babylon; Babylon the Great as well!

Another sign from ancient Mesopotamia are 'lucky' amulets with a 6 x 6 square matrix of numbers from 1 to 36. These were designed such that across and vertically the numbers add to 111 in each row or column. When added together in any one direction; 111 x 6 = 666. In addition the Babylonians referred to 36 constellations, with the sun and moon central to their worship of the starry host. Some even suggest they fostered a direct demonic connection through this adoration of the heavens.

It has also been suggested that known spellings of Nebuchadnezzar, when translated into Hebrew, equate closely to 666. This has been proffered for the 'false prophet' of the last Babylon as well. A Latinised Greek rendering of Muhammad / Mahomet; i.e. 'Maometis' is said to have the value of 666.

Neither of these offerings seem to be concrete enough to be considered firm evidence, but are an interesting discussion point nonetheless. Others have suggested that it is code for Caesar Nero, because he is likely to have been the Roman Emperor, at the time John received the visions. Some analysts have made similar comparisons with other identities from the Holy Roman Empire and Ten Kingdom period.

But it should be remembered, this is the second beast of Revelation 13, the one that comes out of the earth. It is not the beast from the sea, with the seven heads and ten horns characteristic of the Roman Empire, its predecessors and the Ten Kingdoms to follow.

It is true that many of these rulers had reputations that deserved the 666 designator, and were antichrists in their own right. However the number 666 here is explicitly tied to the second beast in this chapter - the last and Eighth Beast Empire.

To the Hebrews the number six was unambiguously seen as the number associated with man. When John talks about the mark of the beast, he also notes it is the number of a man; 666; 'the number of the beast is the number of a man'. In the Old Testament, 666 is also associated with a story of payment in gold. Gold and extreme wealth is a sign of antichrist tendencies.

Regardless of the devotion of individual Muslims and to whom they direct their adoration, Islam (as Babylon the Great) was - and in some countries still is - a monolithic, authoritarian ideology directed at absolute domination of the population.

It has little to do with the worship of the God of Abraham, Isaac and Jacob, except to use the Islamic version of monotheism to subjugate the people. The prophets clearly state; the adoration is directed toward the leadership – the Caliphate or 'Man' represented by the 666. It is not 'Allahu Akbar' that is the centre of praise, but in fact the Caliph and the State. The annual pilgrimage to Mecca (the Hajj) the Kaaba and 'moon' rock, ties the worship of the heavenly bodies to the worship of the Dragon (Satan). Daniel and Revelation through the Son of Man, tie the 'little horn with eyes and a mouth', aka the 666 Beast, to Islam.

To that extent it is simply the last iteration of the 'Animal / Beast Kingdoms' of Daniel's and John's prophecies and follows in the vein of the many kings and emperors of those original empires. Indeed the last iteration is tied to the counterfeit triad of Fascist, Nazi and Communist leaders of more recent times.

Hitler was apparently inspired by the Ottoman annihilation of potential enemies in Asia (notably the Armenians) and Stalin looked on from nearby Georgia. The Nazis were also actively supported by Palestinian and Turkish leaders and by their militias who fought for the Axis armies. These regimes are therefore 'daughters' of the 'Mother of Prostitutes' of Revelation 17.

Caliphates and crusades

Since Islamic rule in the Holy Land and Middle East ran as long as the five previous empires put together, one would reasonably expect that Islam has always had a prominence and an historic presence, in the minds of people in Europe and the West.

Although that would certainly be the case for anyone living in the Middle East, Africa and Asia (and also in Europe during the time of the Crusades and medieval period) it seems in the modern era, our western imaginations have rarely been captured by the ubiquitous impact of the Caliphates. For the most part this will be because our ancestors lived on the periphery of the Islamic Empire, or in fact escaped from it, early on in history.

We have therefore been concerned with more immediate problems within western society, and excising our own demons; the World Wars and the many European conflicts and revolutions that came before, being a central part of those difficulties.

In the west our focus on the Cold War, Communism, the USSR and China have continued to distract us, and therefore we seem to have been largely unaware of the history and issues surrounding Islam, i.e. until more recent sporadic post war terrorism gradually impacted our consciousness.

Nevertheless since the Six Day War in Israel, the Bath party Iraq war with the Ayatollahs of Iran (beginning in the late seventies) followed closely by the Gulf Wars and more recent strife and destruction since the so-called Arab Spring of 2011; events have now reached a pitch that *really* commands our attention.

Millions of refugees have been displaced and terrible destruction has taken place throughout the Middle East. It now seems the American Embassy in Israel is to be relocated to Jerusalem to reflect the actual situation of 50 years; that Jerusalem is indeed the capital of the country. And as a consequence of all this relatively recent turmoil, we find ourselves looking at a very unsettled situation in the region and potentially the world.

This build-up is my personal experience as a westerner, not uninterested in these events, but necessarily focused on the job at hand and going about business as usual. Others will have a different perspective, depending on whether they have had direct involvement, or alternatively no interest at all.

To set the scene then, here is a brief history of the Ecumenical Caliphates since Muhammad:

Rashidun	(Abu Bakr, Umar, etc.)	632-661	Arabia
Hasan ibn Ali	(grandson of Muhammad)	661	Arabia
Umayyad	(Harran based)	661-750	Syria
Abbasid	(Caliphs of Bagdad)	750-1258	Iraq
Abbasid	(Caliphs of Cairo)	1261-1517	Egypt
Ottoman	(Caliphs of Istanbul)	1517-1924	Turkey

The Ecumenical Caliphates are recognised by both Sunni and Shia, and as presented above, have come to their conclusion. However many Muslims seek a return to the old days and momentum has been gathering over the last several decades.

The conflict of earlier times between Europe and Islam has already been briefly addressed in the description of the Crusades to the Holy Land. These began around AD 1100 and continued for approximately 200 years, and there were certainly a few big battles fought and many fortifications and cities were built by the Crusader armies. The Knights Hospitaller and Knights Templar were instrumental in these efforts.

Generally though, the knights were there to provide the civilian participants on pilgrimage, sufficient infrastructure and protection for their needs - and this was achieved with considerable success over the years.

Unfortunately these noble aims eventually degenerated, to the extent that the western pilgrims of the late 13th century were responsible for the sacking of Constantinople. They were composed of militias with specific interests within the western Crusader alliance and this weakened the Byzantine capital, such that it eventually lead to the capture of the city in 1453 by the Ottoman Sultan, Mehmed II.

Nevertheless, the history of Islamic conquest and the number of battles throughout Europe, Africa and Asia, make any offensive military measures by Christian Europe pale into insignificance. Europe was attacked in a pincer movement through the Middle East, Africa, Spain and France on one hand, followed by the conquest of Constantinople and Eastern Europe on the other - eventually advancing to the gates of Vienna itself.

It has been estimated that the total number of defensive battles fought against Islam number over 550. They were against Islamic aggression extending from France to India and beyond.

For a so-called 'religion of peace' there is a lot of history that says otherwise.

Certainly the biblical prophets saw it completely differently in their visions, and those were undoubtedly visions of God.

Of course the Islamists suffered some big defeats in their own right and one such occasion stands out. It occurred centuries after the Seljuk Turkish incursion, when the Mongols of the 13th century sent emissaries to the Islamic State of Khwarezmia. The State covered a considerable portion of Persia, Turkmenistan, Uzbekistan and western Afghanistan.

The Muslim leadership ill-advisedly, but characteristically, killed a Mongol diplomat and sent his remains with his humiliated brethren back to Genghis Khan. The Great Khan was understandably not amused, and promptly arranged an attack on Khwarezmia from three sides. It must rank as one of the most vengeful attacks in history and estimated tens of thousands were slaughtered. It allowed the Mongols access to Baghdad, the Levant and indeed, to Palestine.

Eventually, the Mamluks (mercenaries employed by the Egyptian Abbasid Caliphate after they relocated to Cairo) met the Mongols at Ayn Jalut near Megiddo in the Jezreel Valley.

This was a battle that the Mamluks won and which eventually allowed the Muslims to regain control of their empire.

It is interesting to note, that the leaders of the Crusader Knights were also in difficult circumstances at this time. They lost control of their fortified coastal cities such as Acre (Akko), Sidon and Byblos and were forced to retreat to Cyprus and various Aegean Islands. They therefore tried to collaborate with the Mongols in an attempt to counter their common enemy.

However these negotiations in the late 1200s and early 1300s amounted to nothing.

Incredibly though, the incursion of the Mongols into Europe and the Middle East was almost certainly the vehicle that allowed the Black Death to sweep through western Asia, the Middle East and Europe. When the pandemic arrived in the following century, it virtually swept through overnight and had extraordinarily dire consequences, killing an estimated third of the population. This, and subsequent plagues are surely the subject of the biblical prophecies regarding plague and pestilence, and contributed to the eventual demise of the Byzantine Empire.

All this activity should be seen in the light of other significant movements of people in Western Europe.

Starting before the turn of the millennia, the Norsemen had moved into France and eventually the Mediterranean and were also trading with Constantinople down the Russian rivers.

In 1066, the now French speaking Normans invaded England after being delayed by bad weather. The English king, Harold was already engaged in battle; defeating Vikings who had arrived by ship from the north and were trying to reclaim the Old Norse city of York. Harold's victory celebrations at Stamford Bridge were short lived however, and he was forced to turn his armed forces around and head south to protect London. He met the Norman leader William and his army on the south coast at Hastings.

It all proved too much and England was subject to a French speaking Norman hierarchy for 350 years.

This was a time of castle and cathedral building the like of which had never been seen in England before. It was essentially the same castle building (and contemporaneously) that featured in the Crusaders activity in the Levant; much of it initiated from France, with kings such as Richard the Lionheart for the Normans, together with Frankish and other European monarchs united in defending Christendom and the west.

Knowing this background helps to put the European world in context when addressing the Crusader period. It illustrates that defending Italy, France, Spain and Britain (and Europe in general) against the Muslims was front of mind.

It was a concern before the Crusades and an all-consuming task during the period and for centuries thereafter.

So as the 13th century turned into the 14th, the leaders of the Knights Templar and Knights Hospitaller of St John were losing their mainland strongholds in the Levant, and the economic viability of the Knights Templar was particularly in doubt, as their business model evaporated.

In 1307, Jacques de Molay, the leader of the Knights Templar, returned to France. The French king Philip IV (The Fair) was in deep debt and planning to get his hands on any loot he could find to correct the problem. He already had the Pope where he wanted him, because he had the Vatican transferred to a French location at Avignon.

Significantly, the Knights Templar had become very rich over the years; operating a money transfer business equating to an international bank - together with extensive integrated military escort and accommodation portfolios. This was all too tempting for the French king, who then isolated the Templar leadership from their boss, the Pope, and on their return to France had them arrested and tortured starting in 1307.

Jacques de Molay was in fact crucified on a door (but not killed) at this time, and accused of a number of things, including worshiping an animal head idol, Baphomet (which some Templars associated with mosques).

The Templars were also accused of being unfaithful to the cause, homosexuality and pretty much anything which seemed useful to discredit them and their leader.

The result was the Templar knights were disbanded; Molay and his deputy were burnt at the stake years later, while the king pocketed the Templar fortune. Forty years after Molay's arrival in Paris, the plague took hold and decimated the population of western Asia and Europe. This was in the years 1347-1348.

The date in the middle of the 40 years was AD 1328.

Amazingly, it just happens to be **the precise middle of the second 'time, times and half a time'!**

Now, how significant all this is, I am not completely sure. But certainly, as the forty years between AD 31 and AD 70 were a judgement on Second Temple Judaic administration and the ruling hierarchy, I believe this episode was an indictment on medieval European royalty and government, at a time when the politics of the Ten Kingdoms of prophecy ruled in Europe.

It mimicked the situation which existed in old Jerusalem 1,278 years before. It included the ineptness and failure of Christendom lead by the Papacy to positively influence the royalty of Europe, or protect their own knights. It was also far from the first century Christianity of the Apostles, or for that matter the holiness of the Davidic priests of the First Temple.

All in all, it appears that the judgement which came with the plague era was not entirely undeserved, and in any case, absolutely decimated Europe.

This miserable episode in our Christian heritage came to my attention in the form of a paperback entitled 'The Second Messiah' by Christopher Knight & Robert Lomas.

The book attempts to tie the Masonic movement of London (which officially appeared in 1717 as the Grand Lodge) to an earlier Scottish tradition centred on the Sinclair family and their Rosslyn Chapel. A further connection is then suggested to France and the Templar movement, and the implication is that some Templar knights escaped to Scotland in an effort to elude capture by Phillip the Fair.

It is an interesting hypothesis; one that is likely to be based on solid history. And there is little doubt that many would have attempted to escape from France during this time. There is also the connection with the stonemasons, who were a significant part of the medieval world and the Norman kings particularly.

Nevertheless, while the authors' attempts to tie Masonic tradition back to earlier history are commendable, the book's tone is no friend of the Christian faith.

The authors' findings aside, the idea of a Second Messiah (or at least a prominent religious figure, who was treated nearly as abominably as the Jewish Messiah 1,278 years earlier) was intriguing in the extreme; a figure who may have experienced the anointing of the Holy Spirit as well. When the biblical timeline outlined here became clear, and was obviously an exact fit in the middle of the second 'three and a half times', it suggested to me the proposition that I have put forward above.

Now Jacques de Molay and his story did not play out in Jerusalem, which is the city at the centre of our prophetic timeline. That is a slight concern, and it may be that there is another story based in the Holy Land, which I have not discovered.

Molay was almost certainly wrongly accused; he was tortured and finally put to death at the stake. At the end of the forty years after his arrest, the worst plague of the medieval period struck and is therefore surely a biblical timeline marker.

The fact that Molay was the leader of the Knights Templar who tended to be drawn from royal and aristocratic families (some of whom, may have had Jewish ancestry) makes his story the more valid as an echo of the first heresy in Jerusalem.

That the Templars were unable to re-enter Jerusalem because of overwhelming Islamic forces, also supports the claim.

So Christendom *also* is burdened with a dark history, and this midpoint episode illustrates how debased the elite and high society had become in Europe. Individuals will often be corrupted, no matter how much they have benefited from the very best teaching and example set forth by their faith and society.

The second 'three and a half times' also included centuries of anti-Semitism, as the Jews were blamed for the sins of their fathers. There was reason enough to recognise the continuing denial by religious Jews of the identity of Jesus. Too, that Arabian Jews sowed the seed for Islam, by teaching others to deny Jesus was the Messiah as they did themselves. But demonising them for their blindness to the extent that happened, was not exactly the right of mere human beings, Christian or not.

Even the likes of otherwise great men such as Martin Luther were caught up in the name calling and active discrimination.

Unfortunately the sins of the Kingdoms did not stop there. Once the colonial age was truly under way, there were specific positives brought to the lands which were occupied. The great commission of preaching the Gospel to the entire world was one. European law, administration and medicine and the new knowledge of science, engineering and technology that went with the advent of pre-industrialisation, were others.

On the down side, there were cynical attempts to rob people of their land and livelihood through war and unscrupulous business dealings. One of the most abominable in the English speaking world, was replacing silver with opium as currency in the China tea trade. It left the Chinese to pick up the pieces of destroyed lives.

There was also the introduction of European diseases to native populations. Most of this was unintentional, but unbelievably there were instances where plagues were introduced deliberately to decimate native populations. Sadly, the overt exploitation of individuals eventually culminated in the Atlantic slave trade; an act borrowed from previous generations and in particular from the known example of the Islamic Caliphates.

However it beggars belief, that all this was implemented at the same time Christian missionaries were endeavouring to bring the love and goodwill message to the same people.

There was no justification for such appalling behaviour and therefore the Ten Kings 'receive authority as kings along with the beast' and 'give their power and authority to the beast' as it is described in Revelation 17 - not only during their time in Jerusalem - but long after they had vacated the city.

The Apostle John does not specifically mention these issues, but I suggest they are likely examples of 'beast' activity through the later centuries of the second 3.5 'times'. They are just not highlighted by the Jerusalem centric prophecies.

Now there is a possible identification of tribes and nations throughout all this history, which I barely touched on earlier. It is simply a question of whether there is an association between the Ten Tribes of Israel and the Ten Kingdoms of prophecy.

Figure 2: A map of The Holy Land in the time of Jesus (Yeshua) with inserts from later times. (a) The Star of David at Capharnaum on Galilee ca. AD 350. (b) The fallen columns of Beit She'an (Scythopolis) a result of the AD 749 earthquake. (c) Modern day Jerusalem.

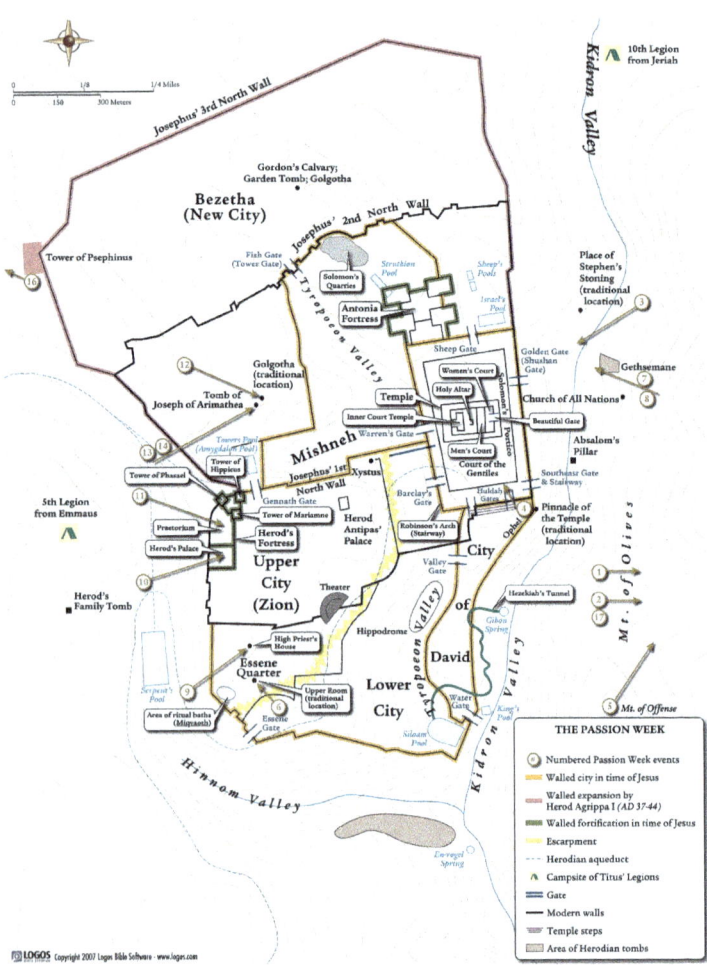

Figure 3: The Old City of Jerusalem with the various walls built throughout the ages. The position of the original City of David is outlined toward the south-east.

Plate 15: Above, an alleyway in the Old City with security on patrol.
Plate 16: Looking north-east to the Western Wall with the Dome of the Rock in the background.

Plate 17: The Jerusalem British War Cemetery on Mount Scopus with graves of young men from every corner of the Empire, including Australia; casualties of the Great War and the liberation of Palestine.

Plate 18: Gethsemane at the base of the Mount of Olives in the Kidron Valley. The olive trees are believed to be about 1800 years old.

Plate 19: The Library of Celsus at Ephesus in today's Turkey. Ephesus is considered to be the largest Greco-Roman archaeological site in the world.

Plate 20: A quiet corner of 'Mary's House', a reconstructed building on the foundations of an earlier one. It is situated on the mountainside behind Ephesus and may have been where Mary lived in later life.

Plate 21: Above; the massive Thutmose III obelisk in Istanbul carved from Aswan granite. **Plate 22:** Below; the Hagia Sophia, the 6th century basilica dedicated to the Holy Wisdom (Holy Spirit) where the 'man of iniquity' sat as God.

Plate 23: The entry to the Holocaust memorial in Jerusalem. The inscription reads: 'I will put my breath into you and you shall live again, and I will set you upon your own soil...' (Ezekiel 37:14)

Plate 24: One of the many plaques in memory of the those who perished in the Shoah.

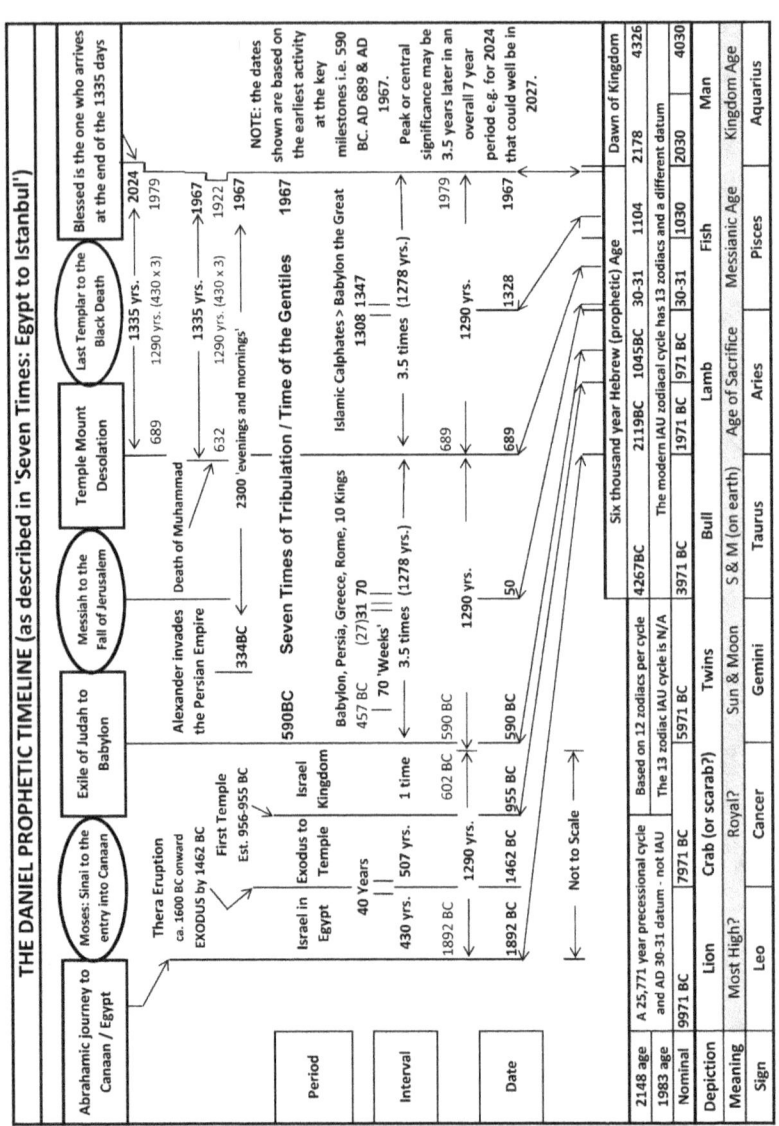

Figure 4: The Seven Times prophetic chart of Daniel, together with other intersecting timelines and the retro 1290 'day' cycles back to Abraham.

There appears to be at least a circumstantial case for an identification here, and some readers may have already recognised this. The seven churches in Anatolia may be a clue, and I have suggested that the three kingdoms that were overcome by Islam were very broadly Egypt (other Africans and Canaan), Assyria (Syria, eastern Anatolia) and Mesopotamia. Alternatively, Assyria and Mesopotamia should be grouped, and Western Anatolia and the Bosporus region becomes the third.

Either way, this would suggest that the equivalent of seven tribes emigrated to Europe and that they ended up as the ruling elites of the western European nations - or some other broad identification, or fulfillment with those nations is implied. If so, this would parallel the earlier experience in Parthia, where Ephraimite and Jewish leaders are believed to have had great influence in that Empire after their earlier captivities / exiles.

Returning to the prophetic word; in Daniel 12 we have a masterful overview of the whole 'seven times' of tribulation and Jacob's trouble, and in fact an extension to the timeline.

However a casual reading is not enough to reveal the full impact of this chapter. Verses 1 to 4 begin with a brief look toward the great tribulation period and its ending, noting a great increase in the search for knowledge. This is easily interpreted as the driver for our industrial and post-industrial age.

Daniel then sees one of the figures beside the river asking of his visionary host, 'how long until all is fulfilled' and receives the reply 'it will be for 'a time, times and half a time' when the power of the holy people has been finally broken'.

I do not think that the answer to the question was one Daniel would have been particularly expecting. I am pretty sure he would have been more interested in the timing of the wonderful things he could see in his vision relating to 'knowledge'. Things like the trappings of our modern world (cars, mighty ships, aircraft, television, computers, medicines etc.) which we now have, and for the most part appreciate.

He would probably have been no different to the rest of us, in being mesmerised by the baubles of our modern industrial and technological age. Instead, he learnt that it would be 3.5 times until the holy people are broken (in AD 689). That is; it would all begin when the abomination of desolation of the Temple Mount occurred.

Daniel 12 in its entirety:

"At that time Michael, the great prince who protects your people, will arise. **There will be a time of distress such as has not happened from the beginning of nations until then.** *But at that time your people—everyone whose name is found written in the book—will be delivered. [2] Multitudes who sleep in the dust of the earth will awake: some to everlasting life, others to shame and everlasting contempt. [3] Those who are wise[a] will shine like the brightness of the heavens, and those who lead many to righteousness, like the stars forever and ever. [4] But you, Daniel, roll up and seal the words of the scroll until the time of the end.* **Many will go here and there to increase knowledge.***"*

[5] Then I, Daniel, looked, and there before me stood two others, one on this bank of the river and one on the opposite bank. [6] One of them said to the man clothed in linen, who was above the waters of the river, "How long will it be before these astonishing things are fulfilled?"

[7] The man clothed in linen, who was above the waters of the river, lifted his right hand and his left hand toward heaven, and I heard him swear by him who lives forever, saying, **"It will be for a time, times and half a time.**[b] *When the power of the holy people has been finally broken, all these things will be completed."*

[8] I heard, but I did not understand. So I asked, "My lord, what will the outcome of all this be?"

[9] He replied, "Go your way, Daniel, because the words are rolled up and sealed until the time of the end. [10] Many will be purified, made spotless and refined, but the wicked will continue to be wicked. None of the wicked will understand, but those who are wise will understand. **[11] "From the time that the daily sacrifice is abolished (and the abomination that causes desolation is set up), there will be 1,290 days.** *[12] Blessed is the one who* **waits for and reaches the end of the 1,335 days.** *[13] "As for you, go your way till the end. You will rest, and then at the end of the days you will rise to receive your allotted inheritance."* (NIV)

We see in verse 11 that from the time the daily sacrifice is abolished and the desolation of the Temple, there will be 1,290 days. This is not between the sacrifice being abolished *to* the desolation, but rather, from the sacrifice being abolished *and* the desolation - *to a future unspecified event*. Finally, there is a further period to the 1,335 'day' where we are told that 'blessed is the one who waits for and reaches the end of the 1,335 days'.

So the first 3.5 times is measured off and then further 1,290 and 1,335 year milestones follow, i.e. a total of 57 years beyond the second 1,278 year period when Jerusalem was finally reunited from under Islam in 1967.

So some simple arithmetic tells us that the **$1,290^{th}$ year was 1979 and the $1,335^{th}$ year will be 2024.**

That result will be affected slightly by the 0.68^{th} of a year (8 months) beyond 2,556 years, which a more exact determination of the 'seven times' would bring. Importantly, if it is spread equally it should not change the actual Gregorian calendar year.

Before we explore the significance of this extended timeline of Daniel, there is one more thing that I hinted at earlier; that there was a fourth timeline which I discovered pointing to 1967.

In my family there are two anniversaries around the timing of the Six Day War. They just happen to be the 5th of June and the 10th of June which are the beginning and ending days of the conflict. The significance of this, is that the dates for the war are firmly embedded in my mind.

So when I was recently researching Muhammad and his birth ca. 570 and death in AD 632, I was surprised to see that he died on the 8th of June - and thought (with mild amusement) - how the heck did the Israelis manage to rig a war around that?

In fact I thought it rather extraordinary, and because of the coincidence with 1967, checked out the numbers.

From 632 to 1967 there are exactly 1,335 years!

This is just another one of those alignments which is beyond coincidence to me and there is a significant date in the 1290th year as well.

It was in 1922 that the British who controlled Egypt to that year, agreed to Egyptian independence. At the same time they were appointed through the League of Nations to oversee the mandate of Palestine and agreed to that responsibility also.

So here is a secondary Israel-Egypt marker and again the words of Daniel's Angelic Being apply; they ring in our ears, 'blessed is the one who waits for and reaches the end of the 1,335th day'.

The end of the age

Now I should emphasise, that I believe this is only a secondary fulfilment of the 1,335 day prophecy. The words of Daniel 12:11 are quite explicit. We should measure the time interval from when the daily sacrifice is abolished and the abomination is set up; that is in AD 689. However Muhammad was the initiator, the false prophet of Revelation 19 and the first of many. It seems appropriate that a secondary 1,335 day timeline begins with his demise, just 57 years before AD 689.

Fifty years on from 1967 the saga continues; a whole jubilee gone and the Dome of the Rock still stands as a Muslim rallying point. Tension everywhere, targeted missile attacks in Syria and the Russian Federation thwarting moves in the UN Security Council to bring perpetrators of chemical attacks to justice.

Since the 'Arab Spring' of 2011, we have several hundred thousand dead in Syria through the activities of the hardline Assad government and various rebel groups, including the rise of Islamic State / ISIL. This in concert with those attempting to counter the atrocities and terrorism, mounted by these groups and bring some sense of security to the innocents affected.

The Islamic State in Syria and northern Iraq, driven by fundamentalist Salafist doctrine is on the retreat. However it is still a potential threat as part of a wider movement and the question is - will it rise again?

Iran has militia in Iraq and is building military bases in Syria. This backing also extends to the Hezbollah in Lebanon. Meanwhile Russian air support is being provided to both the Syrian Assad regime and its Shia supporters. With known and probable further support from other 'kings of the east' (including the Far East) there is a scenario developing which fits the Armageddon description of Ezekiel and Revelation 16.

Turkey (currently a NATO member) is also in northern Syria and working with Al Qaeda and other Sunni groups to oppose the Kurds. The Kurds currently happen to be US and Israeli allies against ISIL. The Turks are increasingly looking to further their advantage, while looking warily at the Shia coalition, now moving over the Euphrates and into Syria.

There could be a turning point if the seemingly impossible happens and the Sunni and Shia forces form a coalition. If that disturbing proposition comes to pass and a coalition of Islamic forces head from the north and east against their common enemy Israel, it would surely threaten an Armageddon war.

In this respect, it needs to be recognised that both the Iranian Shia and the Turkish Sunni are all 'kings of the east'. The difference is; that the Iranian influence west of the Euphrates River is relatively recent in terms of our overall timeline. The Turks, on the other hand, have been established in Anatolia for centuries, since departing their homeland in Asian Turkmenistan. In other words the Euphrates 'dried up' for them a long time ago.

I suggested earlier, that the last beast kingdom could be identified with Arabia where Islam originated. This was based on the 'beast out of the earth' description of the latter verses of Revelation 13. I also indicated that another direct identification can be made, through the 'two horns as a lamb' description in the same chapter.

We have already learned that in the 13th century, the Khwarezmian Empire was destroyed by the Mongols, who subsequently invaded Anatolia on their westward drive. This would have accelerated the Turkish migration westward, which began with the Seljuk Turks entering Anatolia in numbers, around the 11th century.

As 'kings of the east' the Iranian and Iraqi Shia, as well as the Turkish Sunni, were once part of the Mesopotamian and central Asian scene that was the Abbasid Caliphate. However before these events - and even before the Sassanid kingdom before them - we know the same central Asian region was the heartland of the Achaemenid, Medo-Persian Empire of Cyrus.

This is the famous Medo-Persian Empire which was covered in earlier chapters, and the self-same empire described as a 'two horned ram' by none other than the Prophet Daniel. Because the head of the ram is also symbolic of sun and moon worship; emblems of Babylon and Islam, this finishes the historic circle and makes the photo identity match complete!

With that sorted out, it is now time to address the prophecies of the Ezekiel 38-39 invasion narrative of Gog and Magog (or Gog of Magog). There is also the Zechariah 14 vision of war and earthquake, and 'the signs in the sky' paraphrased by Jesus from Isaiah 13 and 34. The latter include the darkening of the sun and moon, the 'stars' falling and the shaking of heavenly bodies. Revelation 16 which has something of a similar narrative will be addressed in the next chapter.

In the case of the prophecies of the Tanakh (OT) particularly, there is a need not to confuse references to earlier wars with those yet to come. Anything in fact from the Judah exile through the 'seven times' not associated with the 'last days'. As a corollary we also need to establish whether a prophecy covers a few short years, or whether it covers many hundreds or thousands of years. The description in Ezekiel 38 and 39:

- Gog and Magog as the principle adversaries of Israel
- A confederacy including Meshech, Tubal, Persia, Cush, Put, Gomer and Beth Togarmah invade from the north
- They invade Israel where people who have been living in fear are now gathered within unwalled villages
- Sheba and Dedan (Arabian Peninsula) and the Tarshish merchants (Western Europe) are concerned with the intent of Gog and Magog
- The wrath of the Lord will be such, that there will be an unnaturally great earthquake causing mountains to fall
- Every man's sword will be against his brother
- Torrents of rain, hailstones and burning sulphur will fall on the invading troops
- There will be such a bloodbath, that the people of Israel will take seven months to make any impact on the clean-up and recover enough fuel for seven years supply
- Many will die in the Valley of Hamon Gog (the multitudes of Gog and Magog) in Jordan

Zechariah 14 makes the following points:

- The gathering of the nations to Jerusalem
- Half of the city to go into exile

- The Lord will fight against the nations and stand on the Mount of Olives which will split in two
- The land south of Jerusalem subsides and the people flee from the earthquake, water will flow out of Jerusalem
- On the day of the Lord the normal evening and morning will not occur
- A plague will strike the troops fighting against Israel and their flesh, eyes and tongues will rot while they are still standing
- Animals will suffer the same fate and the enemy will attack each other
- The wealth of the surrounding nations will be collected and the survivors will go every year to worship the King

Apart from the first two points of the Zechariah story, which appear to be a summary of the original gathering of the 'beast' nations against Israel - and the exile / escape and dispersion of the inhabitants out of the land - the remainder describes the war and earthquake at the end of the age; the Great Day of the Lord.

The earth subsiding to the south of Jerusalem, sounds like a literal event lining up with Ezekiel and Revelation descriptions. The water flowing out of the Temple Mount is repeated in Ezekiel 47 and may be literal, although a spiritual or political interpretation cannot be dismissed. Certainly there is a political divide between the Israelis and the Palestinian Arabs.

Zechariah's description also focuses more on the events in Jerusalem, but is otherwise comparable with the Ezekiel narrative in terms of war, earthquake and the judgement, of the opposing nations. The plagues and devastation are primarily visited on the enemies of Jerusalem, and accord with Revelation 16. However, because of the geographic proximity to the Babylon the Great (Gog and Magog) hordes attacking Israel, we see some disruption and damage to Judea and Jerusalem itself.

Other prophecies which I believe tie directly into our time, are Psalm 83 and prophecies against Assyria and Damascus such as Isaiah 17. Presented here is Psalm 83:5-8

'With one mind they plot together; they form an alliance against you – the tents of Edom and the Ishmaelites, of Moab and the Hagrites,

Byblos, Ammon and Amalek, Philistia with the people of Tyre. Even Assyria has joined them....' (NIV)

These immediate neighbours of Israel have been a constant thorn in the side over many years. This has been especially the case, since the resurrection of the State of Israel starting from World War I. It includes the resistance against Jewish migration to Israel itself, support for the genocide during World War II, the 1947-48 war, the 1967 Six Day war through to the 1973 Yom Kippur war, continuing strife in 1979-1982 against Hezbollah and the air war with Syria. All this; with continuing internal Palestinian unrest and terrorism, highlight the period.

In regard to the Isaiah 17 oracle and other similar pronouncements, we see that Damascus will one day lie in a heap and remain desolated. This is hard to accept, given Damascus is one of the longest continuously occupied cities on earth. My sense here is that if the city is not totalled by war (as seems likely at present) then it will be destroyed by an 'Act of God'; namely the same earthquake that severely damages Jerusalem.

Essentially then, both Ezekiel and Zechariah are referring to the Day of the Lord, at the end of the 'seven times'. This, aligns with the chapters in Revelation, which are tied directly to the destruction of Babylon the Great and that apocalyptic time.

The Sunnis of the Arabian Peninsula (Dedan and Sheba) together with the merchants of Tarshish (the West) will be in fear of the northern adversaries, particularly Persia (Iran) and there is some of that tension and outright war already evident. It includes the ongoing conflict in Yemen. Iraq and possibly other elements in Africa will side with the northern confederacy.

There is another prophecy to be found in Revelation 20, where we see the use of the terms 'Gog' and 'Magog' to describe a conflict late in the 'Messianic' age with the northern kings. This period is referred to as a time when Satan is initially bound for a 'thousand years'.

This description is not easily linked with the Ezekiel or Zechariah episodes, but has the curious reference to Gog and Magog coming from all points of the earth, to engage in battle against God's holy people. In this sense, it is reminiscent of the recent rush of ethnic Syrians and Islamic associates from around the world, to actively engage and fight with the Islamic State in Syria and Iraq. They have been labelled as 'foreign fighters' by both the locals in the Middle East and the nations abroad, and as far as I am aware, this is the only place in the New Testament where the terms Gog and Magog are used.

This passage has been commonly interpreted as a period beyond the end of the 'seven times' timeline and our current age and to be a 'thousand years to come'. However this 'millennial reign' is intriguing because it can also be seen to be in parallel with our current age. I will therefore leave further discussion to be grouped with references in Revelation to the 'seal' and 'trumpet' prophecies and further commentary on Israel.

Regardless of how we interpret Revelation 20, I believe we can proceed confidently in the knowledge, that we are very close to the events closing the 1,335 year period of Daniel.

It means we can now focus on more immediate and remarkable things to do with the end of the age.

Above I mentioned 1922 which was the 1,290th year on the secondary 1,335 year timeline from the death of Muhammad in 632 AD to 1967. The 1,290 year (or 'day' as Daniel describes it) is a peculiar and interesting number.

When we looked at the history of the Hebrews, we found they lived 430 years in Egypt including Canaan. We also found our basis for biblical prophetic interpretation in the 'seven times' in Leviticus. But then there was the 'day for a year' principle in Ezekiel, where he acted out a play to demonstrate to a disbelieving public, that Jerusalem was about to be sacked.

This was where he lay on his left side for 390 days for the sins of Israel and right side for 40 days for the sins of Judah.

I have left the big sums for last; **suffice to say 390 + 40 = 430;** and the next part of the equation **is 430 x 3 = 1,290.**

What does this mean? Well of course it is our Lessons 8. In short, 430 is a significant number for *all* the Tribes of Israel, but especially in the context of the connection to Egypt and the 430 years they spent there, from Abraham to Moses.

So the 1,290th year after the desolation of the Temple is beyond 1967 and is in fact 1979.

It just so happens, that in 1979, Egypt was the first of Israel's neighbours to officially recognise it as a nation state.

It was a most unlikely occurrence, given the post 1967-1973 political climate and was initiated by Egypt's then President, Anwar Sadat. For his trouble he was assassinated by fundamentalist army officers on the 6th October, 1981.

The period 1979 to 1982 (which included the assassination) was also unsettled and extended as in previous 'seven times' markers. It concluded with war at Israel's borders and Palestinian terror attacks within. In 1982 there was another Lebanese conflict, and air operations against Syria and Iraq completed the events of this milestone.

The significance of 1979 as the 1,290th 'day' is another reason I am confident in continuing the timeline to AD 2024; the 1,335th 'day' of Daniel.

There is also no other nation or empire in the region, which has had greater ties to the Holy Land than Egypt. It has been a love / hate relationship with memorable highs and extreme lows, but one which nevertheless endures and survives with considerable mutual respect.

And despite the concerns I have expressed in previous chapters, over some pieces of scripture and their historical accuracy (oh ye of little faith) I also find it absolutely amazing, how prophetic details provided from Jewish prophecies given hundreds of years apart, tie in so accurately and seamlessly.

This surely shows the stamp, through the Holy Spirit, of the Author of our Salvation, the awesome Numberer and Namer!

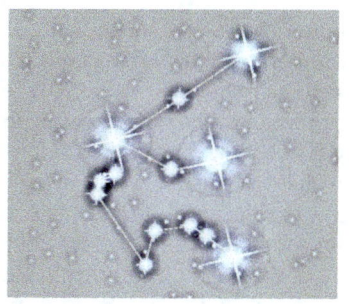

SIGN OF A MAN

CHAPTER 13

The Water Carrier

"At that time the Sign of the Son of Man will appear in the sky and all the nations of the earth will mourn. They will see the Son of Man coming on the clouds of the sky, with power and great glory."
JESUS OF NAZARETH

The 'fig tree' buds

There are many prophecies in the Holy Bible and yet I have focussed only on the writings of Leviticus, Ezekiel, Daniel, Zechariah and Ezra in the Old Testament, and mostly the Revelation to St John in the New. The reason is simple; this message is about the plan of God and when you have a plan, you have to have a timetable and milestones along the way. There is no way to avoid it; this is 'date setting' and decried by many. But surely the prophets have not been told about 'times' and 'days' for this valuable information to be ignored.

In fact the prophecies and promises go way back to Abraham, Isaac and Jacob (Israel) and his sons, but where Judah and the sons of Joseph (Ephraim and Manasseh) feature strongly.

However it was a messy start for Joseph; Judah was given the promise of royalty, while Joseph was sold into slavery.

In Egypt, Joseph had to work diligently to prosper. He had to use all his talents wisely to make an impression and eventually it paid off. He was ultimately responsible for saving his family from starvation and the whole of Egypt from famine and disaster. I believe this pattern is seen throughout the Seven Times, and is consistent with that ancient prophecy of Jacob.

In this regard, the Genesis 48:15-16 blessing to Ephraim and Manasseh by their grandfather Jacob is both wonderful and compelling:

> *"May the God before whom my fathers*
> *Abraham and Isaac walked faithfully,*
> *the God who has been my shepherd*
> *all my life to this day,*
>
> *the Angel who has delivered me from all harm*
> *- may he bless these boys.*
>
> *May they be called by my name*
> *and the names of my fathers*
> *Abraham and Isaac,*
> *and may they increase greatly on the earth".*

I stated earlier that it is not my purpose to argue a case for which nations may, or may not be identified with the Ten Tribe Diaspora of Ephraim, or how we should view Jewish and Israel identity in the West. Certainly there has been an embedded 'elite' which has shaped western identity, vision and inventiveness. However it has already been noted, how the British Empire was instrumental in re-establishing Israel as the land of the Jews and assisted the vision of the Zionists.

In addition, the later incarnation of people in the New World of North and South America (notably the United States of America) have been crucial in maintaining the independence of modern Israel. It is still an invaluable contribution when aid is offered remotely, or the mere existence of western nations deters potential aggressors.

Now one thing is for sure; with so many Jews by faith living in the Americas, the Ezekiel prophecy of the 'dry bones' and 'the two sticks becoming one' have been fulfilled in a huge way. I like to think that out of the ashes of the Twin Tower carnage in New York in 2001, we have the very sign that Ezekiel spoke of; a single stick. The One World Tower reminds us that Israel, the people, wherever they are, and whatever they call themselves, are back together. Even more significantly, the 'two sticks' continue to be joined in the land (eretz) Israel itself; and that has always been the prophetic focus.

In three of the gospels; Matthew, Mark and Luke, we encounter Jesus' Olivet speech to his disciples. After he outlines the destruction coming to the Temple, they questioned him about his dire predictions and the further events that will unfold through to the end of the age. Some of his speech comes from quoting previous prophets and Daniel and the 'abomination that causes desolation' is mentioned specifically. Other portions are new and without doubt come from his personal prophetic knowledge.

It is a summary of the next 2,000 years of world and Hebrew history and I am sure it overwhelmed his followers at the time. Nevertheless the message has been faithfully recorded for posterity, to absorb and understand as best we can.

Here are a few of the prophetic signs mentioned in Matthew 24 and expected at the very end of the age, together with others from throughout the New Testament:

- Wars and rumours of wars; nation will rise against nation
- Lawlessness will increase; persecution of believers will escalate
- There will be earthquakes, famine, disease and plague
- There will be an increase of false prophets and false teachers
- The love of many will grow cold; the masses will believe falsehoods
- The gospel will be preached to the entire world
- The State of Israel will be resurrected in the Land
- There will be a cry of peace, when there is no intention of peace
- If those days had not been cut short, no one would survive
- There will be signs in the sky and the 'sign of the son of man' will appear
- Scoffers will mock all these end time signs and disregard them

Looking at history from our perspective today, with two World Wars behind us where an estimated 80 million people died, well might we ask are those wars in Bible prophecy? Indeed, with 6 million Jews and other minorities losing their lives in the Holocaust alone - where can we find that time of terror specifically described in the Jewish scriptures?

The numbers are truly horrifying, although they did occur in the industrial age, where both the technology and population was such, as to almost guarantee the scale of the slaughter.

Perhaps that is our problem today. The knowledge of the calamities of more recent times, has overshadowed the trials throughout the centuries. We have been disconnected from what went before, because time, a blurred record, geography and lack of interest has not helped in our education. In fact, now that we have our timeline and understand that it did not finish in Jerusalem until 1967, we can see that the great destruction of the World Wars was right at the end of the period.

In the same way that the Seven Times of tribulation kicked off with 70 years of suffering and dispossession during the Babylonian exile, so at the end of the period, 70 years of terror occurred. Only then was the curtain closed on the Seven Times, with the Six Day and Yom Kippur wars in Israel.

There are other points of comparison and contrast as well. The Babylonian exile was divided into two; the oppression and difficulties before the Persian takeover and the hope that blossomed thereafter. In the case of the last 70 years of the tribulation, the Shoah of World War II bisected that period as well. Here the 'final solution' became the final horror, and yet a new world and new opportunities beckoned for both the survivors and the rest of the world, when the conflict closed.

This picture will be reinforced shortly in a study of Revelation 11, where we see the death and resurrection of the two witnesses (the olive trees) which represent the nation.

Now, despite all the problems with prophetic exegesis in the Christian world and the multitude of diverse views expressed, there is one single prophetic figure of speech which rides above all the signs above. It represents the most amazing occurrence against all odds. Nearly 2,000 years after the words were spoken; it shines as a beacon of prophecy. It is 'the fig tree buds' as a symbol of the modern resurrection of the State of Israel.

Given the resistance up to this year (2017) against any recognition of Jerusalem as the capital of Israel (a fait accompli of fifty years) and knowing the difficulties and opposition of the past, it is a sheer miracle that it happened at all.

The sixth seal

There are some chapters in Revelation that I had no intention of addressing, because they are either peripheral to the main story of the 'beast' or 'animal' empires, or they describe sequences of events that have no numbers, timelines or milestones. Since I have relied on inspiration with regard to timelines as the basis for this manuscript and have limited knowledge outside of that, it was hard to see how addressing such chapters would add anything useful to the story.

However I have come to the view, that some of the content I was going to sidestep, needs to be reconciled with the 'time of the Gentiles' narrative. In essence, the subject matter I am referring to is presented in a broadly defined way and with a modicum of symbolism. It lacks those comforting numbers to which I have become accustomed, but could be helpful in other ways.

Revelation may have been completed and compiled before AD 70 and the siege of Jerusalem, although the Apostle John passed away some decades later in AD 98. It means the extent of his involvement with the final composition is rather uncertain and the text we know may have been collated by others. If so, it could account for unusual aspects of the sequences described.

The chapters I will address are Revelation 6, 8, 9 and 10, which describe 'seals' that herald events in a broad brush way. They appear to cover the whole 'messianic' or church age and are momentous eras, dramatic events, or cataclysmic happenings. They seem to parallel the timeline we already know.

This sequence of significant eras and events, also concludes with the same apocalyptic end-time destruction described in Ezekiel 38 and 39, Zechariah 14 and Revelation 16. In other words, Israel will be the pivot, but the scope inclusive of the broader Middle East and the worldwide view of the day.

With that in mind, the relationship between the seals and trumpets is that the seventh seal is the last one in the sequence, but embodies all seven trumpet events. It gives the impression that events are occurring rapidly in these 'latter days' and therefore greater detail is being supplied for the end of the age.

Trumpets 5, 6 and 7 are also called the three 'woes', indicating that the trauma escalates as well.

The seals begin with the 'four horsemen of the apocalypse', and are introduced by four 'living creatures' who are also mentioned in chapters 4 and 5. These are the four camps of Israel lead by the tribes Judah, Dan, Ephraim and Reuben in the order they are described in Numbers 2. Their banners are the lion, eagle, ox (bullock) and man respectively. The horses are also of different colours and they are apparently in chronological order, although only in the sense of their initial appearance:

o   The first is a white horse whose rider holds a bow and is given a crown and rides out as a conqueror
o   The second is a red horse whose rider is given power to take peace from the earth. He holds a large sword
o   The third seal is a black horse with a rider holding a pair of scales. A voice is heard saying 'a quart of wheat for a day's wages, and three quarts of barley for a day's wages and do not damage the oil and the wine'
o   The fourth is a pale horse whose rider is named Death and Hades. He is given power to kill a fourth of the earth by sword, famine, plague and wild beasts

A common historicist interpretation is that the four horsemen depict four (roughly) consecutive eras, of the pagan Roman Empire in the early centuries AD. This is a Rome centred interpretation, justified in part because it intersects with the history of Israel. It begins with 'victorious' Rome AD 98-180, also known as the Nerva-Antonine Dynasty, which is identified as a bowman riding a white horse.

The red horse and rider are often likened to the Roman Civil War era of AD 192-284; a period where 32 Emperors and 27 Pretenders, resulted in war and confusion. The third horseman has been associated with Emperor Caracalla (Marcus Aurelius), but his reign was a short AD 211-217 and within the previous AD 192-284 period.

I am more comfortable with an alternate view; one that suggests the first of the four horsemen depicts the Messiah lead revival out of Jerusalem that became the early Christian Church. The description in chapter six of the white horseman is not as vivid as that of the Messiah described in Revelation 19, but there are certainly comparisons to be made. The bowman then becomes the one who launches the gospel to the world and this moves the focus back to Jerusalem. This interpretation then becomes a parallel one to the four chariots of Zechariah 6:1-8.

The rider on the white horse is followed by eras of war, famine and then a pale horse which represents multiple disasters. Here I accept that the earlier wars (the red horseman) will be those of the Roman era including the Roman Civil War, but will largely be conflict throughout the known world between Rome and its many rivals, with a special focus on the Near East.

The famine period likely begins in the 6th century, when extreme weather events were caused by natural disasters driven by volcanic eruptions in the North Atlantic. The period resulted in missed summers and crops to fail throughout Europe and around the world - and a pandemic that killed many people in Constantinople, Alexandria and other widely dispersed cities.

This is followed by the fourth horseman, who becomes active as early as the 7th century with war and aggression commonplace in the Holy Land (Sassanids, Islam) and around the Mediterranean and Europe. Wars, famine and plague of one sort or another were an integral part of the medieval age; and the invasion of the Muslims, Norse, and Mongols directly impacted Israel, the Middle East and Europe. This parallels the period of 'great' tribulation. 'Knights in shining armour' might be the image we prefer to mark the medieval period, but life was anything but a picnic for the general population, where an estimated one third died from disease alone. This state of affairs continues to the Black Death of the early 14th century and beyond to subsequent wars, famine and epidemics thereafter.

The horses and horsemen are then followed by the fifth seal, which describes the saints who have been martyred, or passed away previously in serving their Lord. They are now clamouring for justice. They are the ones described as reigning with Christ in heaven until the resurrection and judgement day.

Chronologically this seal could be placed anywhere before all the saints have been gathered and so fits naturally anywhere before the events of the end of the age. This fifth seal vision of the saints is recounted in detail in Revelation chapter 7 and 14 as the 144,000 from every tribe of Israel, together with a further 'great multitude' of 'white robed' saints from every nation.

If there is any difference between the 'seal' description and the later ones, it will be its focus on those that have passed away early in the tribulation. It confirms that those who have gone before are with the Lord, while waiting for the final wrap-up.

Up to this point the sequential nature of the seals (guarantees) is only loosely defined, because the first horseman (the Messianic revival) and the fifth seal with saints waiting patiently are really a picture of the whole age – the gospel continues to be spread and the saints continue to gather and wait. This could also be said of the other seals; war, famine and plague.

However the sixth seal opens with a great earthquake that is seemingly yet to occur; the sun turns black, the moon red and the stars in the sky fall to earth. The sky recedes like a scroll and mountains and islands are moved from their places. In this frightening scenario everybody, both high born and low, hide in caves - and according to Ezra - revert to a Stone Age type of existence. This picture in the sixth seal is so dramatic that I (and I am sure others) naturally assume it is still future; and in fact is the self-same earthquake event described in other passages associated with the apocalyptic, end-time Armageddon war. Sure, earthquakes continue as well, but the one described here is the primary weapon of the Almighty against the end-time enemy.

Furthermore, in Revelation chapters 6 to 9 it becomes apparent that the sixth trumpet (2nd woe) - which is seemingly further along the Revelation timeline than the sixth seal - is unambiguously the Armageddon enemy, with their fiery red, dark blue and sulphur-yellow clad army. This confrontation is apparently shaping up in Syria right now and so this seal / trumpet sequence seems a little at odds with Ezekiel and Revelation 16.

This is simply because the end-of-the-age earthquake is being described in the Sixth Seal, rather than in the trumpet sequence of the Seventh Seal where one would expect to find it.

It also appears that the events of trumpets 1 to 4 have already been fulfilled, or are underway. To be blunt, when we compare this sequence of events with that from Revelation 16 and the 'seven bowls' of judgement described therein - or indeed with the Ezekiel 38 and 39 passages - it gradually dawns that the events in chapters 6 to 9 seem a little out of sequence. Even if the 'trumpet' sequence has not begun, why would this humungous and unique event be included as a Sixth Seal, rather than part of the Seventh?

Given these events are occurring rapidly and appear intertwined to a large degree, this might not be significant on its own. It could simply be suggested that the sixth seal *as well as* the seventh, should be grouped together and that the monumental earthquake event described as the sixth seal in Revelation 6:12-17 and the 'great day of the wrath of the Lamb' are part of the final destruction and judgement of the last Beast Kingdom.

Put slightly differently, perhaps the 6th seal battles, signs in the sky and natural calamities begin much earlier than we might imagine; in fact even centuries earlier than the present. It potentially means they could include the World Wars and the earlier 7th seal trumpet events. This would be consistent with the idea of a specific opening for each seal, but then continuing to the climax at the end of the age. If so, it behoves us to look further at the trumpets within the seventh seal, and see if there is evidence to suggest whether they have already begun. We might even spot a half hour break (Rev 8:1) before the trumpets begin.

The fifth trumpet (1st woe) I believe, is one which lends strong support for the 'already begun and happening' case.

To support this view, the fifth trumpet (1st woe) is apparently describing modern weaponry, which has been in use for well over one hundred years in our technological age. The wars of our industrial age and that of any over the next few years, have been described by prophets such as Ezekiel, who talk about 'living things' in action. The living things are variously described as cherubim (figures adorning structures and furniture in ancient Babylon) locusts, scorpions, horses, lions or some other familiar, but recognisable object or animal of the prophet's day.

In other words, I have a distinct sense the 'living things' describe war machines of our industrial age that the prophets see in their visions, i.e. tanks, rocket launchers, attack aircraft, helicopter gunships and the like - and often in formation. They may also be unmanned robotic vehicles, including drones. All modern vehicles are 'living' to the prophets, because they are self-propelled, autonomous and moving.

Now while there is seemingly a strong case for this interpretation, it is also possible that some images might be describing alien, or angelic beings and vehicles of unknown origin.

It is also instructive to track down some of the earlier references, e.g. Ezekiel 1 and 10 and imagine trying to describe modern weapons and firepower that you saw in a vision - and to attempt to do this without the benefit of 20th or 21st century words. Indeed, without an understanding of what these machines are, or how they operate. And so in Revelation 9:1-10 (the 5th trumpet / 1st woe) there is just such a description of locusts which have the 'sting' of scorpions, and we are also told that they will torment people for 'five months'.

By using the 'day for a year' prophetic principle with which we are now thoroughly familiar, 5 months is about 150 years. The question is, when should we start a count; from World War 1 - perhaps sometime late in the 19th century?

This uncertainty may extend the biblical timeline out beyond 2030-31, although in my view, it is most likely that the 'five months' (150 years or 3 jubilees) 1st woe is a further pointer to Daniel's $1,335^{th}$ 'day'. It may therefore be meant to be linked to the jubilee years; 1867, 1917, 1967 and 2017.

Given the Daniel timeline is the basis for this prophetic analysis, there is every likelihood that the 150 year period begins ca. 1867 when the *first terrifying industrial age mechanical monsters of war* were developed; weapons such as the ironclad warships which were utilised in the American Civil War. They were dramatically different from the sailing ships of earlier periods, even those of the Crimean War a decade earlier.

The take home message here, is this is happening in our day, and some of the trumpets (particularly No. 5) overlap the rest. Seal 6 and trumpet 6 appear closely connected, but with a 'half hour' (50 year) break between the Napoleonic Wars and 1867.

If this is the case, we have arrived at the end of the 'five month' period right now (2017) and have an extension of seven years until the beginning of Daniel's 1335th 'day' and 2024. Of course, not everything is strictly sequential and some elements are spread out over time more than others. The full sequence of 'trumpet' judgements within the 'seventh seal' is:

1. A third of the earth is burned up, a third of trees, and all green grass destroyed from' hail and fire'
2. Something like a huge mountain is thrown into the sea and a third of it turns to blood, a third of all sea creatures die and a third of ships are destroyed
3. A great star falls from the sky on a third of the rivers and springs and the name of the star is Wormwood and a third of the waters turn bitter and many die
4. A third of the sun is struck, a third of the moon and a third of the stars, so that a third turn dark (obscured?)
5. The sun and sky are darkened by the smoke from the Abyss and out of the smoke come locusts and scorpions which are given power to torment people for 5 months (the 1st woe)
6. The four angels are released at the Euphrates to kill a third of mankind. They have myriads of troops and horses with heads of lions and stings in their tails – again descriptions of modern weapons (the 2nd woe)
7. The mystery of God is accomplished, but the seventh trumpet is sealed up and not written down (the 3rd woe)

I accept that these trumpet events are yet to be fulfilled completely, but also acknowledge that the impending end of the Seven Times timeline, together with the horror of the 20th century wars, suggests that much of the above has already occurred. My tentative view is therefore; that trumpets 1 to 4 have occurred, or are in progress and trumpet 5 (as described above) is drawing to a close:

o A third of the earth being burned up is a probable reference to the World Wars and all wars since. Both wars were catastrophic and the destruction could well be a picture of Europe and the Pacific (including Egypt, Syria and Israel) and easily covered a third of the globe
o A third of ships sunk are another feature pointing particularly at World War II, and this trumpet may also include the general destruction around the world

- A third of all sea creatures dying is surely attributable to a combination of war, ocean and river pollution, unsustainable development and over-fishing
- The reference to 'wormwood' is highly likely to be of nuclear warfare, nuclear accidents and atomic testing. It has been noted that the name Chernobyl is a subspecies of mugwort plant and means wormwood in Russian
- A 'third of the sun, moon and stars being struck' I suggest is a reference to the smoke of burning cities and oil installations. It could date from the World Wars and certainly depicts more recent events in the Middle East. The imagery may also relate to the deaths of the saints (the stars) and the antichrist armies (sun and moon)

This last dot point has been specifically linked to the First Gulf war of 1990-1991 and the burning oil (from the abyss) to the installations which were such a feature of the Kuwait war. This, by at least one commentator who has noted that the war lasted a month or two longer than a literal 5 months and that helicopter gunships (locusts) were a major part of that conflict.

I tend to agree that the vision that John saw could have been from an episode like this, and that attack helicopters - a major feature of war since Vietnam - are almost certainly the 'locusts' he saw. In addition, four months after the end of the war, Mt. Pinatubo on Luzon Island in the Philippines exploded causing a massive amount of ash to circle around the world. This event cut light dramatically for months, and in northern Australia sunrises and sunsets changed from orange to a deep magenta, like some primeval harbinger of doom - an unusually pretty sight which masked the uncertainty and apprehension.

If a literal 5 months rather than 150 years is intended, I would expect that it would be a Day of Wrath scenario where the four angels are released at the Euphrates (Armageddon). It would make the trumpets an expanded version of the 6th trumpet (2nd woe) Day of Wrath events. This seems illogical and all I can say in the absence of any specific inspiration; 'I think not'.

Now it may be that the 5th trumpet will be a template of sorts for the events of the 6th, albeit the 6th trumpet (Day of Wrath) will be compressed into seven years. At the very least, the 150 year period will be a precursor of what is to come when the bowls are poured out on Israel's enemies.

Regardless; the 'seals and trumpets' are general prophecies for Israel and the nations, but also include the 'bowls of wrath' which are specifically poured out on the inhabitants of the last Beast Empire. This suggests that numerical imagery has been prioritized to reinforce the message. The 6th seal, 6th trumpet and 6 general bowls of wrath up to and including Armageddon, describe the fall of the satanic Antichrist, Gog and Magog forces whose number is 666. The allusion to the dimming of the stars (the Saints) and the 'sun and moon' (the Islamic powers) is even more compelling and refers to the gradual demise of the Ottoman Empire from well before World War 1. This, as the influence of the Caliphate waned and Islam was driven from Europe and the Holy Land at enormous human cost.

The fate of the Whore City of Mystery Babylon will be discussed shortly, but in regard to our Sixth Seal conundrum, I think the answer goes something like this: earthquakes and wars have always occurred, but this last calamitous sequence is so destructive and demanding, that it is like nothing this earth ever experienced in earlier historic times. The 'falling stars' are also real world objects, whether aircraft and satellites, or indeed meteorites, asteroids or comets and they initiate the mega-disaster.

It means the impact of the events toward the end of the age are such, that the sixth seal has been treated as a singular event and 'seal' in its own right. However the tide turned against the Gog and Magog forces well prior to the trumpet sequence.

Woes and wonders

Before analysing the 'trumpets' and 'woes' further, it is instructive to introduce Revelation 11 which describes events in Israel and the topic of the 'two witnesses' in detail. We see a '2nd and a 3rd woe' described in this chapter as well, with the inference that there is also a 1st woe, although it is not specifically numbered.

In the earlier verses of Revelation 11 we find a description of 'two witnesses'. Who are these two witnesses; are they individuals, or are they imagery with a broader interpretation?

In fact chapter 11 acknowledges in summary what we have been considering; that the Gentiles will trample the Holy City for 42 months (verse 2). It then says; verse 3:

'And I will give power to my two witnesses, and they will prophecy for 1,260 days, clothed in sackcloth'. (NIV)

So in just two verses we see the use of two different ways of describing a 'time, times and half a time', i.e. 42 months and 1,260 'days'. In each case of course, it means 1,278 years, or one half of the tribulation. These descriptions and numbers now come from the Koine Greek and explains the new terminology.

During this time, the two witnesses prophecy in sackcloth because of their often desperate plight. The witnesses nevertheless prophecy with great power and impact, and this I believe speaks to the determination and faithfulness of the saints and the power of God available through prayer. It demonstrates the profound sentiment articulated by Alfred Lord Tennyson; 'More things are wrought by prayer than this world dreams of'.

Now if the reader is under any illusion that the period described in verses 2 and 3 is *only* a literal three and a half years, it must be dispelled here. Verse 4 expands on their identity:

'They are the two olive trees and the two lampstands, and they stand before the Lord of the earth'. (NIV)

From verse 5 we find that the two olive trees are identities with supernatural protection in former times. If anyone harms them they will die, they can shut up the rain at will and send every kind of plague on their foes. They appear to be regular super heroes. Nevertheless, we later learn that the two witnesses, 'the two olive trees and candlesticks' are killed and lie dead in Jerusalem for three and a half days. We are then told people from every nation gather around and celebrate their demise.

So the narrative first describes how the two witnesses are overpowered and killed in Jerusalem and then follows up with; for 'three and a half days' men from every people, tribe, language and nation 'view their bodies and refuse them burial'. But then things go terribly pear-shaped for the onlookers, who have shown absolutely no compassion whatsoever toward the witnesses - everything is suddenly and shockingly reversed:

'But after the three and a half days the breath of life from God entered them, and they stood on their feet, and terror struck those who saw them' Revelation 11:11 (NIV)

Given the two timescales of forty two months (1,260 'days') and 'three and a half days', who are we to understand these two witnesses to be? And does 'three and a half days' mean the same as 'three and a half times' (or 1,278 years) as defined in this commentary in verses 2 and 3? Alternatively, is the much shorter period of 'three and a half' days in verses 9-11, meant to be interpreted according to the 'day for a year' rule?

It is pretty obvious the description of the two witnesses is stylised and figurative to a considerable degree. Therefore we must be cautious in our evaluation and not take the imagery too literally. We can be all 'fundamentalist' and stand on our own narrow interpretation of prophecy, or we can follow the prophets and the Lord's guidance and realise there is code here for years, and sometimes there is imagery instead of bald facts.

So 'when they arise' is a spectacular metaphor for the resurrection of the State and Tribes of Israel in their own land. When it occurs, everybody - and no doubt particularly their neighbours – are shocked and filled with fear.

Addendum: *In fact the vision appears relatively simple, but originally I assumed that the 1,260 'day' / 42 months was the second half of the tribulation,* **when in fact it is the first half.**

The verses that really give it away are 1 and 2 and the words: 'Go and measure the Temple of God and the altar, and count the worshippers there. But exclude the outer court; do not measure it, because it has been given to the Gentiles.'

This is typical of the first half of Jerusalem's tribulation, where the Jewish priests still had control of the altar (and Holy of Holies before it was destroyed) but not after the abomination of desolation, when worship was banned and most believers had already fled the Holy Land ('lay 'dead' in the streets').

This resurrection of Israel in their own land was prophesied in Ezekiel 37 with almost identical language: 'This is what the Sovereign Lord says to these bones 'I will make breath enter you, and you will come to life'.' If we then consider the history of Israel and the way the kingdom was split into two after Solomon's time, the prophecy becomes easier to interpret. The two olive trees are of course Judah and the Ten Tribes and the two candlesticks are their dedication to faith and worship.

The Revelation 11 vision has other support as well. It follows the vision of Zechariah 4, which describes Israel as a lampstand with seven lamps and two olive trees on either side.

The two candlesticks can also be viewed as two separate expressions of the faith, although it is not a necessary interpretation. Still, it has been expressed as such in some circles and I offer it as a corollary to the 'two olive trees' interpretation. One faith system is traditional, Mt Sinai and Torah centred, the other Messiah and first century focussed. In their pure form they are both based on the Holy Spirit (Shechinah) experience and worship of God. Of course the reality in both cases is far from perfect, but a broad distinction is recognised by at least some scholars who suggest that Judaism is the faith of the Jewish Diaspora, whereas Christianity is broadly the faith of the more numerous Ten Tribes and their associates.

Now at this point I have gotten a little ahead of myself, because not only have we been told that the 'two witnesses' will prophecy in sackcloth in the first half of the tribulation, but in verse 7; 'when they have finished their testimony, the beast that ascends from the bottomless pit will *'come up from the Abyss and overpower them...'* This begins in the second half of the tribulation, but because the witnesses are rendered as 'dead' for three and a half 'days' instead of 3.5 times /1,260 'days', is this the first hint that there are two intertwined interpretations here?

That 'the beast' ascends from the 'abyss' at the Islamic 'abomination', but that Satan is 'unloosed' in all his fury at the end of the age? Previously, I have been open about my admiration for the prophet Ezra, and sure enough, he also addresses this sequence of events. His prophetic timeline starts ca. 420 – 400 BC as 2 Esdras 7 indicates, and begins with the comment that 400 years will elapse until the Messiah appears. He then describes some of the trials in the end-time, where the 'world will be turned back to primeval silence over seven days'

There is little doubt that Ezra's the 'world will be turned back to primeval silence' is a description of the 6th seal and 6th

trumpet. It lines up with the 2nd woe in Revelation 11; the earthquake that hits Jerusalem where 7,000 residents die. This fixes Ezra, the trumpet and seals and the Revelation 11 earthquake as the same, or contemporaneous 2nd woe events.

From this point on, we are told by Ezra that 'the kingdoms of the world have become the kingdom of our Lord and of his Christ'. His next statement 'the end of the age, the rescue of believers and the final judgement of all', ties in with the 7th trumpet (3rd woe) of the trumpet sequence. This is the one which is sealed up and not revealed in Revelation chapter 10.

Significantly the sealed up 3rd woe *is* revealed in Revelation 11:15, and the whole 'wrap up' picture is dramatically described in Revelation 14 as well:

'The seventh angel sounded his trumpet, and there were loud voices in heaven, which said:
The kingdom of the world has become the kingdom of our Lord and of his Messiah, and he will reign forever and ever.' (NIV)

It all logically follows the Armageddon war and Jerusalem earthquake (2nd woe) after which, God's temple in heaven is opened - it is the final judgement. And so the 3rd woe is finally described and confirmed toward the end of Revelation 11 - and it is exactly as it is portrayed in Ezra's vivid sequence of events.

Revelation 11 only makes mention of the 2nd and 3rd woes by number, implying there was a 1st woe. I believe that 'woe' is addressed in the previous section, as the 5th trumpet.

Moving on, I suggest that the three and a half 'days' can be seen to refer to the whole of the second 1,278 years as an overarching fulfilment of the prophecy. This is because it states that the witnesses lie dead in Jerusalem, and few of those who comprise the two witnesses, existed in Jerusalem under the Caliphates. So the 'resurrection' description is very appropriate and this interpretation then follows the earlier verses of the chapter (verses 1-3) where the whole of the second 'time, times and half a time' of Islamic domination is implied. In this case, a 'day' in Revelation 11:11 then describes a 'time' and not a year.

Nevertheless it is tempting to think, that this part of the prophecy also follows the regular 'day for a year' rule as well and describes when Satan is released from the 'abyss'. It would then sadly, *represent the Holocaust and the horrific three and a half year period* of the mid 1940s during World War II.

This interpretation then parallels trumpet 1, but also trumpet 5 (the 1st woe) where sophisticated industrial age military hardware, appears at the beginning of the 20th century. This latter identification, then gives direct one for one correspondence to a close degree, between the three woes of the 'trumpets' in Revelation 8, 9 and 10 and the three woes of Israel in Revelation 11.

It means: what has evaded scrutiny in the past, has now been revealed. That **the Holocaust is very much a part of the prophetically described events in the Revelation to Saint John!**

If dual interpretations of a prophetic 'day' are required to do justice to this revelation, it is indeed something special.

So at the end of God's appointed time of tribulation, we are told they 'go up to heaven in a cloud'. Then the remnant of these two kingdoms of old 'come to life' in modern Israel and rejuvenate a neglected land. We also know they have brought blessing worldwide through the greater Diaspora. Revelation 11 therefore describes the tribulation, the dire end in World War 2, the Jerusalem earthquake and the final Judgement.

Of course, the Holocaust did not actually happen in Jerusalem and so it follows that both the second 3.5 'times' of Islamic control *and* the ordinary 'day for a year' prophetic rule, are both required to fully appreciate Revelation 11. However Babylon the Great is described as the 'Mother of Prostitutes' and Nazism, Fascism and Communism were surely her daughters at the end of the tribulation.

As we close in on the end-of-the-age prophecies, it is also appropriate to address the prophecy of Revelation 16. This chapter ultimately introduces the Whore City of Babylon the Great and the 8th kingdom of Revelation 17 and 18. It is therefore desirable to discover whether Revelation 16 is also a parallel account to the others, or a more condensed vision of the end-time. The seven afflictions outlined in the chapter, are usually described as the 'Seven Bowls of God's Wrath' or 'Vials' which are heaped on those that 'have the mark of the beast':

- They will be afflicted with ugly and painful sores
- The sea turns to blood
- The rivers and springs of fresh water turn to blood
- The people are scorched by intense heat from the sun
- Yet the 'beast' kingdom is plunged into darkness
- The Euphrates River 'dries up' so the kings of the east can invade Syria
- The 'great city' is split and Babylon the Great destroyed by an earthquake where even islands vanish

Given that Babylon the Great is the principal entity addressed in the last 'wrath', it follows from previous analysis, that the bowls of wrath are primarily directed at the last 'beast' kingdom, Islam. It will therefore include all the entities attempting a resurrection of the Caliphate, or supporting its ideology.

But where does this sequence of disasters begin? Does it approximately follow the 'seals' and start early in the age; say when Jerusalem was finally overcome by Caliph Umar leading up to the 'abomination of desolation'? Or is it closer to the trumpets which I suggested began prior to World War 1? Or indeed, is the sequence restricted to the seven short years of the 'Day of Wrath' according to Lesson 10 and the definition of a prophetic milestone? In fact, the latter definition was one I admitted might also apply to the 'trumpet' sequence and it is quite possible that multiple interpretations apply anyway.

So although it is possible that the first 'wrath' also applies to the 'Black Death' plague of 1348, and the Euphrates drying up to the 11-14th century invasion of Turkey - or indeed some epidemic prior to, or associated with World War 1 (the Spanish Flu Epidemic for instance) I am inclined to believe that the 'bowls' are primarily meant to cover the seven years of the end-of-the-age 'Day of the Lord'. In fact the 'Day of the Lord' is also described as the 'Day of Wrath' elsewhere and therefore the *'wrath'* terminology tells us all we really need to know.

These are judgements aimed at those with the 'mark of the beast' and it seems the 'bowls' are now at our doorstep in the Middle East and about to be unleashed. The 'Euphrates drying up' then describes the Shia invasion that ends in Israel and at

Hamon Gog in Jordan, together with their confederates who are north and east of the Euphrates and moving west this very hour.

Apart from Ezra, there is also support for a seven year Day of Wrath in Ezekiel 39:1-16. This is a prophecy against Gog and its fate in Jordan, where Israel will have enough fuel for *seven years* from the spoils of Gog's destruction. Now this would seem to be seven years toward the end (or after) the actual Day of Wrath, but goes on to say that Israel will also take *seven months* to bury the dead and thereafter will still be looking for bones and cleaning up the horror left behind. This 'seven months' may, or may not be literal, but if not, will probably follow some prophetic fraction of the Day of Wrath period.

Last but not least, the Revelation 11 interpretation above is very likely to impact the way we view Revelation 20. In fact Revelation 11 definitely *does* shed light on chapter 20, but the reverse is also true. I have already hinted at this comparison and it is now time to deal with it conclusively. The choice is between interpreting the chapter as a future millennium, or as a parallel version of our current age - including the 3.5 times of the Babylon the Great occupation of Israel.

I will present Revelation 20 in a similar manner to the summaries in the previous chapters, but implore the reader to follow the chapter in as many translations as possible. The difficulties presented by the phrasing will then become apparent:

- The dragon (Satan / Devil) is bound for a thousand years. After that he is released for a *short time*
- Those that have been beheaded and not worshipped the beast, reign with Christ for the thousand years
- The rest of the dead come to life at the end of the thousand years
- Satan is loosed from his pit / prison at the end of the thousand years and gathers Gog and Magog to battle
- They come from all over the earth and march against the saints and the beloved city, where fire consumes them
- The dead are judged according to their deeds, and if not in the 'book of life' are thrown into the lake of fire with the beast and false prophet

My journey with this chapter has been tortuous. It is a very difficult chapter to interpret unambiguously and has been a vexed question for many. To reiterate, the 'millennial' period in chapter 20 has been described as a time of peace where the Messiah reigns on earth after the end of our current age; meaning after the Seven Times of tribulation and the Second Coming. There are several passages in the O.T. / Tanakh which are used to support this view of a future 'millennium', rather than a parallel fulfilment of the latter part of the Seven Times and Church Age. They are images of peace; such as the lamb lying down with the lion and other heavenly promises. However these prophecies **never suggest** that this time is 1,000 years in length, or is followed by Satan's release **after** a time of relative peace.

In fact it makes no sense as part of the Redemption Plan!

Fortunately, others believe the '1,000 years' described is concurrent with our present age and broadly corresponds with the Church (Pisces / Messianic) Age, or part thereof, and I believe the Crusader theologians were of this opinion. There also appears to be no direct support for the concept of a future 1,000 years in the New Testament, other than what we find in Revelation 20. In some versions of the New Testament, Revelation 20 begins with 'And' and others, 'Then'. However both 'And' and 'Then' can simply be read as meaning the next vision John was given; one that summarises the previous visions he received.

Revelation 19 begins where chapter 18 leaves off, and that is after the destruction of the Whore City at the end of the age.

When we study the ending of both Revelation 19 and 20, we see that in the first case it focuses on the destruction of the 'beast' and 'false prophet', while in the second (chapter 20) the focus is on the destruction of those whose names are not found in the 'book of life' - *together with* the 'beast' *and* the 'false prophet'. In both instances they meet a fiery end and therefore, this is the next indication (after the 'three woes' correspondence) that the verses toward the end of Revelation 20 (after the 1,000 year period is finished) are a parallel account with the last verses of both Revelation 19 and Revelation 11.

Of course not everything has been fulfilled, but equally, we are not talking about a short period, but an Age.

However it has to be said, Revelation 20:4-5 could easily be seen to give a different point of view, at least superficially:

4.'I saw thrones on which were seated those who had been given authority to judge. And I saw the souls who had been beheaded because of their testimony for Jesus and because of the word of God. They had not worshipped the beast or his image and had not received his mark on their foreheads or hands. They came to life and reigned with Christ a thousand years.

5. (The rest of the dead did not come to life until the thousand years were ended.) This is the first resurrection.' (NIV)

While I have not had any dramatic inspiration for my views on Revelation 20, I am absolutely convinced that I have been lead to see Revelation 20 as a fulfillment of all, or part of the Church Holy Spirit Age. Having said that, a short discussion on which 1,000 years is being referred to and whether there is a coincidence of prophetic periods (or indeed what type of resurrection is being described) is appropriate. Therefore a question to be asked is: Are the resurrections referred to in Revelation 20, bodily resurrections, national resurrections (as with Rev 11) spiritual (as people pass away) or figurative only?

Some people are taught 'soul sleep', but Jesus' parable of the Rich Man and Lazarus in 'Hades' hints at the souls of the deceased waiting for their ultimate fate. Parables are sometimes dismissed as being fictional, but others believe they were examples of the day that most people knew of. However this parable is not something that anyone other than a prophet could possibly know about. Fortunately the 5th Seal of Revelation 6:9-11 vividly describes souls waiting in the heavenlies during the tribulation (and our age) and thus gives support to the reality of the parable and the consciousness of the saints after death.

In regard to the national resurrection of the nation of Israel as described in Revelation 11; this does not appear to fit as an alternative way of describing the first resurrection of Revelation 20, because of the timing. (The resurrection of the nation would have to be 1,000 years before Gog and Magog appear.)

Nevertheless it does not preclude the second 3.5 times of their tribulation and subsequent 'resurrection' being in parallel with the 1,000 years we are focussed on.

The souls who are beheaded could also include the saints of the Old Testament period leading up to the crucifixion and the resurrection. That means they may have been resurrected (or at least have passed away and now live with the Lord).

This is hinted at in Matthew 27:51-53, but in line with the concept that Jesus was the first resurrected and from that time onwards, those that pass away have followed. Verse 5 immediately follows with: 'the rest of the dead did not come to life until the thousand years were ended', or similar rendering, confirming that the resurrection period is constrained to the 1,000 years and that most (or all) from the Pisces Messianic Age will not receive their incorruptible bodies until Jesus returns.

In verses 1 to 3 we learn that the dragon is 'bound for a thousand years' and thereafter is let loose for a short while to deceive the nations. The imagery in Revelation 11:7-10 is similar, where we are told 'the beast that comes up from the Abyss will attack them (the two witnesses) and overpower and kill them'. However, as previously noted, the verses of Revelation 11:7-9 appear to be more about the 'beast' Islamic conquest of Jerusalem, than an end of the age prophecy. So although the 'beast', 'serpent', 'dragon' and Satan are identified as the same entity in some contexts, here in Revelation 20 we are specifically told it is the Devil / Satan rising from the Abyss.

It suggests that Satan is being released at the very end of the Seven Times in line with the suggested three and a half 'day' secondary Holocaust interpretation of Revelation 11:8-11.

He is also released for a 'short time' only, in a similar fashion to the 'little while' and 'one hour' that the Ten Kings reign in Jerusalem after the fall of Rome (Revelation 17). That period was in the vicinity of 100 years and so in this instance could be seen to begin as early as the 20th century. So although 'the thousand years' could be seen as shorthand for 1,278 years (the second 3.5 times) it is likely referring to the *whole* 2,000 years.

It means despite tribulation, wars, famines and diseases (the seals) the **Lord reigns over the earth** *and Satan is bound* ***throughout the Church Age / Holy Spirit Age / Messianic Age of Pisces of our Lord Jesus Christ (Yeshua Mashiach);*** *this, as people are saved and refined through the trials of the age.*

This analysis reinforces the view that *the thousand years* is in parallel with the Church Age and corresponds to it. Together, Revelation 11 and 20 imply, that despite 'tribulation' and the occupation of the Holy Land and Middle East by the 'beast', that Satan's shackles are gradually loosened beginning early in the Islamic era....but not thrown off completely until the end of the Seven Times. At that time he is figuratively 'released' by the angel and allowed to do his absolute worst.

This was during the Holocaust and it hurts even to meditate on that diabolical time in World War II.

In the decades since, there have been all sorts of wars and any number of genocides throughout the world. It begins drawing to a close when the 'foreign fighters' i.e. Gog and Magog (Revelation 20:8) are deceived and gather in the Levant against the saints. *We have seen this unfold before our eyes in the form of **Islamic State / ISIL** and the prophecies indicate this will gather momentum on a national scale in the immediate future.*

We are then told that the fate of Gog and Magog (those with 'the mark of the beast') are sealed and they are destroyed by 'fire from heaven'. The Devil that deceived them is then thrown into the lake of fire with the 'beast and the false prophet'.

In summary; verses 7 to 15 of Revelation 20 are very similar to the final destruction in Revelation 16, 17 and 18 and lines up with the judgement prophecy of Revelation 19 as well. All these prophecies end in the 'day of wrath' or the judgement thereafter and tell a similar story to Ezekiel, Zechariah, Esdras and Revelation 19; experience teaching that if an episode is important, it will be emphasised in a number of prophecies in scripture.

In addition, the crucial prophetic monologue by Jesus, who as the Son of Man orchestrated all the prophecies in scripture, gives support to this view.

In Jesus' Olivet discourse in the gospels, he begins with the imminent destruction in Jerusalem, mentions the abomination of desolation of the Temple Mount part way through, and ends his description at the end of the age when the Son of Man returns and his sign is seen in the sky. Those events are then followed seamlessly by the rapture and warning of impending judgement in the following parable.

He specifically mentions the abomination of desolation of AD 689, which is the start of the last tribulation period, and so you would expect a mention of another millennial age and second judgement event, at a landmark time in the future. That is, if there was still one to come. **But there is absolutely nothing.**

In contrast, the picture painted later in Revelation 21 of the New Jerusalem, is a picture of the resurrection promises of the afterlife and is unambiguously placed after the earlier events.

In the end, the reason for closely comparing the passages above regarding an Armageddon war, apocalyptic earthquake and the resurrection, is to ensure with as much confidence as possible, that the events that are yet to be fulfilled are indeed on our doorstep and not centuries away.

Celebration in the sky

I have discussed the measurement of the ages using the precession of the earth, and how the rising of a new zodiac just before dawn was, and is, an indication of the passage of time over approximately 2,000 years. We also discovered that from approximately 12,000 years ago the signs were Leo, Cancer, Gemini, Taurus, Aries, and Pisces. It is also fairly common knowledge in Christian circles, that the letters which make up Pisces, 'the fish' are an acronym in the Koine Greek language for 'Jesus Christ, God's Son, Saviour'.

This is not the make believe world of astrology, but the real world of astronomy. It was the way the ancients measured long periods of time, but also a method, whereby God has kept his purposes and message in the mind of his chosen. We have also deduced from this, that when the Bible prophets talk about an 'age', they are referring to a zodiacal age and that most understood this including Jesus. Certainly Ezra counted the ages that way and in fact, divided the age in which he lived (Aries) up into 12ths and indeed a half of a twelfth. I have also suggested the Hebrews may have used the method before the Greek period and as far back as Melchizedek. In Matthew 24:30 Jesus said:

'At that time **the sign of the Son of Man** will appear in the sky, and all the nations of the earth will mourn. They will see the Son of Man coming on the clouds of the sky, with power and great glory'. (NIV)

This passage introduces the next age, the sign of the Son of Man. Of course it can only be the Sign that comes after Pisces, and is therefore ***The Water Carrier, Aquarius.***

It has been noted that many, in fact most of the visions of Daniel and John, encompass huge amounts of time and are rarely based on a single event. In Revelation 12 we have another overarching summary of Israel's journey. It has elements of the preceding chapter 11 and covers similar ground to Jesus' Olivet discourse. The difference is that it is presented in pictorial form.

We see the image of a woman (the bride) representing the people of Israel, and all grafted in believers from every nation on earth. This promise for Gentile believers of being 'grafted' *into* Israel was recognised by the first century Apostles when they met in Jerusalem. The Apostle Paul came all the way back from Galatia to participate and the consensus was; 'if the Shechinah has fallen on the Goyim, who are we to exclude them?'

We also see that the narrative describes the whole age, from the time of the giving of the prophecy in the first century, through the great Diaspora thereafter. It is also a rather beautiful picture, even though we understand that it describes very troubling times and a flight from the danger at home, to the wilderness of the four corners. Revelation 12:

' A great sign appeared in heaven: a woman clothed with the sun, with the moon under her feet and a crown of twelve stars on her head.
*² She was pregnant and cried out in pain as she was about to give birth. ³ Then another sign appeared in heaven: an enormous red dragon with seven heads and ten horns and seven crowns on its heads. ⁴ Its tail swept a third of the stars out of the sky and flung them to the earth. The dragon stood in front of the woman who was about to give birth, so that it might devour her child the moment he was born. ⁵ She gave birth to a son, a male child, who "will rule all the nations with an iron sceptre."[a] And her child was snatched up to God and to his throne. ⁶ The woman fled into the wilderness to a place prepared for her by God, where she might be taken care of for **1,260 days**.*

And:

¹³ When the dragon saw that he had been hurled to the earth, he pursued the woman who had given birth to the male child. ¹⁴ The woman was given the two wings of a great eagle, so that she might fly to the *place prepared for her in the wilderness, where she would be taken care of for a **time, times and half a time**, out of the serpent's reach. ¹⁵ Then from his mouth the serpent spewed water like a river, to overtake the woman and sweep her away with the torrent. ¹⁶ But the earth helped the woman by opening its mouth and swallowing the river that the dragon had spewed out of his mouth. ¹⁷ Then the dragon was enraged at the woman and went off to wage war against the rest of her offspring—those who keep God's commands and hold fast their testimony about Jesus.* (NIV)

There are some things about the image that should provoke a glimmer of understanding, now that we have studied so many visions.

The Woman is clothed with the sun, but has the moon under her feet. She has a crown of 12 stars on her head and she is about to give birth. In this image we can interpret the sun as the righteousness of God lighting the way for the people of Israel, no matter where they are, or what they have been tasked to do. The twelve stars represent the tribes of Israel, and similar imagery describing the family is first seen in Genesis 37:9-11, where Joseph describes a dream, where his parents and siblings (the sun, moon and stars) bowed down before him.

The 'moon under her feet', represents Israel's struggle and eventual overcoming of all that is Babylonian, including the idolatrous moon god and all that the Statue of Nebuchadnezzar, symbolises.

The red dragon represents Lucifer, the Devil, who from the beginning in the Garden, has been intent on preventing the final salvation of the nation and the world. The dragon also symbolises the beast kingdoms of Revelation, which are described as having seven heads (or crowns) and ten horns. It means that in this image, the Serpent personifies the oppressive empires *from the first three and a half 'times'*.

He tries to destroy the woman, and also devour the male child she gives birth too, before fleeing into the desert for 1,260 'days'. The desire of the red dragon is to remove the last shred of allegiance to the Almighty, our Abba (Father), God of Creation and substitute the worship of 'the Man'; the anti-messiah represented by the number 666 - and ultimately Satan himself.

This scene therefore begins back in the days when the Empires which constitute the composite 'beast' of Revelation 13:1-10 existed – Daniel's kingdoms between 590 BC and 689 AD.

As a consequence of these identifications, and because the woman flees to the 'wilderness' for a complete 'time, times and half a time', we can be sure that this picture begins in the first 'time, times and half a time', but describes all the second and the great Israelite diaspora around the world.

It is noteworthy that this prophecy pinpoints **the appearance of the Jewish Messiah** to the first 'three and a half times'.

And of course the second 1,278 years is the time when Jerusalem was occupied by the Islamic Caliphates.

Although there are many countries that could be referred to as the 'wilderness'; including much of western and northern Europe in the early Celtic days, I think it fairly obvious that the great worldwide colonising days of discovery from 1492 onward, were the last wave and the most expansive. This period began when the Ottoman Caliphate was at its height and when Constantinople had finally been overcome. Therefore the description in chapter 12 is extremely apt and was previewed in verses in the Old Testament such as Micah 5:1-5.

This was the Habiru on the move again. Always searching, discovering; if not new lands, then new knowledge and inventions; science, technology, new ways of doing old things – never happy with the status quo, when something better might just be around the corner. But also exercising the old skills on larger scales; pastoralists, farmers, miners, water managers, resource explorers and gem dealers, physicians and teachers.

Is it too much to imagine, that another sign we have been given is that the largest of the new nations of the wilderness is symbolised by the Bald Eagle ('a great eagle' in the prophecy)? And the Woman representing Israel, also the Statue of Liberty? And are there unrelated signs which may, or may not be explicitly seen in scripture, which herald the end of the age, or the advent of the new?

And what, if any signs have we seen in the sky?

Between the Six Day War in 1967 and the Yom Kippur War of 1973, some momentous events were being enacted on the world stage outside of Israel. The Vietnam War was a protracted conflict brought on by the fear of a communist 'domino' effect in Asia, and part of the threat experienced since the closing of WW II. Many mistakes were made and many losses felt, but at the end of it all was a desire to halt the spread of the horrors seen elsewhere under Godless regimes.

But there was also the promise of new discovery and the first space exploration outside of earth's gravitational embrace. And so in the early 1960s, President Kennedy energised the nation on the back of its rocket program, to attempt to land man on the moon. That there were seven Apollo missions to the moon and six were totally successful (and even the technically troubled Apollo 13 mission brought its astronauts home safely) is I suggest, a testament to miracle.

The fact we can look back over almost 50 years and see that it is yet to be repeated (and that there are more than a few who disbelieve it ever happened at all) makes the whole Apollo program stand out as a divine, preordained marker. It is the closest thing to real life fantasy that my generation ever experienced. The blinding speed of advances in rocket and space technology throughout the sixties and the minimal computing power available to underpin the program, absolutely shows this was no ordinary accomplishment. Of course for America it was 'one small step for a man and one giant leap for mankind' and a genuine source of pride. But was it also the Lord's celebration in the sky at the end of the Seven Times of trial and judgement?

A celebration not only for the Jew, but also the Tribes and Goyim? Would it have been so dramatically successful if there was not the blessing from above? With the Apollo Missions accomplished and the moon under the astronaut's feet, how could there be any doubt for the believer?

The moon of Babylon was at last subject to a greater Israel; the free and democratic west. More than this, the idolatrous sun and moon emblems of Babylon the Great were cast down in the Holy Land, after that other miracle; the Six Day War.

CHAPTER 14

Whore on the Water

'This calls for a mind with wisdom. The seven heads are seven hills on which the woman sits. They are also seven kings.'
ONE of the SEVEN ANGELS

Back to the future

I will now build on previous subject matter, with further prophecies defining the main players at the end of the age. The 'bowls of wrath' poured out on the last Beast Empire in Revelation 16 have already been reviewed and now we continue with Revelation 17. This chapter gives us a picture of a great city at the head of the final 'beast' empire:

"One of the seven angels who had the seven bowls came and said to me, "Come, I will show you the punishment of the great prostitute, who sits by many waters. ² With her the kings of the earth committed adultery, and the inhabitants of the earth were intoxicated with the wine of her adulteries."

³ Then the angel carried me away in the Spirit into a wilderness. There I saw a woman sitting on a scarlet beast that was covered with blasphemous names and had seven heads and ten horns. ⁴ The woman was dressed in purple and scarlet, and was glittering with gold, precious stones and pearls. She held a golden cup in her hand, filled with abominable things and the filth of her adulteries. ⁵ The name written on her forehead was a mystery:

> **BABYLON THE GREAT**
> THE MOTHER OF PROSTITUTES
> AND OF THE ABOMINATIONS OF THE EARTH

⁶ I saw that the woman was drunk with the blood of God's holy people, the blood of those who bore testimony to Jesus.
When I saw her, I was greatly astonished. ⁷ Then the angel said to me: "Why are you astonished? I will explain to you the mystery of the woman and of the beast she rides, which has the seven heads and ten

horns. ⁸ *The beast, which you saw, **once was, now is not, and yet will come up out of the Abyss** and go to its destruction. The inhabitants of the earth whose names have not been written in the book of life from the creation of the world will be astonished when they see the beast, because it **once was, now is not, and yet will come.***

⁹ *"This calls for a mind with wisdom. The seven heads are seven hills on which the woman sits.* ¹⁰ ***They are also seven kings. Five have fallen, one is, the other has not yet come**; but when he does come, he must remain for only a little while.* ¹¹ *The **beast who once was, and now is not, is an eighth king.** He belongs to the seven and is going to his destruction.*

¹² *"The ten horns you saw are ten kings who have not yet received a kingdom, but who for one hour will receive authority as kings along with the beast.* ¹³ *They have one purpose and will give their power and authority to the beast.* ¹⁴ *They will wage war against the Lamb, but the Lamb will triumph over them because he is Lord of lords and King of kings—and with him will be his called, chosen and faithful followers."*

¹⁵ *Then the angel said to me, "The waters you saw, where the prostitute sits, are peoples, multitudes, nations and languages.* ¹⁶ ***The beast and the ten horns you saw will hate the prostitute.** They will bring her to ruin and leave her naked; they will eat her flesh and burn her with fire.* ¹⁷ *For God has put it into their hearts to accomplish his purpose by agreeing to hand over to the beast their royal authority, until God's words are fulfilled.* ¹⁸ *The woman you saw is the great city that rules over the kings of the earth."* (NIV)

This is another of the more difficult biblical prophetic visions to interpret. It compares to other passages which can only be unravelled with hindsight.

However the vision is even more shrouded, because the fate of the Whore City described here is still future. It means the location of this new and bigger Babylon, as Babylon the Great is not at all clear and needs to be established.

While our familiarity with all the imagery of the various beast kingdoms seems to point to 'Babylon, Persia, Greece, Rome, Ten Kingdoms' and their various capitals, the question of the identity of the Whore (or prostitute) City which rides the former empires remains a mystery.

Here are the salient points:

- The 'ten horns' are the kingdoms following the breakup of the Roman Empire and become the seventh 'head'.
- The Harlot city and eighth kingdom are going to their destruction by way of an enormous earthquake
- The description 'once was, now is not, and yet will come' in verse 8, is ultimately referring to the resurrected eighth 'beast' kingdom and also repeated in verse 11
- The 'beast' is an eighth king, but in John's time five kingdoms had passed, the sixth remained (which has to be Rome) and a seventh will exist for a 'little while'
- The 'ten horns' (kings) will give their power to the 'beast' and will wage war against the 'Lamb'
- The 'beast' and 'ten horns' will hate the Prostitute City of Mystery Babylon and eventually destroy her

Can the eighth king here be the beast kingdom which comes out of the earth in Revelation 13:11-18; the beast that mimics the power of the beast from the sea, and could be rightly described as a Whore 'riding on' that beast?

Absolutely! Mind you, if it was not for the earlier analysis of Daniel's visions and their interpretations, together with the revelation of the identity of the two beasts of Revelation 13, it would be nigh on impossible to know the identity and location of Mystery Babylon with any certainty.

But as established earlier, the 'beast out of the sea' can be interpreted in exactly the same manner as Daniel's statue of a man in Daniel 2, or indeed his many beasts in Daniel 7. It corresponds to the time of the early empires.

Unfortunately there is a reference in Revelation 17 which seems at odds with my previous assessment of chapter 13 - at least superficially - and this needs to be cleaned up. It also appears that a trap has been deliberately set for the foolish, and in all honesty, I have to put my hand up and say 'it fooled me'. Indeed we are explicitly warned in verse 9 that this passage is difficult and 'calls for a mind with wisdom'.

The verse 8 description appears to be saying that the 'beast' which the Whore is riding (as a whole) ' ***once was, now is not,*** *and yet will come'.* It repeats this twice, as if to make very sure that it is stuck in our heads and cemented in our minds.

Verse 11 repeats the message of verse 8, saying again '*The beast who once was, and now is not,* **is an eighth king. He belongs to the seven...**'. This is very significant, because it adds that the beast in question, belongs to the previous seven and so is not the composite beast that verse 8 suggests. It is simply one of the seven which comes back as the 8th king.

In the previous verse; verse 10, we learn that; *'The seven heads are seven hills on which the woman sits. They are also seven kings.* **Five have fallen, one is, the other has not yet come**; *but when he does come, he must remain for a little while'*. Here again, individual beast kingdoms are being described, rather than the whole composite beast and it explains that the prophecy is referring to a line of consecutive kings or empires. It is also clear, that they are the seven heads and ten horns on which the 'Mother of Prostitutes' is riding in verse 3.

Given these heads and horns have the same symbolism that is seen in Revelation 12 and 13, they must represent the same kingdoms. It follows that the 8th king is previously identified with one of the seven kings from the composite beast, but no longer exists when the prophecy was given. Since we know pagan Rome existed in the first century, verse 10 excludes Rome as the final and eighth beast kingdom and by inference, confirms that Rome is the sixth kingdom and current empire.

So verse 8 seems to be suggesting; that the composite beast (all seven kingdoms) no longer exists at time of writing and will subsequently be destroyed, but is at odds with verse 10 and also with verse 11 and 'he belongs to the seven'.

This is something of a conundrum, but the odds are stacking up against verse 8 being a literal proposition. So, is there more clarity to be had here?

If the 'beast' referred to in verse 8 is the beast represented by the Whore City (as verse 11 suggests) then this would not be a problem, but the phrasing of verse 8 does not make this clear.

However, this is where wisdom kicks in. Identifying any individual animal (beast) is the same as identifying a human. The part of the body *that most represents the entire individual is the head, unique body shape, size or special characteristic.* It is how we identify all creatures, *including those who were the beast kings.* If we then go to Revelation 13 and the beast out of the sea, we have a picture of an animal with a 'mouth like that of a lion' which is also a mouth that 'blasphemes God and slanders his name'. This sounds a little like the little horn of Daniel 7, but also represents the original Babylon kingdom, on which Babylon the Great is based. The kingdom where the philosophical roots for the new Babylon are to be found.

The lion's mouth actually began with the neo-Assyrian Empire, out of which ancient Babylon emerged. But it is also directly linked to Babylon by the Daniel prophecy where the wings are ripped off the statue (Daniel 7:4). It therefore identifies Babylon as the kingdom which was fatally wounded and which becomes the 8th beast kingdom, Babylon the Great.

This should not surprise; **plainly, the name says it all.**

This emphasises that verse 8 has been written in an obscure fashion, and needs to be viewed in the light of the Revelation 13:1-10 description.

If it had been made crystal clear in order for anyone to understand, it would have read: '*The **(mouth)** of the beast, which you saw, once was, now is not, and yet will come up out of the Abyss and go to its destruction'.* This would then have singled out *the specific kingdom* in the 'beast out of the sea' sequence (Babylon) and align seamlessly with verse 10.

Normally the 'head' would be the most recognisable body part as in Daniel 2, but unfortunately when you have a seven headed beast it becomes a more difficult identification.

The upshot of all this, is that prophetically, *the original Babylon and the eighth 'beast' kingdom 'Babylon the Great',* are treated as one entity and not two.

It has also been determined, that the Ten Kingdoms could be said to attain some prominence from as early as the sacking of Rome by Visigoth King Alaric in AD 410, or at least by AD 476 when Rome fell to the Germanic King Odoacer. Importantly though, under Emperor Constantine the empire and its power had already transferred to the east and Byzantium in the 4th century. This is of great prophetic significance.

Another question arises in regard to the terminology used to describe the short reign of the Ten Kings in Jerusalem - and how to date this. We are told the seventh 'head' must 'remain for a little while' in verse 10 and for 'one hour' in verse 12, and I have determined that this period somehow aligns with the short Eastern Roman (Byzantine) rule in the city.

If we compare an approximate 100 year period to the total 2,556 years of the Seven Times and *express the period in terms of a fraction of a day,* then it could be validly called 'one hour'.

I realise this sounds a little crazy, but it does not violate the general rule of prophetic periods of time. By default, Constantinople automatically morphs from the Roman Empire into the Byzantine, about 'one hour' before Islam appears in Jerusalem.

Furthermore, in AD 538 when the Bishop of Rome was designated *the* leader of the church under the Justinian Decree, it becomes plain that this general time could be seen to mark the beginning of the REAL split between the western Roman and eastern Orthodox churches. It can also be seen as the transition from a Roman Empire to a Ten King identity, and this is around the time when Emperor Justinian passed away in AD 565.

This is seemingly at odds with the understanding of the citizens of Constantinople, who back in the day, considered the Eastern Roman Empire lasted from 330 to 1453. Neither is it completely consistent with the view of modern historians.

It is therefore a big call on my part, but Justinian was the last Latin speaking Emperor, despite the official language changing in 620, and the officially recognised church split occurring as late as the 11th century. The Constantinople citizens may have imagined they were still the mighty empire of yore to the very end, but they were day-dreaming. Modern historians have a retrospective which reflects the reality more closely; their renaming of the Empire as 'Byzantine', unwittingly follows the biblical identity far more closely.

Accordingly, I have a sense, that this is the way history was portrayed to the prophets of Judah.

Whatever historical and dating option best fits the Ten Kingdom period in Jerusalem, verses 12–14 say the Ten Kings give their power to the beast. The question is which beast? Is it the beast with seven heads of Revelation 17 (the one the Harlot is riding) and the composite beast of Revelation 13? Alternatively, is it meant to be the 8th beast represented by the Harlot?

There is little doubt it is meant to refer *primarily* to the beast she was riding on, before the Islamic takeover of Jerusalem. That is the image portrayed and the one the Ten Kingdoms conclude as the seventh head. After the transition is fully implemented, Babylon the Great then 'rides on' the composite beast of Revelation 13:1-10, because it follows the earlier Empires and occupies Jerusalem. Constantinople then continues alongside the Islamic Empire until it is finally overcome by the Ottoman Caliphate in AD 1453 - and this occurred well after being kicked out of Syria and the Holy Land in the 7th century. As the power shifted to the Caliphates, compromises were made (beginning in Jerusalem) and therefore during the early Islamic period, the Ten Kings kowtowed to the Caliphates as well.

The way this saga unfolds; suggests that if the Byzantines were not the last of the three early kingdoms to be toppled in Daniel 7 (the ones which were suppressed) then they were directly in the firing line after the first three had fallen to Islam.

The fact that the Ten Kingdoms never achieve any supremacy again until the end of the age (when Daniel's 'Stone' Kingdom is implemented) is complementary to this picture and not at odds with it. The Kingdoms emulate the previous beast empires, by persecuting the saints in Jerusalem via an ambivalent Byzantine Empire and the Sassanid assault on the city.

They continue to occupy the land, and fail in their duty-of-care in respect to the future security of the people or the city.

Moreover it is in total agreement with our Jerusalem-centric prophetic worldview and is backed up by the last point in the list above, which shows the changing state of affairs in the relationship of the Ten Kingdoms with the Eighth Beast Empire.

Eventually, when the Caliphate has weakened at the end of the age, they finally turn and destroy the last Beast Empire, and this occurred dramatically in World War I, with unfinished business up to 1967 in Jerusalem and military action ever since.

The Ten Kingdoms are prominent *at the close of both 3.5 times of prophecy.* In the first instance for a short while in Jerusalem prior to the year 689 and then again at the close of the second 3.5 times and the end of the age. It means that there is no conflict between Revelation 17 and anything in Daniel 2.

Now there is one thing I have not addressed completely and needs further clarification. It is the question of the identity of *all* the seven empires that precede the Eighth Beast in Revelation 17. The visions of Daniel only speak of the lineage of empires commencing with Babylon at the beginning of the Seven Times timeline. They amount to only five in number, if we include the Ten Kingdoms as a single entity.

We have also seen in both Revelation 12 and 13 a mention of the 'seven heads' or 'crowns' (kingdoms / empires) which are part of the symbolism and back up the chapter 17 story. We also know that the 'seven heads' are empires which existed prior to AD 689 during the first 3.5 'times'.

After the discussion above, and much other analysis previously, it is clear that the seven 'heads' of chapter 12 (the dragon), chapter 13 (the beast out of the sea) and the Whore pictorial of chapter 17 are consecutive empires. They are also the same as the 'kingdoms' described later in Revelation 17. This is a major contribution to our understanding and was alluded to earlier; because now we know that unlike the Ten Kingdoms, the 'seven heads' are sequential. They are not contemporaneous and we already know the identity of five of them.

In addition, whereas the Roman Empire is the **4th beast kingdom of Daniel** (albeit with the Ten Kings) it has become apparent, that Rome is *also* **the 6th kingdom of Revelation 17.**

This is a vital link and allows us *to synchronise the two sequences* described by Daniel and John. There is also no mention in Daniel of 'seven heads' or kingdoms at all, and certainly not as heads or crowns on the 4th Roman beast. The Reformers and subsequent historicist commentators have promoted the idea, that the 'beast out of the sea' in Revelation 13 is a sequence of kingdoms associated with the Roman, or post pagan Rome period. I suspect they have not made this error deliberately, but have wrongly assumed that it must also be supported by Daniel.

That is simply not the case. However, with five known kingdoms (empires) in Daniel's beast sequence, the question that remains is the identity of the other two 'kingdoms' of Revelation and where they fit chronologically.

In order to put Daniel's empires in context, there is a strong case to be made for *starting the sequence with Egypt under Thutmose III*. In fact it is a watertight case based on confirmed history and another personal revelation I will share.

At its height around 1450 BC, Thutmose's empire extended into Syria in the north and over the Euphrates. It also extended along the southern coast of Anatolia. This all took place after the Thera eruption occurred and the Exodus drama was over - or drawing to its close. If we start there with our Jerusalem centric view of history; we have Egypt (which becomes the first Empire to subjugate Israel) followed by the neo-Assyrian Empire attack and conquest of the Ten Tribes ca. 722 BC and the siege of Jerusalem under Sennacherib in 701 BC. It is after these two empires, that we come to Daniel's sequence of Babylon, Medo-Persia, Greece, Rome and the Ten Kingdoms.

That completes the seven 'kingdoms' or 'heads' that existed prior to the rise of the Islamic Caliphates. It is a sequence which unexpectedly begins with the Exodus and Moses saga and ties all subsequent history and symbolism together.

Abraham and Senusret

There is one more unexpected and pretty much inexplicable experience of mine, which points back to Egypt as the start of the Revelation 17 empires.

I was out with family watching a super-hero movie (I make no claim to being a cinematic connoisseur) and eating some popcorn as you do. It was well after I expected any new material to be brought to my attention for this book. In fact, I already knew that I had been blessed with far more insights than I ever expected, and beginning to correct my numerous textual errors.

I also had a debilitating staph infection contracted after a surgical procedure and was far from well and needed a break.

While I was by myself, I seemed to get a nudge indicating there was more to go and that my job was unfinished. So instead of going back into the cinema, I decided to grab a coffee and some raisin toast.

I was having some quiet time by myself, albeit in a corner of a shopping mall, when something came to mind that I had wrestled with earlier. It was those numbers again - and specifically about 430 being multiplied by 3 to give the Daniel 1,290 'day' milestone. A milestone that was peculiar to the relationship between Israel and Egypt.

Now why anyone would think this way in the circumstances, I really have no idea. It just seemed that I was meant to know something special. At any rate, it dawned on me that although I understood the derivation of the 430, I had no idea of the significance of the number 3. It was simply that 3 x 430 equals 1,290. So I went over the whole thing in my mind again, starting with the most recent event in 1979.

This was the year that Egypt officially recognised Israel as a nation state and of course 1,290 years before that, was the central year for the whole Seven Times; AD 689 and the 'abomination of desolation'. It occurred to me that there was an Egypt-Israel association here as well. After all, this is where so many Jews and Christians were expelled from the Holy Land, or were willingly on the run. By then the Muslims were also overrunning Egypt, however there was really nowhere else for anyone to flee from Jerusalem, *but* Egypt. The reason for this, was that the northern Umayyad Caliphate from Syria and Anatolia was now in control. This effectively blocked the exit to the north.

In addition, Jesus' reference in Matthew 24 to 'fleeing into the mountains if you live in Judea', 'how dreadful it will be in those days for pregnant women' or 'pray your flight will not take place in winter or on the Sabbath' and 'for there will be great distress unequalled from the beginning of the world' are verses referring to the 'abomination of desolation' of Jerusalem and its aftermath – not the earlier destruction under the Romans. This is easy to gloss over without a timeline, but becomes blindingly obvious, when we note the placement of these phrases directly after the reference to the 'abomination of Daniel'.

So then my mind began to roll again and turned to the 1,290 years before that, which is 602 BC. And of course that is back to our previous discussion on the kings of Judah; Jehoiachin, Jehoiakim and Zedekiah who were the ones who provoked the

Babylonians in the first place. It resulted in the siege and sacking of Jerusalem in 588-586 BC, and that was most definitely because the Jews were negotiating with the Egyptians for military aid or intervention.

Suddenly I realised I had the origins of my number three (3); 1979 was the third episode where Egypt and Israel were especially drawn together. In the first case Egypt was a potential military partner and subsequently an escape route. In the second, a place of refuge once more - while in 1979, Egypt was reaching out after the 1967 conflict and finally recognising the Israelis as neighbours.

It was just as well I had gone for the mug option and bought a *large* coffee, because that begged the next question. Although I had three dates, I had not gone back three ages. What if I went back a further 1,290 years; all the way back to 1892 BC?

It only slowly seeped into my consciousness, but I just knew in my heart of hearts, that this would mark the first time that the Hebrews went down to Egypt. This date *had* to mark the start of the original 430 years, when Abraham left Harran in the northern Levant and entered Canaan and by default, Egypt!

It began to dawn that this was a remarkable insight; the beginning of another incredible timeline. Not technically the Seven Times one, but one that is directly linked and intimately connected with it; a timeline beginning with Abraham and Egypt and the Apiru / Hyksos Asiatic saga that started it all.

It also begins in the reign of Pharaoh Hkakheperre Senusret II (Sesostris II) of the Egyptian 12th Dynasty, in the late Middle Kingdom. At least he seems the most likely candidate, given the best accepted dates I can find. The reign of the next pharaoh, Senusret III, who is considered the most powerful of all Middle Kingdom pharaohs, began in 1878 BC and concluded in 1860 BC. So depending on when Abraham actually visited Egypt, he could also be a candidate for the famous meeting described in Genesis 12. We are told that Abraham was an old man when he left the north, so a Senusret king is pretty much a given.

This also reminds me of the Egyptian writer Manetho's otherwise strange family association of Senusret (Sesostris) with Ramses II of the much later 19th Dynasty. We only have fragments of his history, so one wonders whether the original Egyptian documents were implying a genealogical descent of the New Kingdom pharaoh, from the pharaoh that Abraham met. Any suggested direct association between Moses and Ramses would have been incorrect, but a possible family link between Senusret and the New Kingdom pharaohs, becomes more understandable in the context of Abraham's initial journey and his ancestral connection to Moses.

Anyway this was all rather overwhelming; I was really not expecting another lesson in Egyptology and Bible prophecy on that particular Saturday afternoon – a lesson providing a whole new timeline and more historic date setting!

Significantly, from 1892 BC we can also calculate another biblical date for the Exodus. By adding **the 430 years** the Abrahamic family lived in Canaan-Egypt, we arrive at **1462 BC**.

I described earlier, how others have previously counted back from the best known dates for the David-Solomon kingdom, using the 480 years of 1 Kings 6:1. This was discussed in the Exodus Cataclysm chapter and a date of 1450-1448 BC was noted as a best estimate.

This is only 12-14 years later than 1462 BC and given the unknowns in the derivation of the date, a good match.

We have also determined that in all our milestones throughout the Seven Times timeline, that a seven year period for complete implementation of each milestone event is implied - even warranted. Given this new timeline ties into the Seven Times, there is no reason to believe Abraham's original migration was any different (probably with some family visits home in that period). The rule should also apply for the Exodus event.

If we use our nominal 7 year period then, this tail-end period of the Exodus would have occurred 1462-1455 BC. This draws even closer to the dating derived from the biblical genealogical timeline and the 480 years from Solomon; 1450-1448 BC.

This period is all within the overall reign of Thutmose III (1479–1425 BC) who came to the throne as a child. It means 1462 BC is also within the co-regnal period of his aunt, Queen

Hatshepsut who reigned 1473-1458 BC. That period is also not historically certain and it may be possible that Thutmose seized power earlier. Regardless, a seven year (1462-1455 BC) window fits the transition of power like a glove.

I emphasise once again, that I believe all indications point to a late Second Intermediate Period / early New Kingdom dating as the *beginning* of the Exodus / Thera eruption period. This is the conclusion that I outlined previously, given all the scientific data and Egyptian history available.

In fact, an early Thera eruption date also ties in with Tjebnutjer's (Manetho's) other seemingly unlikely suggestion, that there were only seven generations between Abraham and the Exodus. Over the nominal 430 years from Abraham to Thutmose, the seven generations make little sense, unless 70 year lifetimes are used. If we use an initial eruption date of 1620 BC and measure the years back to the end of the reign of Senusret II (1878 BC), there are only 258 years and Manetho's seven generations begins to look much more plausible.

Even if the period turns out to be a little longer and lines up with the reign of Ahmose, it will still be within reason and provide more backing for a longer migration out of Egypt.

It also seems highly likely, that another contingent (and perhaps the largest) left when they knew that Thutmose III was about to assume kingship as an adult. His reputation as an ambitious leader and warrior (and likely tyrant) would have driven this refugee event, at a time when Queen Hatshepsut was losing her influence and security was no longer guaranteed in the land.

Once this transition of power occurred, the situation in Egypt and surrounds would have descended into active warfare and the Thutmose III lead Egyptian occupation of Beit She'an in northern Israel ca. 1450 BC (with attacks on Kadesh and points north) would have been the norm.

In addition, there are potentially similarities in the way in which each of the three consecutive 1,290 year periods unfold. It is likely that if we examine each 430 year segment separately, we would find even more patterns within the whole timeline. Indeed, the 430 years from 1032 BC to 602 BC would be a significant one, covering an extended period of the Kingdom of

Israel; one that would encompass King David's reign completely and the early Kingdom period of Saul as well.

Suffice to say; given the shock of receiving this revelation at such a late date, I will leave any further analysis for another day and for those who have a better grasp of Israel history.

Now this whole new timeline is based on a number; 1,290 and its components 430 and 3, together with no specific information on how to interpret the 1,290 'days' in the Daniel 12 prophecy. To go so much further, and attempt to reverse engineer a timeline purportedly to show the beginning of Abraham's entry into Canaan and Egypt - and extract a new date for the Exodus - will seem wildly speculative and optimistic.

All I can say in defence of this extraordinary tale, is that the initial 430 years in Egypt is the first crucial clue. It is reinforced in Ezekiel 4 in regard to Israel and its numeric identity - and when you lay out the Seven Times timeline - the significance of the 1,290 year milestones in regard to the Egypt-Israel relationship becomes apparent.

Furthermore, not only have I had an interest in prophetic numbers (however intermittent) throughout my life, but the Lord has blessed me in my own personal journey, by way of signs through numbers - and I shared one previously in regard to the dating of the Six Day War. If he wanted to bring to my attention a new prophetic insight, there would be no better way.

So Egypt certainly satisfies the criteria for the Revelation 12, 13 and 17 'seven heads' pictorial *and* also the first empire of the Revelation 17 sequence of kingdoms.

Ever since the Egyptian Exodus, the seven empires (followed by the Islamic Empire) invaded and controlled Judea, or seriously threatened Jerusalem.

I am now confident that this sequence of empires is the one John's vision portrays. Obviously other possibilities could, and have been proposed. However all the prophecies go back to the Son of Man who revealed them, who as our Messiah preached in 'the great city', which he figuratively called 'Sodom and Egypt'. That city is the city of Jerusalem and everything is subject to a timeline based on Jerusalem - not Greenwich Mean Time or any other time.

The confusion for prophecy analysts trying to unravel the scripture is; which 'beast' is which?

What has now been confirmed with confidence, is that the Eighth Beast was not Rome or the Ten Kingdoms, but a resurrected Babylon under successive Islamic Caliphates. Given the firm foundation of the Seven Times of prophetic history and the identity of the empires, it is now imperative to move forward and identify the Whore City of Mystery Babylon as well.

In some ways the puzzle parallels the experience of our own lives. In the midst of our journey it is not always easy to see why we are going through the ups and downs that we do; neither the character building hard times, or the times of satisfaction and success. But in the twilight of our years and in retrospect, if we have been true to ourselves and our calling, the pathway we have trodden begins to make sense.

It is no longer the mystery it once was, when we could not see the wood for the trees.

Mystery Babylon

As the Lord unexpectedly brought prophetic timing issues to my attention after decades of complete neglect, the revelations were at first dramatic, then as I realised I should write them down, they began to flow, sometimes intermittently and cautiously with much study and referencing, at other times as fast as I could type. In fact, writing seemed to be an act of faith that the Lord used to reveal more.

But occasionally the flow stopped; stubborn, unyielding, grinding, writer's block. At the end there was an interpretation regarding Mystery Babylon and the Whore City which I was so concerned about, I was paralysed. I had no idea of its identity.

I knew that the interpretation I was familiar with over many years was incorrect. I was sure that Babylon the Great was *not* the Holy Roman Empire, or the Whore City the Vatican and Rome. But I needed to be sure. And if Rome was not the Whore City, then which great city could it be?

Despite much trepidation, I had long come to the conclusion that Islam was the final 'beast' of Revelation 13 and that Muhammad and his line of Ayatollahs, Muftis and Imams were the

'False Prophet'. Also that the Caliphs and Sultans were the head of the 'beast'. So I was sure that it was indeed the Islamic Empire in its many guises and Caliphates over 1,278 years. But which city was the 'Whore' of Revelation chapters 17 and 18?

I paused one morning and prayed specifically over this question. Not even kneeling, but nevertheless yearningly and deliberately. As I mentioned earlier, I was chronically ill, my body poisoned by a super bug and I was finding it difficult to sleep and therefore to write. It was early November 2017, and I felt I might not be able to complete this manuscript, without more specific inspiration. The identity of the City was critical.

Forty-eight hours later I was again thinking about the city on seven hills. The obvious one was Rome, but Jerusalem is also built on seven hills. I knew Mecca was said to be built on seven hills as well. This meant all three cities most associated with the largest monotheistic faiths – Christianity, Judaism and Islam were all built on seven hills. But the description in Revelation 18 did not seem to apply. We are told it is a massive port city, standing as it were, at the centre of the world. A city where all the nations have access and cannot stay away. A city with which the nations feel they *must* trade; where the spoils are just so lucrative, much too hard to resist and totally addictive.

In fact a place that could be truly called the 'Gate of God', as the Babylonians called their original city on the Euphrates.

There was another passage of scripture that I had never quite got a handle on. It was the 2nd Epistle to the Thessalonians chapter 2. Here the Apostle Paul is talking about the 'man of lawlessness' to come. 2 Thessalonians 2:

"Concerning the coming of our Lord Jesus Christ and our being gathered to him, we ask you, brother and sisters, 2 not to become easily unsettled or alarmed by the teaching allegedly from us – whether a prophecy or by word of mouth or by letter - asserting that the day of the Lord has already come. 3 Don't let anyone deceive you in any way, for that day will not come until the rebellion occurs and the man of lawlessness is revealed; the man doomed to destruction. 4He will oppose and will exalt himself over everything that is called God or is worshipped, so that he sets himself up in God's temple, proclaiming himself to be God. (NIV)

This man of lawlessness / sin / perdition, seemed very much like another description of the evil empire and its leader(s) of Daniel's prophecies. But this 'antichrist' figure could have been one of many vile leaders and despots within the great empires, or indeed, those who attempted to resurrect them in some form or other. Men like Charlemagne, Napoleon, Mussolini, Hitler and Stalin come to mind as European examples of leaders who might fit the description.

But why was Paul addressing the people of Thessalonica? After all, this prophecy and its timeline would still be true if it was referring to the Jerusalem Temple. It simply says that Jesus will not return until the Antichrist appears - and that will not happen until the restrainer is removed and the rebellion occurs.

If Rome and the Vatican were the Whore City of Revelation 17 and 18 (as just about everyone has taught except the Pope) why did Paul address the folk of Thessalonica, when it would have been far more appropriate to write to the Church of Rome?

It was something of a puzzle, and somewhat similar to the question of the Seven Churches of Revelation, and why they were chosen to be the subject of the early chapters of Revelation. There were dozens of synagogues / churches throughout the Levant and Mediterranean, that could have been used as examples to teach spiritual and moral principles. Why were these seven churches in western Asia Minor the ones selected?

We learned from 2 Esdras, that this was where many of the Ten Tribes folk fled out of Assyria. So that is obviously of key significance, but was there something more? Could it be that the church folk of Thessalonica in Europe and the church assemblies in Asia, were both being warned of things in the future that particularly applied to them?

I was drawn to reading the history of the final assault by the Muslims on Constantinople, and some things came into focus. Thessalonica was one of the last cities to fall, before the fall of Constantinople itself. Constantinople was a virtual island in the end, with Ottoman forces already occupying the land well into Greece, the Balkans and today's Bulgaria. The city was completely surrounded and had to be supplied by ship from the west, right up to the end.

On the 29th May 1453 the city finally fell to a force lead by an Ottoman Sultan, Mehmed the Conqueror.

The picture began to gel. Somehow Paul, Silas and Timothy already knew, that the Thessalonians were one day going to be in the middle of an onslaught, by the 'king who will exalt and magnify himself'. This was the 'ruler' symbolising the Beast Empire and its leadership, depicted by Daniel and John. It seems they wanted to reassure the locals, that this event was a long way off and anything they heard to the contrary was false.

I do not know how they knew; perhaps Paul or one of his colleagues had a vision, perhaps they had been in contact with John in Ephesus, Patmos or the region and he enlightened them. But it made sense. Something was going to happen in this part of the world. It was just not going to happen anytime soon.

I then reopened some on-line posts I had checked out before; one in particular caught my eye. It included a map of western Anatolia and the Seven Churches of Revelation, but also the Bosporus and adjacent parts of Europe, including Thessalonica. I then reflected on my holiday, and the location of the island of Patmos offshore south eastern Turkey. It had not been far away from my coastal drive from Ephesus back to Izmir.

I did not get to see it, but knew in my mind's eye, that Patmos was tucked out of view, behind the much larger island of Samos which lay before me.

It is a beautiful, rugged and scenic coastline, reminiscent of some parts of Australia's Pacific south coast, and for some reason I felt rather at home in this part of the world. Perhaps it was the landscape, or a couple of people I met. Perhaps in part, because I remembered that my paternal grandmother had mitochondrial DNA linked to this part of the world.

As I meditated on these things, everything became very personal; I was thinking what would the Apostle John have felt? What did he see in his vision? Was his location in this part of the world significant in understanding the identity of the city?

I then had the distinct leading to check to see if the City of Istanbul was built on seven hills. I do not know exactly why I had the thought, but even a rather dull witted person begins to sense, when something extraordinary is being unveiled before his eyes.

From a western perspective and trading as we do with so many huge port cities in the Americas, Europe and Asia, Istanbul does not readily come to mind as the Whore City of Babylon the Great. But consider; the city of Constantinople was selected by Constantine for its location and as a replacement for Rome - including its seven hills. He would have noted its fabulous potential as a sea trading location, with access to the Black Sea through the Bosporus and the many countries in Asia and Eastern Europe, reliant on that trade. We know that even the Scandinavians traded down the Russian rivers with the city, and Islamic artefacts found in the Baltic region are evidence of this.

The city is located in a central northern point of the Eastern Mediterranean. It sits like a Queen in close proximity to the mighty kingdoms of yore and the Red Sea and Indian Ocean are not far away. There is direct access to the Levant and Egypt, all the Mediterranean nations of North Africa and Europe - and through the Gates of Gibraltar to 'all the nations of the earth'.

And that is not all; it links Europe and Asia and sits on the old Silk Road route. It pipes oil and gas from the Asian and Middle Eastern petroleum fields, as well as all manner of tradeable produce. It is unique in a way that describes true uniqueness. When you really think about it, it is hard to find superlatives which do not apply to the city and its location.

Memories of my visit reinforced this. Flying into Ataturk Airport, I saw a city of high rise accommodation apartments, a towering business district, signature buildings, bridges and motorways. Ultra-modern infrastructure and ancient historic buildings were all blended into an amazing metropolis, and a wonderful sight. Even at height it was impressive - with a magnificent view from the Sea of Marmara all the way north to the Black Sea in one stunning backdrop. I did not doubt it was a city of over 14 million people.

On the ground and driving into town along the coastal motorway, hundreds of bulk carriers, container ships and oil tankers were standing out to sea. I was familiar with Australian bulk coal and iron ore ports, with dozens of ships waiting patiently offshore for their turn at the terminals - but this was something else. And then there was the horn shaped water body within the southern entrance to the Bosporus; the Golden Horn.

The little horn of Daniel comes to mind; more symbolism, little clues or coincidence?

Yes, this city as Byzantium, became part of the Roman Empire and under Constantine, exploded on to the international scene. It became the Byzantine (Eastern) Empire with opulence beyond measure. But....when it fell to the Muslims in 1453 and the Ottoman Caliphate rose to its ultimate splendour (and decadence) under the Caliphs and their vast entourage, we begin to see the description that applies in Revelation and elsewhere.

They had envied their Mesopotamian and European predecessors for so long. Now *they* could live the dream and possess the world; in fact all that the earlier beast empires had to offer and more. It was now *their* turn to rule like the great empires of yore; Egypt of the pharaohs, mighty Assyria, Nebuchadnezzar's Babylon, the King of Kings of Persia, and the Greek 'gods' who ruled all the way to India and Egypt after them. But even more than that, they were now taking over the city which demonstrated Rome's wealth, power and glory at its very peak.

This was the final iteration of Islam as an Empire. Fuelled by trade, unlimited slave power, splendid infrastructure like the Hagia Sophia, the port facilities, Roman roads and aqueducts; more than enough on which to build a superpower. In fact when Paul in Thessalonians describes the man of lawlessness 'so that he sets himself up in God's temple', I began to wonder if it really *was* the Temple in Jerusalem that Paul saw in his visions. What if it was the future Hagia Sophia of Constantinople / Istanbul? Young's Literal Translation convinces me:

'The son of destruction who is opposing and is raising himself up above all called God or worshipped, so that he in the sanctuary of God as God hath sat down, shewing himself off that he is God' (YLT)

It is a mighty church building, still standing, despite countless earthquakes. It has been used as both a Christian place of worship and a mosque for centuries. The foundation stones of Constantine's original sanctuary apparently still lie under the massive 6th century Basilica. The more I dwell on it, the more I think the Hagia Sophia *is* the place of worship that is being described. Hagia Sophia means 'Holy Wisdom', i.e. Holy Spirit. It is a church dedicated to the Holy Spirit of Almighty God.

This explains so much about Mystery Babylon and why it was condemned by the prophets; about the pretence of worshipping our Creator and Saviour God. It is all about the 'Man' and self-aggrandisement to the detriment of millions.

If the passage was referring primarily to the Temple Mount, as many translators suppose, then it really does not fit. The Temple Mount was razed of all previous structures; all the bits and pieces of the Second Temple and other temporary buildings that may have existed before 689, were knocked over or buried. From there the Islamic builders started from scratch.

The Temple grounds would also have to be regarded as the equivalent of a building. This would really be stretching the truth, even though the grounds to the south, were extended as a manmade platform in Herodian times. Jerusalem also has no relationship to Thessalonica (or Constantinople) whereas Constantinople and the Hagia Sophia are geographically one.

So it is obviously appropriate, that this episode occurs in the very heart of the Whore City, Babylon the Great. We see the Sultan simply making himself at home in the pre-eminent and most magnificent church building of its day - and sits as God.

Fancy that! A church sanctuary built in the 6th century, recorded by prophetic vision in the New Testament! In fact, the only Christian Church building specifically mentioned in the Bible at all!

On that day in Istanbul, when I gazed around me within that amazing edifice, little did I realise that I was soaking up an experience out of the pages of the Bible - and that the Lord led me there to complete this picture for you...

And there is one thing more; a rather enlightening discussion surrounding the Koine Greek word used for 'temple' or 'sanctuary' in Thessalonians and elsewhere.

It is the word 'naos'.

Naos is often contrasted with another word for temple used in the New Testament; that of 'hieron'.

Now some commentators distinguish 'naos' from 'hieron', by suggesting that 'hieron' is used for the building, whereas 'naos' is limited to spiritual imagery.

In this context, 2 Thessalonians 2 is portrayed as a story of the spiritual battle between God and Satan in the heavens. They suggest it is not an actual sanctuary and that from earlier usage in the New Testament, naos (at most) refers to the temple of our bodies and minds, and our struggle with temptation.

It is true that naos is used for the temple of our bodies in the NT, but others have pointed out that naos is also used for structures where the Holy Spirit was expected to dwell and be manifest - and indeed this is the case. For the Temple in Jerusalem, hieron is used to cover all the buildings of the Temple complex, whereas naos is used only for the Holy of Holies - where God manifested himself to the High Priests.

Therefore a Christian sanctuary where the Holy Spirit is welcome, is well and truly a building which can be described as a 'naos'. Furthermore, since Hagia Sophia means Holy Wisdom (a variant of Holy Spirit) it therefore has added meaning. It may well signify that 'naos' was used to reflect the actual name of the sanctuary in the prophecy!

So the Ottoman Caliphate was a power that was to last over 400 years, but would cap off over 1,278 years of Islamic dominance in the Holy Land, the Middle East, Egypt, Africa, Asia and Europe; indeed for a 'time, times and half a time'. We can now understand how the Caliphate was built on the infrastructure of the 'beasts' of the previous regimes. And why the Whore City rides on the back of the beast, in the prophetic imagery.

Istanbul began as a colony emerging from fishing villages and ferry points on the Bosporus. The known history begins in the 7th century BC and is probably Thracian-Illyrian related. It was later called Byzantium / Byzantion by the Greeks, when they arrived in the region. After the early days - where some Egyptian presence was felt on the south coast of Asia Minor and an Assyrian influence prevailed later - Persia, Greece, Rome and the Ten Kingdoms all owned a presence at the site.

To my mind, the riddle has been solved. We do not have to propose a new port built by the Saudis in the Red Sea, or any other city - ever. We do not have to pretend that a land locked city like Jerusalem or Mecca, meets the description any longer. In fact we do not have to continue the pretence that Rome was a port either, but acknowledge that over the years, the Romans built large ports on the coast to expand the trading capacity of the city itself. They had to, the Tiber was simply never large or navigable enough, to make Rome the port of the Empire.

On the other hand Constantinople (Istanbul) on the Bosporus, astride both the Black Sea and the Mediterranean through the Sea of Marmara and the Dardanelles, *has* fulfilled the role of a super port for seventeen hundred years. It also fulfils the prophetic description of the Whore, Mystery Babylon over hundreds of years. It has the runs on the board, so to speak.

It also highlights the sub-title of this work; 'Egypt to Istanbul'. It is the city which is central to the last iteration of the Islamic Empire – the Ottoman Caliphate.

The emblematic obelisk of Thutmose III positioned near the Hagia Sophia and Blue Mosque, echoes the synergies and history that tie the Empires together. It ties the first of the 'seven heads' (Egypt) with the Sixth, Rome, and its Eastern Empire - and thereby to the Seventh (the Ten Kingdoms). And since the obelisk and city still stand, they tie directly to the Eighth World Empire. The span is from the Exodus and the Law, until the Apocalypse and Judgement; from Moses to the end of the Age.

What I find truly inexplicable on reflection; is how the true identity of the Whore City has been hidden for so long.

I do not know why this is so, but have the distinct impression, that we have all been purposely blinded by something that surely is in plain sight. There has to be some greater purpose.

In hindsight though, it is obvious that AD 476 and the focus on the demise of Rome by Odoacer is partly responsible. What has been forgotten is that Rome was no longer the capital of the Roman Empire, because Constantinople had taken its place.

Following on from these astounding things, it is now appropriate to provide some excerpts from Revelation 18. They describe the city and its final fate:

"'Fallen! Fallen is Babylon the Great!'
She has become a dwelling for demons

and a haunt for every impure spirit,
  a haunt for every unclean bird,
  a haunt for every unclean and detestable animal.
³ For all the nations have drunk
  the maddening wine of her adulteries.
The kings of the earth committed adultery with her,
  and the merchants of the earth grew rich from her excessive luxuries." ⁴ Then I heard another voice from heaven say:
"'Come out of her, my people,
  so that you will not share in her sins,
  so that you will not receive any of her plagues;
⁵ for her sins are piled up to heaven,
  and God has remembered her crimes.

And:

¹¹ "The merchants of the earth will weep and mourn over her because no one buys their cargoes anymore— ¹² cargoes of gold, silver, precious stones and pearls; fine linen, purple, silk and scarlet cloth; every sort of citron wood, and articles of every kind made of ivory, costly wood, bronze, iron and marble; ¹³ cargoes of cinnamon and spice, of incense, myrrh and frankincense, of wine and olive oil, of fine flour and wheat; cattle and sheep; horses and carriages; and human beings sold as slaves.

Then:

²¹ Then a mighty angel picked up a boulder the size of a large millstone and threw it into the sea, and said:
"With such violence
  the great city of Babylon will be thrown down,
  never to be found again.
²² The music of harpists and musicians, pipers and trumpeters,
  will never be heard in you again.

And finally:

Your merchants were the world's important people.
  By your magic spell all the nations were led astray.
²⁴ In her was found the blood of prophets and of God's holy people,
  of all who have been slaughtered on the earth." (NIV)

    The vision that John saw and wrote about in Revelation 16, and whose plagues were listed earlier, is found immediately prior to the description of the City in chapters 17 and 18.

The chapter describes the judgements coming to Istanbul and the region around, and identifies the residents as 'the people who had the mark of the beast and worshipped his image'.

We have support for our identification of Mystery Babylon, from extra-biblical sources as well. The prophecies are confirmed because of the vulnerability of Turkey and Istanbul to earthquake. The city sits just 20 kilometres north of the North Anatolian Fault complex, which crosses the Sea of Marmara.

The fault system is a strike-slip fault, similar to the San Andreas Fault system in California, and even on major fault lines, earthquake movement does not happen often. However when there is slippage at one location, it is only a matter of time before there is a ripple effect and the slip propagates along adjacent sections to relieve tension. The movement is inexorable and the longer the system remains locked, the more violent the next 'quake will be. It is a bomb waiting to explode.

So to a student of geoscience, this vulnerability of Turkey to earthquake along the east-west, North Anatolian Fault line is obvious. It also links up in the east, with the northern boundary of the Arabian Continental Plate along the Zagros Mountains of Iran - and through the East Anatolian Fault, with the western and southern boundaries of the plate along the Afro-Syrian Rift.

The background driver is continental collision; with the African Plate forced under and against the European and Anatolian Plates. It is a largely hidden line of volcanic activity from west to east; Sicily through Italy to Santorini Island, and it is caused by subduction and melting of crustal rock.

The uncertain boundary conditions along the north-south Afro-Syrian Rift (which runs along the Beqaa Valley, Jordan River, Dead Sea and Gulf of Aqaba) also potentially links the destruction in Turkey, with the earthquake predicted for Jerusalem. In East Africa rifting is continuing and there may be pressure on Arabia from beneath the Red Sea.

It is also possible, that this monumental event could be triggered by a bolide impact of some sort - this, given the description in Revelation. Verse 21 describes the imagery of an angel throwing a boulder into the sea. However if a trigger is required to initiate such a cataclysmic event, nuclear explosions might also be sufficient and Revelation 17:16-17 suggests something of that nature. On the other hand, the imagery may simply be emphasising the immense destruction by earthquake and fire, that would result from a conventional earthquake event.

I used the term 'uncertain boundary conditions' above, because although the plate boundaries and rift valley are well studied, I suspect the political unrest and outright war in recent times, has interrupted rigorous seismic monitoring in Syria and Iraq. However in Turkey, the earthquake history along the North Anatolian fault line is well known and has been seen to progress westerly toward Istanbul over the last century.

In fact seismologists have an expectation that a big earth movement will occur in the next twenty years or so. And because the North Anatolian Fault complex passes through the sea just a few kilometres south of Istanbul, it could just explode as if an angel had thrown a boulder into it.

This is frightening confirmation of our biblical prophecies; prophecies which describe islands disappearing and the surface of the earth being rearranged and unrecognisable in places.

Pondering on this, my first thoughts were that the islands of the Aegean were in the direct firing line and most at risk. But then it became obvious that every boundary of the Arabian Plate - including the cities in the Zagros Mountains - could be affected. This might easily include the Arab Emirate States along the Persian Gulf, with their monumental, glitzy, high rise towers and fancy, manmade islands built on sand.

A picture of concrete piles poking out of the sea came to mind, and then it dawned that, not only could the Burj Khalifa come down, but this whole catastrophe will be the Almighty's judgement on 1,400 years of Islamic deception and abuse.

Revelation 16 gives a vivid portrayal of this destruction and chaos, including the collapse of the 'cities of the Gentile nations' and the disasters offshore. It suggests the possibility of a wider fulfillment of the Babylon the Great prophecies, perhaps even damaging the Caliphate capitals of yesteryear from Arabia to Harran, Baghdad and Cairo. It means there will almost certainly be some destruction at other Islamic centres of power as well. Indeed, much of the western world today matches the description used of Jerusalem itself; 'Sodom and Egypt' as it was described by the Lord - and you would wonder why this sin would not be addressed and judged in the same hour. That is, in the absence of serious repentance.

If so there could well be punishment reserved for such moral decay and unfaithfulness. And just as Jerusalem will not entirely escape being damaged as the Temple Mount is cleared of the 'abomination' and prepared for the future, so it could be that a clean-out may occur on a much larger scale. This end-of-the-age earthquake could affect many nations in the region and way beyond, with tsunamis at sea an obvious and predictable threat. But there will also be terrible disruption to trade, and the oil from the Middle East will surely dry up.

Now it must be said, that visions dramatically portray the ferocity of events, and (as we have seen) often pinpoint specific places that will be utterly levelled. However visions do not always give the reader an appreciation of the areal extent of events, or how long the terror will last. Given this qualification, my description here could be way over the top, or indeed the very opposite, with devastation occurring on a global scale. It could also transpire that the earthquake will only affect the Eastern Mediterranean and western edge of the Arabian plate.

So how will these events unfold? Will it all be tied to Daniel's 1,335[th] day? At one point, I began to assume that most of the apocalyptic things may have already happened. The World Wars, the Six Day War, Vietnam War and the Iraq Wars being end of the age examples. It seemed the current turmoil in the Middle East might simply be the final wind-down and demise of hardline Islam. I also assumed that the 1,335[th] day (2024) would be the absolute end year of all prophetic events.

However the concept of a seven year pattern for each prophetic milestone, gradually became impossible to ignore and re-emphasised the Armageddon climax. The historical and scriptural support was *too* strong and 2024 as the 'beginning of the end' seemed more likely. It meant that the following seven years (as outlined in prophetic Lesson 10) were almost certain to be part of the 2024 year prophecy. Moreover, there seemed to be direct associations and descriptions in several places in scripture, with dramatic events of war and very specific earthquake activity in the Middle East which had yet to occur.

Although Daniel tells us very little, other prophetic utterances give a fuller picture. Ezekiel 39:15 talks about an event in the Valley of Hamon-Gog, where 'all Israel' buries five-sixths of the army of Gog and Magog. It is apparently a valley near Dhiban (Dibon) in Jordan. This event may also be described in Isaiah 63, where the Lord returns from Edom (Bozrah) after dealing with his enemies. However, I am now totally convinced that the Bozrah victory is the one where British Empire forces defeated the Turkish and German military on the Gaza-Beersheva front in 1917; the battle that lead to the liberation of Jerusalem.

In addition, Ezekiel 38 and Revelation 16 refer to the expected apocalyptic, Armageddon conflict on the mountains of northern Israel, using similar imagery. Given, the current horrific scenes coming out of Iraq and Syria would seemingly indicate a precursor conflict has begun, it is tempting to suggest that the descriptions of events at 'Hamon-Gog' and 'Armageddon' are more generic, than literal, and symbolic of the wider conflict. Frighteningly, I believe they will be horrifyingly real disasters, with immense destruction coming upon the invaders.

Hamon Gog in Jordan will involve the slaughter of whole armies of Shia (and perhaps Sunni) Muslims, while the battle on the northern mountains of Israel and the Jezreel Valley is vividly described by Ezekiel. In the latter case, the judgement is directed at the Chief Prince of Meshech and Tubal; the Turkish head of the massed armies of Turkey, Persia, Cush, Put, Gomer and Beth Togarmah:

'I will turn you around and **put hooks in your jaws** and bring you out with your whole army….all **your troops and the many nations with you** will….advance like a storm….the fish of the sea, the birds of the air….every creature that moves along the ground, and **all the people on the face of the earth** will tremble at my presence….**the mountains will be overturned**, the cliffs will crumble and every wall will fall to the ground….I will **summon a sword against Gog** on all my mountains….every man's sword will be against his brother….**and I will execute judgement…...**'.

The scene is beyond disturbing and almost unimaginable! Massive attacks on Israel will come from the north and east, with probably little respite elsewhere. As the northern alliance grows, the influence we already see with Iranian Shia in Iraq and southern Syria might be compromised somewhat, and a more southerly (and direct) east to west route into Jordan will develop near Dhiban. So whether we look at earlier wars from the 20th century, or the 2011 'Arab Spring' uprising as a 'beginning of sorrows', the current, ongoing devastation in Iraq and Syria, appears to be heralding the end time apocalypse.

I also encountered a post on social media, talking about the last 7 years from 2011 until now in mid 2018. It used the imagery of the new moon peeping through as a sliver of crescent, and eventually blossoming into the full moon. It paralleled my own journey from immediately after my visit to Egypt and the Eastern Mediterranean. But when I saw the imagery being put forward as a 7 year journey from crescent moon to full moon, I thought, 'hang on, it takes 14 days (and a bit) to get from new moon to full moon'. And of course no one knew that better than Moses as he planned the Exodus.

So here is another strand, as it were, in the cable supporting our prophetic bridge. Using the day for a year guide, we have 14 years from the Arab Spring in 2011, until 2024 (inclusive) and the end of Daniel's timeline.

My sense now is that much, if not all, will occur within the 10 years from now in 2017-18 ramping up to 2024, possibly peaking (or finishing) in 2027 - and concluding 2030-2031. The Daniel prophecy for the 1335th 'day' is paramount and extremely likely to herald the general time of the return of the Lord.

Given this background, I am reluctant to put forward the following proposition, but feel it would be uncourageous and perhaps unfaithful of me not too. I hope it does not contravene the idea of *exact* timing to the day or the hour. History will be the judge, so I just issue a cautionary disclaimer to the reader.

Be watchful and alert and above all be prepared; the Jerusalem earthquake *may* be predictable in the following manner. If you had any doubt before, it should be dispelled by now, and quite plain, that the Lord through the Son of Man, his Angels and his Prophets talk to us through numbers

At the beginning of the Seven Times, the 590 BC marker was three and a half years before Nebuchadnezzar finally subdued Jerusalem during the siege. So if we begin a timeline from three and a half years after 590 BC (which is the mid-point of the Babylonian siege of Jerusalem in 587 BC) and add 1,335 years to this, we come to AD 749 (1,335 - (587-1).

Visiting Israel and Jordan as I did and getting a history lesson along the way, it was impossible not to learn of the massive earthquake of 749 which rocked such a wide area in the Levant around Galilee, the Jordan Valley and Dead Sea region.

Now the year 2024 is the 1,335 year since the desecration and desolation of the Temple in AD 689, which in turn was the mid-point of the Seven Times. But it was only the preparation year for the construction of the Dome of the Rock. If we therefore do the same thing for the second half of the 'seven times', but use estimated peak building activity on the Temple site of 692 and the same 3.5 year offset; we have: 692 + 1335 = 2027.

Some analysts are particularly keen on the idea, that key latter day events will fall on the September-October Hebrew feast days. This is a concept based on the nature of those feasts and the apparent relevance to end time events which they herald; i.e. Trumpets (Yom Teruah / Rosh Hashanah), Day of Atonement (Yom Kippur) and Tabernacles (Sukkot).

It is an interesting proposition. One timing scheme which I investigated would tend to confirm the autumn (fall) feast days, based on historical precedent and a month by month count.

Another model suggests that a June 2027 earthquake event is very possible. A third model implies an event toward the end of the year; actually January 2028. It depends on how the 8 months extra of the 2556.68 year timeline is spread over the Seven Times period - amongst other considerations. A fourth scenario based on the Meton cycles (Appendix 5) could mean an even earlier event in January 2027, or leading up to that year.

A further point of interest, is that in both 1967 and 2027 the new moon which heralds the beginning of the Sacred Mosaic year, occurs in early March around the 10th of the month. Furthermore, these milestones are 60 years apart and could have special meaning according to Jewish notions of the number six.

In fact, this 60 year period exactly mimics the period from AD 689 and the 'desolation' of Jerusalem to the 'quake of AD 749. It therefore ties in *exactly* with the first 1,335 year period.

So can the pattern at the end of the first 3.5 times guide us at the end of the second? I will not speculate further, but the year 2027-28 seems a distinct possibility for a mighty earthquake.

In terms of Istanbul, the earthquake event will most likely happen in a similar timeframe and related to the same widespread plate tectonic and seismic events. At the time of writing I have little insight into when that city will be destroyed. It could happen earlier, and a possible indicator might be that 2024 is 100 years (2 jubilees) since Ataturk declared the Ottoman Caliphate over - and things come in twos when it comes to Islam

However, as Ezekiel 38 describes; this immense catastrophe will surely coincide with the destruction of the Islamic hordes at Armageddon and quite likely with those in the Valley of Hamon Gog. *It means that ultimately, these battles will be the key to timing, whenever they occur.*

Finally, earthquake propagation scenarios along the Anatolian Fault Line, Arabian Plate and Red Sea, could potentially be modelled to some degree or other by seismologists. This might give some guidance as to the most likely sequence of events.

The seven year period will be absolutely pivotal in Jerusalem; a time to avoid because of war and earthquake on one hand, but one that will almost certainly herald the imminent return of the Jewish Messiah, Yeshua – our Lord Jesus.

On the other hand, the Lord himself issues a caution in regard to the timing of 'end of the age' events. In Matthew 24:36 he is quoted as saying:

"But about that day or hour no one knows, not even the angels in heaven, nor the Son, but only the Father'. (NIV)

Exactly which events he is referring to is open to debate; whether the day or hour of his return, or the absolute end of the world. More importantly from a personal perspective, after these words Jesus uses parables to illustrate that we should all get our lives in order and be prepared for his coming.

We therefore need to be wide awake and living the wholesome and productive lives we know we should; being faithful and true in everything we do. To all those who pass away before his coming, it means something a little different; to be prepared before our lives are closed and the silver cord is broken.

In the context of his second coming and to those left alive, he was saying 'Look you have been given all the signs of my coming, and they include the signs embraced in 'the fig tree' imagery; the signs in the sky - stars falling to earth - the sign of the new age - the earthquake that will hit Jerusalem - the timeline of Daniel - *plus* - my precursor warning example of Noah and the Flood (Matthew 24:36-41). But.... as for the day or the hour, that is known only to the Father himself'.

With those words ringing in our ears, it behoves us to be extremely cautious about using the symmetry of history, and any secondary timelines we find, as anything more than a confirmation of the 1335th day milestone itself.

If Jesus does not return somewhere within the Daniel 1335th 'day' period, it is possible that the Exodus timeline from the Passover to Sinai and the giving of the Law is a 'type' for a period to come; a period where all will be fulfilled.

It would obviously still be based on prophetic principles, but the $1,335^{th}$ 'day' could simply be the last warning milestone.

We know that in Moses day, the way the year was judged to *really* have begun, was the ripening of the first fruits. The year was effectively delayed for one or more weeks after Passover, if the grain had not already ripened. Perhaps the Almighty will do the same thing now, in order to allow for a full harvest of souls.

Again, I can only emphasise that this book is not focussed on predicting events and certainly not to the day or hour. It is simply meant to outline the Lord's prophetic timeline and Daniel's $1,335^{th}$ 'day'. I trust that in doing so, it provides everyone the clearest possible view of the very near future.

The key message is that our biblical prophets saw *real* visions of God and have warned us in advance. And as much as we are told that the prophecies were to be shut up until the last days, all indications are, that the last days have arrived.

Epilogue

Some readers may now sense we have moved through a complete circle within our biblical discussion. That from earlier events where we saw the Almighty working through natural disasters, such as the Flood of Genesis and the volcanic eruption of the Exodus, we have now moved to the eschatological end of our journey with the earthquake of the Apocalypse.

Humanly speaking it is really beyond comprehension, that forces which are measurable and reasonably well understood, are at the same time predictable by vision and somehow ultimately manipulated by God.

Even so, this should not be a complete surprise. I have tried to convey a hint of this throughout these pages by mentioning the relatively recent scientific understanding of the sub-atomic world, where mind bending gravitational effects, quantum mechanics and chance theory underpin our reality.

I dare say that this is a small part of the world the Bible calls 'spirit', and is the basis for the comparatively simplistic world we experience with our own biological senses.

As Goethe simply put it, 'The greatest happiness of the thinking man is to have fathomed what can be fathomed and to quietly reverence that which is unfathomable'. There is just so much that we can understand about the world around us, and for our own happiness and mental wellbeing we should quietly reverence the rest - that which is currently beyond us.

There is another circle that is being completed as well. Despite Islam's association with the old city of Babylon and the fact that it arose in Arabia, the satanic association is as much with Turkey as anywhere. In our biblical journey we first met the Adversary in the Garden of Eden. We saw how closely that was linked with sun and moon worship at Harran in eastern Turkey - and that going way back to Gobekli Tepe, there are hints of the birth of the post hunter-gatherer world where our troubles began. This circle continued through the Umayyad Caliphate, which was also based at Harran.

It ends with the destruction of Istanbul, as the sun and moon emblems become 'darkened' and the Final Beast Kingdom of Daniel and John is judged by the Son of Man.

We have learned that the mark of the beast, 666, is the number of the beast and also the number of a man. There seems to be something in common with all the successive empires, their Kings, Caesars, Emperors and Caliphs which is reflected in the statue that Nebuchadnezzar had made. The one he directed everyone to bow down to under pain of death.

It was always about the reign of 'Man' as a living idol. Antichrists all, pseudo-messiahs driven by their own desires and goaded on by false prophets. They all aspired to deity and to be worshipped as God (or as if they were God). The Ottoman Sultans eventually made it happen in the Hagia Sophia.

But the story of the Hebrew prophets is only the story of the main line of this idolatry, as it directly affected Israel and the Middle East. Like a disease, it seems to have spread from the centre of the ancient world to the four corners. Throughout Europe we had various kings, emperors and leaders effectively trying to do the same thing, either on the back of the Roman and Holy Roman Empires, or independently down to recent times.

The three demons of Fascism, Nazism and Communism are reminders of the recent past.

There are despots in many countries today, who still aspire to this exalted status and have installed themselves - or have attempted to install themselves - as leaders for life. There have been many in Africa and Asia, and we know some by name.

We still see this tendency in more open societies, where the masses are often used and abused; denied fair wages, working and living conditions and sometimes affordable health support.

The story of China and the Forbidden City, based on the same Cult of Man with its heavenly symbolism, harems and complete authority over the masses, is a cautionary one indeed. Although this belief system could have risen independently, it is most likely an influence from the west, via the Silk Road and contact with the Middle Eastern empires.

The Qin Dynasty of western China were neighbours of the Macedonian Greeks and the Achaemenid Medo-Persians before them, and the Qin subsequently rose to power and unified the whole country, but with uncompromising force.

So it would not surprise, if it was indeed the influence of the 'beast' kingdoms, that inspired the Qin Emperors.

Modern China and other Asian regimes continue a dangerous dance with the same authoritarian rule. Such 'kings of the east' have certain things in common with militant Islam which may be too easily dismissed. They are usually atheistic or humanist in philosophy - or are monarchies with absolute power.

Unlikely bedfellows at first glance, but they share totalitarian ideals. We should not be surprised, therefore, if they choose to confront the West by using the Islamic 'kings' in their attempts to destroy the freedoms we hold so dear. On the other hand the tide has already turned for Islam, and Muslims worldwide are being targeted in many countries as distrust grows.

All this makes future scenarios difficult to predict and a world where no one can seemingly be sure of their neighbour.

We could also be forgiven for thinking that the emphasis in the Bible is focussed on eliminating idol worship in the traditional sense. In fact we have learned the Seven Times prophecies say otherwise. They point to autocratic god-men as the *real* idols to watch out for, and that the prime enemy is 'man' himself. It also suggests that any ideology that is focussed on self-worship is potentially far more destructive to the individual than we might imagine - and there are many self-absorbed 'isms' today; some of them old, some of them new.

They may exhibit some positives in regard to self-development and self-esteem, but the biblical prophecies indicate that our self-centredness is actually more damaging to ourselves and others than anything else could possibly be. We should therefore not be surprised when atheistic and eastern philosophies lead to 'man' based government, where any notion of a Higher Power and external accountability is denied.

But those of us who were blessed with humble and discerning parents were taught the truth of the matter. Put others first; do not bully, do not coerce. Respect and compassion is the only way. To be truly human and the best we can possibly be, we need to acknowledge the Almighty first and keep ourselves fully grounded as Children of God.

In the context of the theme here, I am also reminded of a beautiful worship song written a few years ago by Bill and Gloria Gaither. It is called 'There's something about that Name'. It sums up the whole history of the world from very ancient times as described by the prophets. It says 'kings and kingdoms will all pass away, but there is something about that Name'.

In a few lines it refers to the mighty empires of yore which have come and gone and the name of the Lord, our Saviour.

Focussing once more on the last Empire, Islam, and end time events; Jesus is acknowledged by their teaching as a prophet who brings a new perspective to the world at his first coming as 'messiah' and is a sign of the impending day of judgement at his second. In one version of their end time scenario, Jesus is a sort of accessory to God's plan. He returns to fight a false messiah (Al-Masih ad-Dajjal) at the end time in order to assist in establishing peace on earth at his return. He is not the central figure in this scheme and is not acknowledged as a divine figure coming in power, but rather as a slave of God.

The Islamic idea of 'messiah' is therefore in name only. His crucifixion and resurrection are denied by Islam, but he nevertheless somehow reappears to herald in the end of the age, before dying a few years later as an ordinary mortal.

In this portrayal, Islamic teaching and imagery paints Jesus as a secondary prophet who is surpassed by most others.

Nonetheless he is acknowledged as a 'John the Baptist' like figure heralding Muhammad. Muhammad is then lifted figuratively to the highest heaven of God in the Hadith and therefore an exalted figure, while a 'Mahdi' identity is a temporal world leader who potentially replaces Jesus in the minds of Muslims.

It is difficult not to simply dismiss this scheme, as a pseudo acknowledgement of the God of Abraham and his Messiah, and an attempt to validate Muhammad as a genuine prophet of God. And although this convoluted and contrived scheme makes little sense to non-Muslims, adherents are in effect brainwashed into believing that Jews and Christians have no future except by being incorporated into Islam.

To most of us, this picture will seem inexplicable, but I believe there is something diabolical underlying it all, which is a threat to Muslims themselves. Just as Sultan Suleiman once walled up the Eastern Gate in the Old City of Jerusalem to prevent the Messiah / Prince of the Covenant entering on his return, so this scheme could be psychologically designed to inhibit Muslims recognising the true Messiah right to the end.

They may be eyewitnesses to his arrival, or see him on television perhaps, but because they have been told he will die as an ordinary man, will disregard future Jerusalem events as incidental. Such a despicable attempt to steal souls at the very last minute can only originate from a satanically clever source.

This is not about flesh and blood alone, but as the Apostle Paul stated: about 'principalities and powers'; powers we know little about and who are no friend of humankind.

In tandem with this Islamic scenario, I see potential parallels with false 'futurist' expectations in Christian prophetic quarters.

Many are expecting an antichrist figure to appear in Jerusalem at the midpoint of a future seven years. However before this can occur, many events outlined in the Old Testament prophecies and all the events of the first three and a half times have to take place. Some expect the first four seals of the Apocalypse to be completed, and other visions such as the woman of Revelation 12 and the events of Revelation 11 and 13 are still seen as future in many minds. The third of mankind killed through disease and pestilence predicted in the four horseman prophecy is part of this. And to top it all off - they are expecting the Temple to be rebuilt so that the antichrist can defile it on his arrival.

This is all madness, and has the effect of putting the seven years of their erroneous 'tribulation' on the 'never, never' and the end-time prophecies of the O.T. and N.T. are no longer parallel and cohesive strands of prophetic history.

So if the Lord returns anywhere near the mid-point of the seven year period I propose in this volume (or the mid-point of some other seemingly significant seven years) it could well be that deceived Christians will also ignore his return.

Like many Christian folk, I have always believed that the following quotes paint a picture of an arrival that cannot be missed: *'As lightning that comes from the east is visible even in the west, so will be the coming of the Son of Man'* and *'they will see the Son of Man coming on the clouds of the sky with power and great glory'* as well as *'he will send his Angels with a loud trumpet call'*. However it may transpire, that there will be incredible confusion surrounding his appearance, and that some aspects could be easily misinterpreted.

Because the Lord will return to Jerusalem, very few will actually see this occur in the flesh, and much will probably play out in the media for most of mankind. It will also occur while war and horrific earthquake rages around the Middle East and may even appear to be the alien invasion and abduction (the rapture) that many dread. Indeed, the episode could also be dismissed as another space adventure or television drama.

In short, the turmoil through natural disaster and war could mask the signs of Divine extraterrestrial intervention.

In fact an extraterrestrial intervention is a good way of looking at this scenario for those who cannot handle the idea of Divine and Angelic beings. However it will be an intervention from *beyond* the highest dimension we could possibly imagine.

Jesus also used a 'carcass and vulture' analogy to describe the chaos; chaos which might involve riots and the media gathering to report anything at all, for a news scoop.

Primarily though, it will be outright war, with 'Gog and Magog' gathering like vultures on all sides.

Under these circumstances the return of the Lord could be dismissed as inconsequential, incomprehensible or simply 'fake news' - and this 'fake news' cry is everywhere today.

Indeed, the events could be interpreted as the appearance of the **'antichrist'**, rather than our **Messiah.**

So, as much as it seems unlikely that anyone could possibly miss the return of the Lord, I feel compelled to suggest caution. In any 24 hour period a third of the world are sleeping, many others are working indoors, or away from media contact, and therefore only a third might be in a position to be paying attention in any meaningful way. It means that every eye may see him – could hardly miss the angelic invasion one way or another – but will they recognise what is actually happening and the identity of the central figure? Perhaps not.

Regardless of the merits of the above, the danger and condemnation that will come in the near future is for all those who promote war, terrorism, violence, theft and deception for their own ends; whether Muslim, Communist, Fascist, Capitalist, or uncommitted Christian and Jew. Anyone in fact, who seeks to join forces with those who are exploiting God's people for their own gain, or who prey on others for whatever reason at all.

I suggest however, and hope earnestly that this does not have to be the judgement of those who are good at heart and whose ego does not get in the way of an acknowledgement of their Creator. That those who are filled with compassion for others and who find the oppression by militant regimes and tyrants anywhere abhorrent, will have an opportunity for life.

In fact, there are anointed Christian evangelists who have conducted missionary journeys to troubled places around the world; places like Pakistan, where a cross section of people have been blessed by the Holy Spirit (the settling and dwelling) and in some instances physically healed. This is a sure sign that the Lord knows his own and is gathering his people to himself.

It is also obvious that only those with the desire for love, peace and goodwill in their hearts can inherit the promises of God's Kingdom. After all, why would anyone select selfish, and unrepentant people for a Kingdom of peace and joy? No one of course, and certainly not the Lord; he is Saviour, he is patient and he yearns for all to follow him, but he is also Judge.

Jesus taught we ought to love the Lord our God and our fellow human beings, even our enemies, i.e. those who hate us.

He also taught that he was the only way to Father God, both through his teaching and through the blessing of the Holy Spirit which is within his authority to give. We are therefore not asked to offer up much if our hearts are already in the right place, but simply confess our sin, turn away from known wrongdoing to the best of our ability and redirect our lives toward the Lord and follow his Messiah (Christ) in every way. Given a genuine yearning to follow and trust the Lord nothing is impossible.

Turning once more to our timeline, we see the year 2030-31 will be the 2,000th anniversary (40 jubilees) of the death and resurrection of our Lord Jesus, according to the understanding in this volume, and the analysis in Appendices 4 and 5.

It is also the end of Daniel's extended Age of the Gentiles and the seven years of his 1335th 'day'. It follows that the Lord, through Daniel, has ordained an extra 57 prophetic years for all of mankind, beyond the end of the Seven Times and 1967. We might therefore ask; do these extra years link or synchronise the 'Age of the Gentiles' with the end of the Age of Pisces?

I believe this is so and one way of considering the period. However using an AD 30-31 starting date for the age (and 13 zodiacal ages, as does the I.A.U.) means the Pisces Age ended in 2011-2012 and we are already there. Alternatively a 2,148 year zodiacal age suggests there is some way to go (in fact to AD 2178) until the 'Sign of the Son of Man' finally appears.

Despite this, I am as certain as I can be, that when Jesus used the term 'appearing', he meant the first glimpse before dawn of the very first stars of the constellations of Aquarius, and that this 'appearing' will align with the Daniel 1,335th 'day' milestone. And while the Hebrew astronomical experience was largely Mesopotamian and based on the number 12, history shows that when Joseph's sons Ephraim and Manasseh became major tribes of Israel, that there were actually 13 tribes, with the 13th being the Levites who were dispersed throughout *all* Israel.

This could be an indication that a nominal zodiacal period is applicable for the prophetic Pisces age, and also harmonise with Jesus' words that the days would be 'cut short'.

Another significant, perhaps more pressing question relating to the end of the age; is whether the current conflict in the Middle East will remain regional or escalate to World War III.

It is obviously difficult to know how the future will unfold outside of the region, but there are still a few short years remaining for a global military disaster to develop. In addition a future conflict could, as always, be exacerbated by threats and misunderstandings. If the situation develops in a similar way to World War 1, where the demise of the Ottoman Empire occurred as a sideline to a much larger conflict, the circumstances would leave the Israelis exposed while their historical allies deal with their own global issues.

I hope, however, that 'all Israel' will continue to observe, and be involved in the Middle East and that the modern state of Israel is not forced to stand alone. At any rate, one hopes and prays that they will have allies in a position to help on the day, and that those allies will make themselves available.

As previously noted, the earthquake and natural disaster aspect could definitely have global reverberations, although the Armageddon events of war and earthquake will be centred in the Middle East and region. The dire description of 2 Esdras 7:30 and the Sixth Seal of Revelation 6 do not bode well, and we should therefore prepare ourselves as Jesus taught, for the 'terrible day of the Lord' - and those who still need too, should 'come out of her my people'.

Finally, the prophets have little to say after the 1,335th 'day' and the prophetic timeline stops dead. I think it apt therefore, that there is still some mystery surrounding the end of the age and our contemporary times.

However they *do* tell us that there will be the Parousia (Second Coming / Appearance) and Rapture of the elect, and this will occur in the 'same hour' that the Jerusalem earthquake strikes. This rescue of believers is essential, because the world will not continue forever, and those in Israel will be saved in front of their enemies. In Matthew 24:35, Jesus simply states:

'Heaven and earth will pass away, but my words will never pass away.' (NIV)

And almost unbelievably, there is a promise and hope of a new life and a new world of joy; where resurrection bodies ensure that pain, disease and death are put to rest and sorrow is no more. Revelation 21:3-4

'And I heard a loud voice from the throne saying, Look! God's dwelling place is now among the people and he will dwell with them. They will be his people, and God himself will be with them and be their God. He will wipe every tear from their eyes. There will be no more death or mourning or crying or pain, for the old order of things has passed away.' (NIV)

When I hurriedly put the itinerary together, for my Egyptian and East Mediterranean adventure prior to the 'Arab Spring', I was focussed on the Exodus drama and the part played by the Santorini eruption.

Because of a couple of books I had read, I was also intrigued about the final fate of Mary, the mother of Jesus, and where she played out the last years of her life. We are told in the Gospels that the Apostle John was entrusted with her safety.

Whether she ended her days in Jerusalem or in the hills behind Ephesus in Turkey - or the more speculative possibility - with Joseph of Arimathea and his mythical links to Britain, were all drivers for the places I visited.

What I did not expect, but have come to know, was that the Lord was gently guiding me around the places with which he wanted me to be familiar. This, before he started to reveal the details of the great Seven Times timeline in this book.

That revelation of the identity of the beastly players, their plots and plans and favourite haunts, was to happen over the next seven years.

Come Lord Jesus.

"To him who is able to keep you from stumbling and to present you before his glorious presence without fault and with great joy – to the only God our Saviour be glory, majesty, power and authority, through Jesus Christ our Lord, before all ages, now and forevermore!" Jude (Y'hudah) 1:24

The year of our Lord 2018

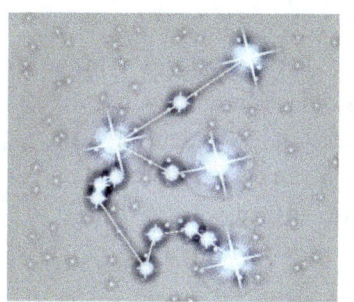

Appendix 1

Lyrics to a rock tune

Glory to His Name

Yes the promised King of Kings is coming back again
It won't be long now, now my friend
Before the Judgement Day, hear what I say
You've got to get together with the Lord today
There's no doubt about it, He's the only way
Glory to His name. Glory to His name.

We read in the scriptures what the prophets said
The world will be filled with fear, fear and dread
In the last days, the end of the age
The Children of Israel go home to stay
Turn to the Lord, without delay
Glory to His name. Glory to His name.

Times are a changin' and a changin' fast
The future is a fadin', fadin' like the past
But Christ will come, the Chosen One
To crush Satan's hordes at Armageddon
It may be getting late, but the battle's won
Glory to His name. Glory to His name.

david 1973

Appendix 2

The false prophet

It is instructive to go over a few passages from the Qur'an in order to see how Islam and the Islamic Caliphates began. Also, the thinking behind some of the Muhammadan pronouncements in regard to Jews, Christians and the Al-Mushrikun (polytheists and anyone who joins other gods to Allah).

Firstly, it has to be acknowledged that there is much in the Qur'an applying to Muslims and their relationship with fellow Muslims, which is perfectly acceptable to Jews and Christians. This is a given, because a good deal of the relational teaching is based on the Holy Bible, or sourced from other religious texts.

The Bible is copied (although not always accurately) with a cursory acknowledgement of those Scriptures. Extant variants of the Qur'an, such as the 7th century Sana'a palimpsest are also problematic. Despite this, we are told that Muhammad received his visions directly through the Archangel Gabriel.

For a society to survive at all, it has to have some semblance of law and order, otherwise it would disintegrate almost immediately. If the system retains some personal focus - such as the role of family - then that is even better. So although the structure may be adopted from another faith system and distorted to a considerable extent, it can look like the real deal to those who have known nothing else. Wrapping up any nastiness embedded within an ideology with copious amounts of relative truth and sweetness, has always been a proven ploy that works.

As is often the case with many ancient texts (and the Bible is not immune in this respect) we sometimes find words which justify war and killing. We even find commands from God *too* kill whole societies. The context of the account may be non-existent, or minimal and the justification questionable, but it is all that some people need to initiate their own antisocial behaviour and violent acts. It is the curse of fundamentalism.

So where does contemporary Islamic terrorism come from? Are the jihadists and 'Islamists' just a modern aberration, with no connection to Islamic teaching of the past? Unfortunately no; they know there is no need to respect non-Muslims and their societies under Qur'anic teaching. Such radicalised people take their Qur'an literally, and any other allegiance they may have is very much secondary.

In the Qur'an it becomes perfectly clear from the start, that anyone outside Muslim society is unwelcome - or if useful, very much a second rate associate of the system. For example in regard to women, it is not acceptable to take a married woman as a wife if she is a Muslim, but that does not apply to married women who are infidels (non-Muslims) and slaves. It opens the door to all manner of gratuitous and perverse behaviour.

Sura 3 (chapter 3 of the Qur'an) states: *'Let not believers (Muslims) take infidels for their friends rather than believers...'* which is fairly innocuous and not dissimilar to exhortations in scripture for believers to avoid being part of the 'world'.

However in the New Testament, 'world' implies pursuing selfish pleasure and bad company to the detriment of your faith, whereas in the Qur'an, it seems to be designed to drive a wedge between Muslims and followers of the Bible. This command is repeated elsewhere in other forms, and is more typically associated with sects that do not allow their adherents any freedom of thought. It leads to the inevitable; *'As for those who become infidels, after having believed, and then increase their infidelity – their repentance shall never be accepted',* Sura 3:84.

So we see at a fairly early stage a clever web being woven to ensnare by propaganda and fear, and why the different brands of Islam see each other as infidels and rivals. Where the beliefs of other Muslims differ from your own interpretation of Muhammad's teaching, they are then seen as worse than other infidels; cannot be forgiven and should be dealt with accordingly.

In the same chapter, Muhammad appears to be writing after an encounter of his militia with the militia of Jews and Christians somewhere in Arabia; a battle which he apparently won and therefore he gloats in Sura 3:10: *'Say to the infidels: ye shall be worsted, and to Hell shall ye be gathered together; and wretched the couch!'* This sentiment is repeated in Sura 98 which ends with: *'They are the worst of creatures'* (referring to Jews and Christians); who are nevertheless referred to as *'people of the book'* in other places.

At this point it becomes apparent that he is now taking it upon himself to punish the 'infidels' for sins real or imagined and is starting to relish that prospect. He attempts to portray his actions as righteous and sanctioned by God.

Despite the obvious aggression of the new regime from the beginning, the rhetoric gets worse and the decline of Muhammad is more akin to that of a despotic warlord than a prophet. In fact Adolf Hitler's hatred of the Jews and the Church developed in a similar way. It is difficult not to avoid the conclusion that in both cases, their psychopathic demeanour gradually worsened, as they lost all sense of reality and their lust for power grew.

In the Hadiths (further teachings) we have 63 – (1767): Jabir bin Abdullah said: *'Umar bin Khattab said he heard the Messenger of Allah [Muhammad] say: 'I shall certainly expel the Jews and Christians from the Arabian Peninsula, until I have only Muslims there''*. In Hadith 7333 there is a similar quote from Umar in regard to a directive he heard from the 'Prophet', *'Most certainly you will fight the Jews, and you will fight them until a rock says: 'O Muslim, here is a Jew, come and kill him''*.

Examples of the more extreme teaching in the Qur'an are found in Sura 9. This chapter is believed by some scholars to be the last chapter which was written. In Sura 9:5 we have: *'and when the sacred months are passed, kill those who join other gods with God wherever you shall find them, with every kind of ambush: but if they convert.... let them go'*.

These threats are based on the Qur'an, rather than Jewish scriptures of course, and there are plenty of other threatening passages with the same message. They leave no one in doubt that ultimately the choice is conversion - or slavery and death.

Consequently escaping, leaving everything behind was usually a better prospect. The choice was to acknowledge Muhammad as *the* Prophet of God and the Qur'an as the definitive word of Allah, or cop the most extreme discrimination imaginable. It is the most onerous requirement in Islam and why at its heart, it is totally incompatible with western society and Judeo-Christianity. And one might well ask why the supposed crime of *not* acknowledging Muhammad as a prophet is a blasphemy offence in hardline Islamic countries anyway?

After all Muhammad was *not* God and this is another sign of the conflict between man and God in the Islamic mind. On one hand Christians are condemned for believing that their Messiah / Christ is Divine, but at the same time Muhammad himself is raised to that august and unique position by slight-of-hand.

It is hypocrisy of an extraordinary kind and began in the seventh century with Muhammad himself.

Personally, I doubt that Muhammad (or his immediate successors) were totally ignorant of the concept of a Messiah visiting the earth as an incarnation of Almighty God. It seems more of an excuse on their part to demonise Christians, simply because they could. It was slander pure and simple. They go to so much trouble throughout the Qur'an to show how awesome and powerful God in fact is, that to then infer that it is impossible for God to appear in a form of his choosing - and equate that with idolatry - becomes hypocritical in the extreme. It also signals Muhammad's own ambition to be seen as 'The Prophet' with Messianic status, despite his denials.

For those who rebel within strict Islamic societies, the consequences can be life threatening. Blasphemy laws can result in imprisonment and likely death at the hands of the mob, or by an Islamic court. Islam is opposed to lying, but once the status of Muhammad and the Imams are questioned, the infidel becomes the liar and very much the target. Often the only hope for victims, is that their case is picked up by the international media.

Now it is very likely, that Jewish and Christian society in Arabia at the turn of the 6th and 7th centuries, was corrupt to some degree or other and guilty of a level of opportunistic behaviour. This cohort would have been a combination of expatriates from the Levant and local Arabs and probably reasonably well off. Muhammad also knew Catholic Christians personally and by all accounts at least one was a relative. The Marian doctrines of the Roman Church were also compromising and the devotion to Mary (adapted from Diana / Artemis of Roman and Greek paganism) a pretty obvious association that could be validly condemned. Given these shortcomings, Muhammad would have used any angst caused by the situation to rally the native polytheistic Arabians - and envy was a likely part of that.

Regardless of provocation, in ancient times the prophets of Israel were given visionary forecasts that clearly show, that those who were to suffer under the Caliphates *would* one day be avenged. Saint John vividly describes Muhammad's own fate.

But where does Islam, the Eighth World Empire, sit today? The Ottoman Caliphate was the last iteration of that monolithic edifice and the last Caliph removed from office in 1924. Although the Caliphate had been in decline for many years, it was only in 1917 that its power was truly lost in the Holy Land.

Since the age of European exploration, many Islamic countries have had a rather modified form of existence. Some such as Malaysia and Indonesia are democracies, or nominally so, and to some extent or other are positive legacies of the colonial era. Pakistan is very much borderline and the prosecution and sentencing to death of Asia Bibi for 'blasphemy' in that country, illustrates the continuing danger to Christians. Nevertheless, these countries with large Muslim populations promote a freer lifestyle for both sexes than the Caliphates before them and Egypt is a notable example. It could well eventuate that this schism between hardline and more temperate nations will grow and that the latter will side with the western Christian nations and survive much of the judgement to come. Despite the positives, they all teeter on the edge of a return to more fundamentalist rule and individual outcomes are uncertain.

For all this, many Islamic nations in the Middle East and region have changed little from earlier times. In fact some have regressed into totalitarian rule of one brand or another and Iran is a prime example. The Emirates with their bright and flashy cities which mesmerise tourists, harbour a world where Muslim women are ultimately *not* free and like Saudi Arabia, require a male guardian no matter who she is, or what her age.

Saudi Arabia also has tentacles that stretch out to many nations, in an attempt to draw free and independent Arabian born women back into the fold. It is also the Islamic nation that leads in Islamic 'evangelism' around the world.

Despite this, the countries on the Arabian Peninsula are vulnerable to the ambitions of the northern 'kings of the east' and are therefore currently aligned with the West, with the mutual benefits that presents. And although seemingly faltering, Islamic State (ISIL / ISIS) has been the most extreme expression of fundamentalist Islam in recent years. It shows how debased the ideology can be, when implemented literally. John's vision portrayed something absolutely horrific and it may well have been the Islamic State foreign fighters committing atrocities.

He saw that there was no hope for these people, because their hearts were turned to evil and their confessed ideology only an excuse for their horrific actions. Therefore the prophecies condemn those who are instrumental in continuing, or resurrecting the 'beast' empires, as an uncaring and hateful form of despotic administration with global ambitions. It condemns those that want the system to persist because of their aspirations of power and wealth at the expense of others. In short, the judgement is for those 'who take the mark of the beast'.

The plight of non-Muslim women within Islamic societies has often been concerning over the years. However it has been truly despicable within Islamic State where torture, rape and sexual slavery ending in death have too often been the result. The plight of the Yazidi women has been particularly heart rending, and this horrific episode illustrates the despicable trickle-down economics of Sharia; where some of the spoils are shared to the male rank and file. And why would we expect anything else where 'man' is worshipped?

However the Apostle John has warned us: Turkey is the one to watch. It may, or may not be the most aggressive, but it will pay for its sin as the last Caliphate Empire.

We have also seen, that the prophets taught that Islam would be a far more subtle system than earlier empires. Because it purports to worship the God of Abraham, Isaac and Jacob it has an appearance of respectability, whereas in fact it uplifts the power of the Caliph, President or Oligarchy. It achieves this so effectively, that the leaders of short-lived atheistic and humanist regimes such as Communism, Fascism, Nazism and Laissez faire Capitalism must gasp at its cleverness.

Nevertheless there are glimpses of hope within Islamic society and the story of Malala Yousafzai - who was shot in the head by the Taliban in 2012 - is one of those episodes. She was advocating a girl's right to education in Pakistan, and after her medical ordeal, became the youngest Nobel Prize winner ever. She freely acknowledges the role of her amazing father.

If the authoritarian political component within Muslim societies is removed and the unsavoury teaching within the Qur'an ignored, then goodwill within the (often minority) Muslim communities in that situation can render the religion benign to a considerable extent. Many are nominal Muslims anyway and secular by disposition. Indeed, there is sometimes more in common between Muslims and conservative Christians in terms of morality, than other, often aggressive secular groups within the 'humanist' and 'progressive' communities.

Remember, for the most part the ancestors of refugees were the ones subjugated by the early Caliphates and they are therefore victims of the system. It is therefore important to engage with refugee Muslims within our society in a friendly and helpful manner. Sport and recreation have an enormous potential for good, as have other communal activities. Ultimately, education and opportunity within our society is vital for them.

Despite the above, we should not be naïve. When Muslims live in the west, we should be aware that a few fundamentalists will feel obliged to sabotage our society any way they can. This is often through illicit activity of one sort or another, rather than overt force or terrorism. They may also threaten fellow Muslims who *do* want to live in peace with their neighbours.

Sadly, some Muslims will also feel obliged to join the fight in the Middle East against Israel in one last push.

Now I have limited this discussion to a few Qur'anic verses to show how it all started and illustrate some of the more violent aspects of the traditions which emerged.

Once understood, it becomes apparent that the hate speech toward 'infidels' in the Qur'an is what drives Islamic extremists to jihad - even against their own. It makes a mockery of anyone who suggests that there is no link between basic Qur'anic teaching and violent Islamic extremism.

Simply put: Islam has been by far the most enduring of the antichrist regimes foretold by the prophets. A regime bent on the destruction of Israel and all that is truly right and of God.

Appendix 3

Unravelling Ezra

Here is an extended discussion of the post exilic sequence of events from the Books of Ezra, 1 Esdras and Nehemiah and other references we find in the Hebrew Tanakh. The point of the study is to determine the best date for the beginning of Daniel's 70 week prophecy pointing to the appearance of the Jewish Messiah.

As an objective starting point, the Persian Achaemenid king list (in so far as it is required) is presented below. It is derived from substantial Persian and linking historical records:

Cyrus the Great	559-529 BC
Conquest of Babylon	537-539 BC
Cambyses I	
Cyrus II	
Cambyses II	525-521 BC
Conquest of Egypt	525 BC
Bardiya (Smerdis)	522-521 BC
Darius I (the Great)	521-486 BC
Temple completed	520 or 516 BC
Xerxes I (the Great)	486-465 BC
Artaxerxes I	464-425 BC
Ezra arrives in Jerusalem	458-457 BC
Walls rebuilt by Nehemiah	444-445 BC

These dates are accurate to within 12 months as far as I can ascertain, except for the 520 / 516 BC date toward the end of the exile period. There are four separate Persian royal decrees proposed by students of prophecy based on the biblical texts, these are:

Decree to rebuild the Temple

Ezra 1: 537 BC

This date is within a year or two of the 1st year of the reign of Cyrus as king of Babylon. Prior to the invasion of Babylon, Cyrus was king of Medo-Persia and the northern and eastern provinces of that kingdom only. 42,360 Jews return to Jerusalem to rebuild the Temple

Ezra 6: ca. 520 BC (2nd year reign of Darius)

This is the time when Darius receives a letter from Tattenai and those opposed to the rebuilding of the Temple and Jerusalem, but confirms Cyrus' original proclamation.

Reviewing this, we find the rebuilding of the Temple was frustrated by the Trans-Euphrates coalition throughout Cyrus' reign. Then a letter is sent by Tattenai to Darius in an attempt to stall construction further. But Darius enquires after Cyrus' original instructions, finds the document in the fortress of Ecbatana in Media, reaffirms them and the Temple is eventually completed on the third day of Adar in the 6th year of the reign of King Darius (Ezra 6:15).

The year 520 BC seems a little too early for Darius and could be as late as 516 BC. This would also closely coincide with 70 years of exile from the end of Nebuchadnezzar's siege of Jerusalem in 586 BC.

Ezra 7: 457 BC (reign of Artaxerxes)

Ezra arrives in Jerusalem in the 5th month of the seventh year of the reign of Artaxerxes (458-457 BC) with hundreds more returning Jews. This occurs after receiving permission and impressive support from the king.

'Artaxerxes, king of kings. To Ezra the priest, a teacher of the Law of the God of Heaven: Greetings

Now I decree that any of the Israelites in my kingdom, including priests and Levites who volunteer to go to Jerusalem with you, may go. (Ezra 7:12-13 NIV)

Ezra chapter 4 seems particularly convoluted, with the first five verses and the last, verse 26, referring to Darius. The intermediate verses 6-8 include references to king Xerxes and Artaxerxes and then another letter written in Aramaic (presumably from the Trans-Euphrates again) led by a commanding officer by name of Rehum and associates to Artaxerxes. The letter causes the king to halt the rebuilding of the city by the Jews and Benjamites.

It emphasises that Jerusalem is a rebellious city, troublesome to kings and provinces and that is why it was destroyed. It includes these words: Ezra 4:16

'We inform the king that if this city is built and its walls are restored, you will be left with nothing in Trans-Euphrates'. (NIV)

And Artaxerxes response, Ezra 4:21

'Now issue an order to these men to stop work, **so that the city will not be rebuilt** *until I so order'* (NIV)

It seems that the Jews were already rebuilding the city. Certainly Rehum, the commander and his Trans-Euphrates people said this was the case. The question is whether it was true and if so, had the Jews been given permission earlier? Certainly with over 42,000 returning originally some rebuilding would have been necessary from the start.

Another valid question is whether the letter above was really addressed to Artaxerxes (Ahasuerus) as stated, or was it a previous king – Darius or Xerxes I? In other words has a scribe made an error, or the record subsequently been misinterpreted?

In fact Xerxes I (reign 486-465 BC) is believed by historians to have also been referred to as Ahasuerus.

This gives rise to the possibility (at least superficially) that the order *not* to rebuild the city could have been given by Xerxes I, who was the king that Esther the Jewess married.

Although the book of Esther does not appear to link directly to the Ezra account, we know from her story that Haman, who was an anti-Jewish senior official of Xerxes, tried to have the Jews killed throughout the kingdom. Esther and her guardian

Mordecai had this attempt at genocide reversed and Haman was hung on the gallows he had prepared for Mordecai.

But is it possible that it was during the Haman period, leading up to the 12th year of the Ahasuerus (but this time Xerxes) reign ca. 474 BC, that the order was given that was later countermanded by the next king, Artaxerxes (also Ahasuerus)?

This could have been as short as a decade before Artaxerxes came to power and tie the Darius-Artaxerxes accounts together.

Nehemiah 2: 444 BC (reign of Artaxerxes)

The backstop to this drawn out saga, is that Nehemiah's later record explicitly says that the city walls and gates were not finally repaired until much later.

In the month of Nissan in the 20th year of King Artaxerxes, Nehemiah asks the king if he can go to Jerusalem. He has learnt that the city walls are still in a state of disrepair and the gates burnt (Nehemiah chapters 1 and 2).

He is given permission to go on a return journey and obtain timber for building and repairs and guaranteed safe transit through the region of the Trans-Euphrates provinces. When he arrives he does a covert inspection of the city and walls to determine how much work is required to restore them.

After the usual scepticism by outsiders, he convinces the Jews in the city to start rebuilding, which they do with enthusiasm. He is all cashed up with money from the king and provinces and the authority to get it done - and an injection of cash is often the catalyst for action.

As the work progresses, the Trans-Euphrates interest, Ammonites and Arabs plot to attack Jerusalem. But the work goes ahead and is completed rapidly. We then find similar statistics and descriptions repeated in the book of Nehemiah, to the ones found in the original Cyrus campaign to Jerusalem to rebuild the Temple. There is more about Ezra, the priest, reading the law, a national confession of sin, the dedication of the finished wall and various reforms to restart a working society.

There is little doubt that this time (445 BC and directly thereafter) is where the city walls and gates were finally completed under Nehemiah and financed with another injection of royal funding, valuables and resources from the provinces and various other donations. But is this decree from Artaxerxes to Nehemiah the one we should use as a start date for Daniel's 70 week prophecy, or were there earlier ones which were not acted on fully, which might apply?

After all the obvious immense opposition from neighbouring states, should not disqualify earlier edicts just because the job remained unfinished. An edict is still an edict.

The historical record from the first book of Esdras (the non-Hebrew version) has similar problems to the book of Ezra. Firstly it perpetuates the apparent chronological mix up with Artaxerxes seemingly reigning before Darius as recorded in chapter 2. It would be easy to suppose the writer has just jumped into the future for comparison sake, but the last verse closes with the following verse. 1 Esdras 2:30

30 Then, when the letter from King Artaxerxes was read, Rehum and Shim'shai the scribe and their associates went in haste to Jerusalem, with horsemen and a multitude in battle array, and began to hinder the builders. And the building of the temple in Jerusalem ceased until the second year of the reign of Darius king of the Persians. (RSV)

This aligns exactly with Ezra 4, but I strongly suspect the first sentence in verse 30 refers to building the city (Artaxerxes and Ezra), while the second sentence refers to the earlier event; Darius, Zerubbabel and the Temple. In fact some translations divide this verse up into two, recognising that the names of two different kings suggests two separate events and times. The possibility that Darius II is meant here, also seems unlikely.

The second problem as I see it, is that in 1 Esdras the first migration of over 42,000 people back to Jerusalem happens in Darius' reign and not that of Cyrus and (for example) Zerubbabel is therefore associated with Darius and not Cyrus.

In Esdras we see a lengthy story where three of Darius' attendants vie for the king's favour by asking some tricky questions of a philosophic nature. Zerubbabel wins the contest and his favour is for the release of his people to rebuild the Temple. Whether the story is actual history or artistic licence remains a moot point, but chapters 5 and 6 of Esdras certainly point to Darius as the king of the first mass move out of exile.

Here we have little choice, but to assume that the Trans-Euphrates opposition of Rehum and Shimshai and their successful appeal to the king to halt proceedings in Jerusalem, in fact happened. It certainly accords with other clues we are given and there does not appear to be another candidate king prior to Darius who would fit the bill.

Cambyses II, who is the only one with a lengthy reign, spent most of his time on the campaign in Egypt, with consequent opportunity for pretenders to usurp the throne.

The version of events in 1 Esdras appears to confirm that with the original Cyrus proclamation, not a lot immediately occurred. There was too much opposition from adjacent states and Darius is the one who accelerates the program, after a review of the original Cyrus proclamation found at Ecbatana.

Fairly obviously, a later scribe attempting to make sense of a historical sequence he is not familiar with, has become confused and tried to paper over things with this concluding remark in both books. Similarities with the description of the kingly authority given to Zerubbabel, Ezra and Nehemiah suggest to me that some doubling up may have occurred here too.

So the walls and gates of Jerusalem were not refurbished until Nehemiah returned to the city with financial support and guarantees from king Artaxerxes starting around 444 BC. This was the second decree from the king (or his predecessor Xerxes) after the work was previously halted at the behest of the Trans-Euphrates coalition under Rehum.

Regardless of the Ahasuerus identities, if the return of the 42,000 happened later under Darius as it is recorded in 1 Esdras, then we have testimony in chapter 4 verses 43, 47-50, 61-63 that both the Temple AND Jerusalem were to be rebuilt. The text emphasises this. For example, Zerubbabel asks Darius:

*43 Then he said to the king, "Remember the vow which you made **to build Jerusalem**, in the day when you became king,*

47 Then Darius the king rose, and kissed him, and wrote letters for him to all the treasurers and governors and generals and satraps, that they should give escort to him and all who were going up with him **to build Jerusalem**. *⁴⁸ And he wrote letters to all the governors in Coelesyria and Phoenic'ia and to those in Leb'anon, to bring cedar timber from Leb'anon to Jerusalem, and to help him **build the city**. ⁴⁹ And he wrote for all the Jews who were going up from his kingdom to Judea, in the interest of their freedom,* **that no officer or satrap or governor or treasurer should forcibly enter their doors***; ⁵⁰ that all the country which they would occupy should be theirs* **without tribute***; that the Idume'ans should give up the villages of the Jews which they held;*

⁶¹ So he took the letters, and went to Babylon and told this to all his brethren.⁶² And they praised the God of their fathers, because he had given them release and permission ⁶³ to go up **and build Jerusalem** *and the temple which is called by his name; and they feasted, with music and rejoicing, for seven days.* (RSV)

So we have Darius on one hand dating around 516 BC (or earlier) and on the other Artaxerxes in 444 BC under Nehemiah's leadership. But it is the proclamation and undertaking between these two which is of specific interest, the one with which Ezra was specifically tasked; the one countermanded in the first instance and then reversed.

Artaxerxes commands Ezra the scribe to take some of the remaining people, priests and Levites to Jerusalem. He is given considerable authority to take whoever he needs, gold and silver from Babylonia, other resources from Syria and Phoenicia and he is not to be taxed by other satraps (regions). He is also told to appoint judges who will judge and teach all those living in the region including the Trans-Euphrates people.

This is quite remarkable and not dissimilar to the respect and authority that was heaped on Zerubbabel's shoulder by Darius above. Although the words do not specifically refer to the walls of Jerusalem, it would seem Ezra's jurisdiction is to extend to the whole region west of the Euphrates and adjacent to the Mediterranean.

Ezra therefore has to be a very strong candidate for the first official Medo-Persian proclamation to rebuild Jerusalem.

Appendix 4

A dating dilemma

In this analysis of the prophetic timeline, I have shown that each milestone can be imagined to extend over several years.

I have suggested seven years is very likely, with the first year key in most cases and that there is scriptural support for the proposition as well. The seven year templates were also carried into the Messianic alignment, where we saw that in the first 3.5 times that AD 50 was the midpoint between AD 30-31 (the death and resurrection of the Lord) and AD 70-71 AD and that Jesus' ministry was seen to be a prophetic half of seven years.

Despite linking that up satisfactorily with Ezra's return to rebuild Jerusalem in 457 BC (via Daniel's 70 'week' prophecy) it has to be admitted that there are many analysts and commentators who support an AD 33 date as the crucifixion and resurrection year. They often refer to the year as 32-33 and therefore two years from the AD 30 or 31 years, I have focussed on in the text. It could therefore be argued that if I had used the midpoint of the nominal seven year period as the key marker, rather than the first year, that this would push the resurrection year along to AD 33-34 and be closer to the view of the supporters of year 33.

However this proposition would make a mess of the Ezra timeline based on using solar years (as I previously presented) and not something I could stomach. It is because of these unresolved matters, that I am going into more detail to address some of the historical data around the year of our Lord's Passion.

If Jesus' ministry began sometime in AD 27, the three and a half years I have proposed could reasonably extend to the Passover of AD 31 but no further. With this in mind, the background for the period in question is the reign in Rome of Tiberius Caesar who lived from 42 BC to AD 37.

He decided on self-retirement on the Isle of Capri in AD 26 and pursued a life of self-indulgence and depravity; something not uncommon amongst the Caesars of the period. Nevertheless Tiberius remained engaged enough to appoint an ex-Captain of the Praetorian Guard as regent; a character by the name of Aelius Sejanus who administered the Empire for five years.

Sejanus was a thoroughly ruthless man, anti-Semitic to boot and eventually began plotting his way to the top by eliminating the competition in Rome any way he could. However a trusted sister-in-law of Tiberius, Antonia, alerted the Emperor to the nefarious goings on in Rome and he promptly orchestrated a public denouncement of Sejanus in the Senate. Sejanus was summarily executed the same day; October 18th AD 31.

Now the connection of this event to Jerusalem (by at least one commentator) follows the logic, that this was the pivotal event, that prompted Pontius Pilate's decision to allow Jesus to be tried by the Jewish Sanhedrin for blasphemy. It was a mock trial to be sure, but whereas Pilate would have once given no quarter to the Jews, he now washes his hands of the affair and lets them proceed. It is a circumstantial case and something of a long bow to draw, but supports the notion that Jesus was crucified after AD 31 and not before October that year.

Pilate was appointed as Prefect of Judea about the same time Sejanus was appointed Regent by Tiberius. It is therefore likely that Sejanus initiated the Pilate move and that Pilate then implemented the Sejanus anti-Jewish policies in Judea.

Both Publius Cornelius Tacitus and Flavius Josephus in their 'Histories' and 'Wars of the Jews' respectively, describe the unique Hebrew concept of a 'hidden' God and also the aggravation of the Jews by Pilate. One such episode occurred when images of Caesar were brought to Jerusalem in order to antagonise the Jews over their aversion to idolatry. The book of Luke hints at even worse abuse (Luke 13:1) where Galilean worshippers had their own blood mingled with religious sacrifices by Pilate; a rather unnecessary anti-Semitic act.

So the argument is that after the 18th of October everything changed; and whereas, while Sejanus was alive anti-Semitism was rife, after that date Pilate's position was in jeopardy and he was far more likely to agree with Jewish demands.

As logical as this is on the surface, Sejanus' demise would have occurred only six months after Passover that year, if my information is correct. The unsettling events in Rome had been occurring for some years, and would have filtered through to the Provinces and Pilate well before the October of AD 31.

In fact we are told Pilate himself was reported to Tiberius for his unnecessarily antagonistic acts to the Jews and reprimanded by him. It means Pilate would have been just as wary of Caesar Tiberius as his appointed representative.

The knowledge that Sejanus was getting rid of rivals at a rapid rate would be extremely worrying and likely to end badly one way or another. As a consequence Pilate was in a terrible quandary, but I suggest he decided to do what he thought Caesar would want and placate the Jewish leadership. They were in his face and further unrest in Judea was not a prospect that would advance his career. His previous aggravation and amusement at their expense, therefore ceased and he washed his hands of the whole business. But this most likely occurred at an earlier Passover, sometime before the 18th of October AD 31.

Run, hop, step and jump

As interesting as the Tiberius Caesar, Aelius Sejanus and Pontius Pilate story is, our attention is now directed to more definitive historical and scientific evidence recorded in the Gospels and other pagan and historic Christian literature:

- The record in the New Testament Gospels of the life of Christ
- Descriptions by early second century pagan and Christian historians relayed by later (mostly) Christian scholars
- The various calendars used in the first century before the introduction of the Julian based AD-BC calendar of Dionysius Exiguus in 525
- Computer generated ancient calendars, full moons and ancient eclipses
- The timing of earthquakes in Judea from recent geological evidence in the Dead Sea

Some of the best sources of information come from historians of the 2nd century, or those who commented on them in later periods and had access to documents no longer available today. Biblical and extra-biblically recorded events and people of interest include the earthquake recorded at the time of Jesus' crucifixion, the darkness that engulfed the land for three hours from midday onward, and anything from archaeo-astronomy which might usefully constrain historical events.

To that end, some of the non-gospel sources are Phlegon, a Greek who wrote during the reign of Hadrian (AD 117-138) and came from Tralles, near Ephesus. Another gentleman from the 1st or 2nd century called Thallus (who is thought to have come from Syria). A third theologian and teacher called Clement of Alexander who also wrote in the 2nd century and Justin Martyr, a Judean Christian apologist who lived about the same time.

Others have studied the 46 years for the Temple reconstruction from Herod the Great's 18th year (John 2:20 and Josephus' 'Antiquities of the Jews') but I will focus on the writers above.

Addendum: *At this point I would like to apologise to readers who picked up a copy of this work in the three years or so after publication and note some differences here.*

After further revision, I realised that some corrections were required and this Appendix is an example. Previously, I aligned the Olympiads incorrectly and found relationships between various ancient calendars very confusing. In fact I still do, but will attempt more precise calendar and 'day of the week' alignments and address the issue of '*after* the 62nd week' of Daniel 9:26.

With that off my chest, the Daniel 70 week prophecy and the Seven Times timeline outlined in the text, tend to point to AD 30 or 31, rather than AD 32 or 33. However with the calendar alignments corrected, the possibilities are spread, to the extent that all four Passover candidates require reassessment.

In fact, computer generated full moon data and Passover dates for the period, were checked out over a much wider range.

Now Phlegon is quoted by many later authors including Eusibius, Jerome, Anastasius and The Chronicon Alexandrinum, while other historians such as Julius Africanus, Maximus and Origen, cite his work. Importantly, Phlegon majored on the Olympiads; their numbering and usefulness as a historical timescale, and events which intersect with them. These quotes are of considerable interest to us, but come in a number of versions.

The ancient Olympic Games occurred at four year intervals as they do now, and any one four year period is called an Olympiad; a run, hop, step and jump so to speak. It was an alternative system to the relative dating generally used in ancient times, based on the reign of kings and emperors. But the Olympiads also augmented calendars based on the sun and moon.

In those days, each Olympic year began at the northern summer solstice rather than January. This loosely aligns with many of the ancient calendars, and the Tiberius ascension date of 19th September which is critical to our period of interest.

Note: *All the calendars are found in dot-point form beginning on page 534 and discussed in detail there.*

Thallus wrote his Histories at an uncertain time, but they are only confirmed in a second century quote by Theophilus writing around AD 180. Clement is thought to have been born in Athens, but after converting to Christianity, he then moved to Alexandria where he wrote and taught ECF notables such as Origen.

George Syncellus, a 9th century monk, quotes the 3rd century Julius Africanus who comments on a Thallus reference. It is a description of an 'eclipse' of the sun as the cause of the 'three hours of darkness' at the time of Jesus' crucifixion; a description which is identical to that of Mark in his Gospel.

The relevant Africanus excerpt:

' "... after the **most dreadful darkness** fell over the whole world, the **rocks were torn apart by an earthquake** and **much of Judea** and the rest of the land was torn down." Thallus calls this darkness an eclipse of the sun in the third book of his Histories, without reason it seems to me. For... how are we to believe that an eclipse happened when the moon was diametrically opposite the sun?'

So Africanus points out that it is impossible for a solar eclipse to be the cause of three hours of darkness during Passover (Easter) because, by definition, the Passover always comes 14-15 days after a new moon and is therefore at the full moon. For an eclipse of the sun to occur, the sun has to be on the same side of the earth as the moon, but behind it, whereas at the full moon, the sun is clearly on the same side of the moon as the earth and shining its light directly on the moon's surface.

Furthermore solar eclipses never last anywhere near three hours; indeed a few minutes at most. The passage nevertheless describes the unusual period of darkness and an earthquake.

Interestingly, Phlegon is also interpreted as repeating this oft repeated fiction in regard to a solar eclipse and Julius Africanus is scathing in his response. However, Phlegon and Thallus may simply be describing the obvious; *that the sun was **obscured for a long time*** and therefore a 'great eclipse' of unknown origin occurred. *The word may have had a broader usage at the time.*

Now Africanus refers to both Thallus and Phlegon (as do other later writers) and of course there are questions about reliability and corruption. Regardless, he and the Jewish Consul Tertullian specifically take the author to task over his 'eclipse' description (Africanus' comment is dripping with sarcasm) but at the same time, fixes the event to Tiberius Caesar.

Despite reliability issues, the two sources seem to be basic historical accounts. Importantly, Phlegon is describing an earthquake and darkness over Bithynia in the far north west of Asia Minor, while Thallus seems to be a lot closer to Israel and Jerusalem and describing a similar earthquake to that of AD 749.

This brings us to the historian Eusibius, who was also known as Eusebius of Caesarea. He was a Christian scholar who became the bishop of Caesarea Maritima around AD 314 and quotes Phlegon verbatim. This narrative is also attested to by Syncellus and Jerome:

> Jesus Christ....underwent his passion **in the 18th year of Tiberius**. Also at that time in another Greek compendium we find an event recorded in these words: *"the sun was eclipsed, Bithynia was struck by an earthquake, and in the city of Nicaea many buildings fell."* All these things happened to occur during the Lord's passion. In fact, Phlegon, too, a distinguished reckoner of Olympiads, wrote more on these events in his 13th book, saying this: *"Now, in the **fourth year of the 202nd Olympiad**, a great eclipse of the sun occurred at the sixth hour [noon] that excelled every other before it, turning the day into such darkness of night that the stars could be seen in heaven, and the earth moved in Bithynia, toppling many buildings in the city of Nicaea."*

There is rather a lot to take in here and it would seem to have similarities to Thallus' description. I will return to this passage from time to time, but will now address the question of the earthquake and the surrounding circumstances.

Because the point of reference for Phlegon's account is today's western Turkey in Bithynia, it means the area is adjacent to Istanbul and uncomfortably close to the North Anatolian Fault zone of Revelation fame. This was outlined in the last chapter of this book and it is therefore no wonder that buildings were toppled in Nicaea. If fire also resulted (which is usually the case in great earthquakes) then the three hours of darkness is not surprising either. However Nicaea is 800 kilometres away from Jerusalem and is not easily linked without an understanding of the plate tectonics and seismicity of the region..

Nicaea was also not far from Phlegon's home town near Ephesus and so we need not doubt the general accuracy of the record, despite the years in between. Having experienced an Australian bushfire myself, and been surrounded by burning suburbs where almost 500 homes were lost, I have an idea of the carnage. One surprising thing about this experience was the complete blackness outdoors that was only interrupted by the occasional fireball. It meant using a torch to see anything at all, for about two hours from 1.30 pm onward.

So to me, the Phlegon description could be explained on the basis of the obvious destruction, except for one very unusual observation - the stars were visible during the day. This seems to rule out the likelihood that smoke or dust was responsible, because that of course, would have affected the sun *and* the stars. Having said that, I have also experienced a situation where a column of black smoke completely obscured the sun, but not the rest of the sky. So perhaps stars might be seen, depending on the circumstances and the position of the observer.

Nevertheless, something else has in all likelihood happened here; probably something extraordinary and rare.

It could have been another solar system object; something large which passed close to the earth on a path toward, or away from the sun. Even then one wonders if it would have been dark enough to allow the stars to be seen in broad daylight, or completely obscure the sun for three hours. However if it was a near miss, it could well have triggered the earthquakes.

I have no answers at this point, but suspect that the explanation will be most interesting and the 'eclipse' likely experienced over thousands of square kilometres. On the other hand, could the destruction experienced in Asia Minor and described by Phlegon been incorporated into the Gospels? After all, there were many Jews in Asia Minor to experience the Nicaean disaster, and we met some in the Seven Churches of Revelation.

Regardless, both Bithynia and the Jordan Valley are earthquake hot spots (and the Bible record from Joshua onward attests to the latter region's susceptibility) and so there is no question that there were earthquakes and tremors during the early 30s in Jerusalem as well as Bithynia.

The evidence is in the exposed sediments of the Dead Sea. Since 1960, the Dead Sea water level has been dropping rapidly as water is used upstream and this has facilitated research work. There are three locations where work has been carried out - or is still underway - and one is an exploratory trench at Ein Gedi. Here the annual varves are dated by counting upward from the varve (seismite) of the well-known 31 BC earthquake recorded by Flavius Josephus. Above this there is a distinct disruption in the range AD 29-33 and one relatively large earthquake occurred, approaching 5.5 to 6 on the Richter scale.

The nature of the varves with their often distinctive summer and winter lithology, unfortunately degrades in some parts of the column. It depends very much on if, and when, the rain arrives in any particular year and this is where the difficulties in identification and counting arise. However C_{14} dating of carbonaceous material continues, along with pollen studies within each varve to establish the years and seasons.

It means there was earthquake activity in both Bithynia and Judea, but the specific year of interest is yet to be established.

Returning to the comments attributed to Phlegon and Thallus above (and the subsequent commentary on their records) we are told that the Bithynian earthquake occurred during the 202 Olympiad, and the Tables provided on page 538 and 539, show that the fourth year of the 202 Olympiad corresponds with AD 32-33. It means the Olympiads are measured from a point 202 Olympiads ago and are traditionally dated from 776 BC.

Aristotle was one Greek scholar who confirms this, and importantly, this is the start date that Phlegon uses as well.

SEVEN TIMES • 529

The Metonic and Callippic cycles that tie into the Olympiads are discussed in Appendix 5, but for now the Tables provided show the Olympic BC-AD years of interest to us.

Note: Callippus began his Callippic Cycles of 76 years (19 Olympiads and 4 Metonic cycles) in the summer of 330-329 BC. It also transpires that *330-329 BC is 9 years into the 24th Metonic cycle dating from 776 BC,* rather than more obviously connecting to the beginning of a Metonic cycle.

Because every Olympiad year begins at the northern summer solstice, it also means that in the Table the Julian / Gregorian years are chopped in half. By counting back in groups of four from Olympiad 202 (AD 29-30 to AD 32-33) the Olympiads can be marked off back to 26-25 BC (Olympiad 188.3) and the beginning of the 5th Callippic cycle. This covers the First Advent of Jesus quite conveniently.

According to Eusebius (his quotes from Phlegon are in italics) the Passion of Christ (the torture and crucifixion) occurred in the 18th year of Tiberius. Notice particularly, *that the 18th year comment is Eusebius' own* and not that of Phlegon.

Interestingly, Eusebius cannot be counting Olympic years from Olympiad 198.2 (AD 14-15) to Olympiad 202.4 (AD 32-33) inclusively, because there are 19 Olympic years (and corresponding Passovers) and not 18. He is therefore using another calendar (or counting convention) and likely options are the actual Tiberius regnal year or the Jewish Civil. It means that the last 6 months of AD 32 are counted in Olympiad 202.4 and the last 3 months of the Civil and regnal years are too.

However the 18th Passover and likely earthquake is in the previous Olympiad, 202.3. If this is correct, it follows that the 202.4 Passover of AD 33 will *not* be a Passover of interest.

Eusebius (in the 4th century) tells us elsewhere, that in the 16th year of Tiberius and also in the 17th year, Jesus Christ, Son of God was preaching his message (throughout Judea and Galilee). He believes that this accords with the account of the earthquake and darkening of the sun by Phlegon, and this scenario will be based on the three Passovers in the Gospel of John.

Those that have read Eusebius' commentary at length, note several issues with his historical text, however we can be certain that Eusibius was aware when Tiberius' reign was officially inaugurated on the 19th September 14, and that this is the first place to start for his interpretation of the Phlegon transcript.

As a Greek historian, Phlegon kept a record of the Olympiads right through to the 229th Olympiad in AD 137, including the overall chronology and associated historical information. He would have had considerable material at his fingertips, including the many historical texts that were held at the Celsus Library in Ephesus. The library was built (or rebuilt) in AD 117 as a monumental tomb for the local Governor and the remains of the building are still inspiring today. Because of Phlegon's considerable historical expertise, a 6th century writer named John Philoponus (Philopon) is one of many to comment on his work:

*'Phlegon also makes mention of the eclipse of the sun as the event which transpired when Christ was put on the cross, and not of any other, is manifest: First, because he says **such an eclipse was not known in times prior; for there is but one natural way of every eclipse of the sun:** for the usual eclipses of the sun happen only at the conjunction of the two luminaries: but the event at the time of Christ the Lord transpired at full moon; which is impossible.....'*

This passage also notes that this was no normal solar eclipse, but also makes it clear why. Philopon then continues:

*'The same thing is proved also from the history of Tiberius Caesar: For Phlegon says, that he began to reign in the **2nd year of the 198th Olympiad**; but that in the **4th year of the 202nd Olympiad** the eclipse has already taken place...'*

Here we see Philopon's reference to the beginning of the Tiberius reign, but also a second version of Phlegon's account where the 4th year is highlighted, as in the Eusebius version.

However it emphasises that '*the eclipse has already taken place*'. The Greek verb describing the 4th year of the 202nd Olympiad (as per the first quote from Eusebius) is apparently written in 'the perfect active infinitive tense' and describes something that has already happened; in this case *before* (or by) the 4th year of Olympiad 202. If Philopon's version of events is correct, and the 'infinitive tense' explanation true, then Phlegon is referring to an event that happened *before* the 4th year.

Perhaps more cautiously; *before the end* of the 4th year, but anyway, this complements the Eusebius version of Phlegon very nicely, because the Passover of 202.3 (AD 32) comes in the second half of the Olympic year and immediately prior to Olympiad 202.4. It therefore looks like an open and shut case for an AD 32 Passover, except that there is a chance that Eusebius used an Egyptian convention, where the early weeks of the Tiberius reign in AD 14 were added to the following year of the Egyptian calendars. This would make his first year an extended one and reduce the regnal count. This method is described later.

In any case, there is a glimmer of hope for an earlier Passover, and as it transpires, we have a third version of Phlegon which again comes to us through Philopon, but this time through Balthesare Corderio in *De Mundi Creatione,* 1630:

'And of this darkness, or rather of the night, Phlegon.... made mention. He says 'in the **second year of the 202nd Olympiad** there was an eclipse of the sun, of a greatness never known before. At the sixth hour of the day it was night, so that the stars in the sky appeared."

This transcript confirms that it was Phlegon, who describes the eclipse of the sun, however there is a dilemma; when did this event occur? Was it the *4th year of the 202nd Olympiad* as the Eusebius version states, *or the 2nd year?* Philopon then comments on the solar eclipse explanation as a matter of course.

Given, I'm still hopeful that the Passover in question might be earlier than Olympiad 202.3 (and AD 32) it remains to explore this 2nd year option. Is it possible that a Tiberius countdown (18 years or something else) could still be valid counting from an earlier time than the official Tiberius' regnal year?

This seems most unlikely until we examine the life of Tiberius leading up to his inauguration. Tiberius had a very turbulent life, with memorable highs and extreme lows, and as a consequence, it seems he was not at all keen to take on the responsibility of Emperor. After heroic victories in Pannonia, Dalmatia and Germania, his personal life deteriorated, and amongst other things, he was forced by Augustus to divorce his first wife Vipsania and marry Augustus' daughter, Julia; a rather unhappy and wanton character. Tiberius' troubles resulted in self exile on Rhodes in 6 BC, but he eventually returned and it is generally believed that around AD 13 his status was made equal to Augustus as a co-Princeps. This was after Augustus' grandsons and heirs had both died.

This co-regnal arrangement meant that if Augustus himself passed away or was incapacitated, there would be no imperial vacancy that a pretender might use to seize power.

Now AD 13 is covered by two Olympic years (197.4 and 198.1) but the historian Suetonius Tranquillus states that it was actually in AD 12 that the co-Princeps was established.

On this basis, it would seem reasonable to assume that Olympiad 197.4 (AD 12-13) covers the first co-regnal year - and this in turn suggests that Olympiad 198.2 (which so obviously is the starting year for the official Tiberius reign) has a genuine rival. It would also seem that Eusebius (in the third and fourth centuries) was unaware of the manuscript that Philopon had access to, and never sighted the Olympiad 202.2 (second year) manuscript option for the Bithynian earthquake.

This new Tiberius' start date begins two years earlier than Olympiad 198.2 and therefore appears to point to AD 29-30 (Olympiad 202.1) using the same numbering logic as before.

An AD 30 or 31 Passover date ties in nicely with Ezra's return in 458-457 BC as well; the likelihood being his contingent returned to Jerusalem in the Autumn of 458 BC and hunkered down, or alternatively, arrived in the spring or summer of 457 BC. At the end of the 69 weeks (483 years) that translates to AD 26-27 and the ministry of John the Baptist likely begins during AD 27, with Jesus' ministry commencing months later.

The physician Luke, who wrote the Gospel and the book of Acts in the 1st century, says that Jesus began his ministry in the 15th year of Tiberius, and this again depends on which Tiberius year we use to begin our count. From the co-regnal year Olympiad 197.4, this points to AD 26-27 (or possibly AD 27-28) to a reasonable degree of certainty, depending on which calendar is used; regnal, Jewish Civil, Julian or Olympic.

If the count is meant to be taken from the Tiberius official inauguration in AD 14-15 (Olympiad 198.2) then the 15th year and the likely Passover candidate will also be deferred by two years. The considerations also depend on conventions used in the first century and this will be examined shortly.

We also know from NASA records, that only a few alignments are possible for Passover and the Crucifixion for our period of interest. This is because we are looking for a full moon and in that regard, NASA has not only gotten man to the moon, but has brought the (full) moons to us!

They also have tables of lunar eclipses going back to the first century and beyond - and lunar eclipses only occur at the full moon. From AD 29 through to AD 33, Passover and the full moons only appear in the month of April, except for AD 31 when it occurred on the 27th March according to the Gregorian calendar. So surely this will nail the year immediately?

Not quite, but it knocks out the 25th of April AD 31 where there was a full moon and partial eclipse, but was not in the month of Nisan (a necessary requirement).

Because we have looked at this problem fairly rigorously and used an Olympiad based table, we can now see that none of the Olympic years 202.1, 202.2 or 202.3 can be eliminated at this stage, with AD 33 also an outside chance. It all rests on opposing testimony and relying to some extent on the Tiberius co-regnal period. It means some sort of picture is emerging in relation to the earthquakes and three hours of darkness, but there has also been too much ambiguity and error in communication. There is simply a desperate need for further solid data.

Calendar confusion

A quick review of the various Macedonian and solar calendars used during the Greek and Roman Empires will now assist.

The Macedonian calendar used in Syria and other parts of Greek controlled Asia, was not exactly the same as the calendars used in European Greece, and many regions and cities in Asia had there own customised Macedonian calendars as well.

However they were all based on the ancient Macedonian lunisolar calendar, which was similar to the old Babylonian. The difference was that each year of the old Babylonian calendar began near the spring (vernal) equinox, whereas the Macedonian began at the autumnal equinox. Significantly, after the Alexander the Great conquest of Asia, the Babylonian calendar converted to an autumn start and the Jewish Civil calendar followed. This meant that most Greek expatriates were using the same calendar, albeit often with different names, orders of the months and sometimes, different start times.

Despite this attempt at standardisation, there was still too much confusion, and the Greeks realised this at the beginning of the Hellenic period in the fifth century BC; in fact back in the days of Herodotus of Halicarnassus. They therefore sorted out much of this, then and there, by adopting the Olympiad numbering and four year count system across their world. Each Olympic year also commenced at the summer solstice, so that the games were a combined New Year and summer celebration. It meant each city of the Greek Diaspora could keep their treasured heritage calendars, but synchronise through the Olympics..

Now, the origins of the Egyptian solar year are pretty much lost in antiquity, and this is the next calendar I will address. It was traditionally linked directly to the Nile flood around July, and also with the appearance of the star Sirius, but somewhere back in the day, it became a mandatory 365 day year.

An inscription from Kom Ombo (south of Thebes and Edfu) only shows 360 days, but the most likely explanation is that the Egyptians did not always record the short intercalary month of 5 extra (epagomenal) days added each year, on their records.

The Egyptian 365 day solar year was also known as the 'wandering' or 'vague' year, because it was without leap years. This needed correction and occurred in Egypt when the Augustus' Alexandrian calendar was introduced in the third decade BC (26 BC). It was based on the Julian year, but started on August 29th (rather than January 1st). As a consequence, the Macedonian lunisolar calendar (introduced to Egypt from about 222 BC by Ptolemy II) was gradually sidelined by the corrected solar year and only tolerated out of political expediency.

In order to provide a comprehensive picture of the calendars used in Asia and Egypt in the first century, here is a summary:

- The Olympiad calendar beginning June 30th, or at a convenient new moon after the summer solstice

- The later version of the Egyptian Civil calendar beginning at the Nile flood and dating traditionally to July 19th
- The Egyptian Alexandrian (subsequently Coptic) calendar aligned with the Julian calendar, but beginning August 29th
- The Macedonian / Jewish Civil lunisolar calendar used throughout Asia and beginning Dios / Tishrei / Sept-Oct
- The modified Egyptian Macedonian luni-solar calendar also beginning each year at the autumnal equinox, but in the first decameron (decan) of October
- The Mosaic lunisolar adapted calendar beginning 1st Nisan
- The beginning of Tiberius Caesar's reign on the 19th September, AD 14 and some very interesting observations concerning Egyptian dating customs
- Our baseline BC-AD retrospective, Gregorian calendar

A visual inspection of the above calendars, shows that the Olympiads and Egyptian solar calendars all commenced a month or two before the Tiberius inauguration date on the 19th September. In fact, the annual start dates were deliberately aligned in an attempt to get over the confusion of so many different calendars. Importantly, they all cover the same Tiberius regnal year *and* the following spring Passover. By contrast, the Egyptian Macedonian luni-solar calendar begins in the first ten days of October and therefore after the 19th September.

From the information I have gained from various sources, the Egyptian Greeks achieved this by using an appropriate start date for their year in the first decameron (ten day week) of the Thoth–Dios month. Exactly how they did this I am still uncertain, but I assume the lunisolar year continued in the normal way and a day selected annually to link to the solar year.

The next thing to digest is that the AD (anno Domini) year of the Lord calendar was only created in AD 525 by Dionysius Exiguus, who lived in the region of Romania-Bulgaria today. He was a Roman Consul and decided that the use of the then, current Diocletian calendar was not exactly appropriate, given Diocletian was a notorious persecutor of Christians. Accordingly, AD 525 directly followed Diocletian 247 and it seems he used the Roman Consular year as the basis for the beginning of his new calendar; a calendar that began on January 1st.

In regard to the calendar start year, he may have simply begun his count based on the statements from the Gospels about the beginning of John the Baptist's ministry. If so, he then used the correspondence of the '15th year of Tiberius' and 'Jesus was about 30 years old' statements to roughly estimate Jesus' birth.

Subsequently the BC count was initiated for older dates, but those involved did not think to use a BC-AD 0 year at the conjunction. This is why New Year's Eve 1999 / January 1st 2000 was one year short of the 2,000 years one would expect.

Chronologists have unscrambled all this to calculate dates for the reign of previous Roman Emperors and all our earlier BC heroes proleptically, and mere mortals are therefore beholden to them as a basis for further analysis. Today we use the Gregorian calendar almost universally, and that is because the Julian year was based on a 365.25 days, whereas the actual solar year is closer to 365.24. For the early 1st century, the computer calculated discrepancy is 2 days with the Julian beginning first, but that cannot be ignored when fixing dates precisely.

I will now focus on N.T. passages describing the last Passover and the Justin Martyr First Apology (ca. AD 155-157):

*'For he was crucified on **the day before that of Saturn**; and on the day after that of Saturn, which is **the day of the Sun, having appeared to his apostles and disciples**...'*

Here Justin Martyr relates the Passover in terms of the Roman calendar, and of course most people assume that anyway.

In fact it may be true, but the Gospels only talk about Sabbaths (and high Sabbaths) and the 1st day of the week, after the manner of the Mosaic calendar. Despite this, Justin Martyr, as a Jew from Nablus, is plainly using the Julian calendar for fixing the Passover and cannot be arbitrarily ignored.

Another point is that the Old Testament confirms in Exodus 12:2-11; Leviticus 23;4-8 and Numbers 28:16-18 that the 14th of Nisan is the Passover, whereas the New Testament passages in Matt 26:17-19; Mark 1412-16; and Luke 22:1, 7-13 could be seen to take a modified view, that the 14th is extended into the 15th and the Feast of Unleavened bread. It would seem that Jesus and his disciples actually followed the Mosaic lead by celebrating the Passover on the eve of the 14th - out of necessity of course, but obvious intent as well.

Being mindful that the Gregorian date will be adrift of the Julian by 1-2 days, we can again explore the years within the 202 Olympiad. Because we still need more supporting evidence, here is a reference from Clement of Alexandria, ca. AD 190:

> "And treating of His passion, with very great accuracy, some say that it took place in the **sixteenth year of Tiberius,** on the **twenty-fifth** of Phamenoth; and others the **twenty-fifth** of Pharmuthi"

This passage is interesting on at least two fronts. Firstly it gives us two possible dates for Passover; either the *25th of the month Phamenoth*, or the *25th month of Pharmuthi* (both Egyptian names). Secondly it gives us another alignment for Tiberius – the *16th year of his reign*. The 16th year appears to conflict with the Eusebius' dating, but **the 18th versus the 16th year** has now been reconciled by the Augustus-Tiberius co-regnal period.

When we study Clement of Alexandria's reference, it becomes obvious that he is reviewing historical commentary based on either the Egyptian Macedonian lunisolar calendar (a calendar which had received some direct Babylonian input in the early Ptolemaic period) or the Egyptian Alexandrian solar calendar with the subsequent Roman Julian leap year reforms.

Both these calendars used the same Egyptians names for their months, and so the question is; which of these calendars were the basis for the days being described? When we investigate either of the possible candidate calendars, we see that Phamenoth and Pharmuthi (the 7th and 8th months) align directly with the Macedonian Greek names of Artemisius and Daesius used elsewhere in the Empire. They were also designed to roughly match the months of Martius (March) and Aprilis (April) of the Julian calendar, beginning on the 1st of January.

The Egyptians also used a fairly unique way of measuring the reigns of their pharaohs. When a new ruler rose to power and the count was restarted, if the first year of the reign of the king or pharaoh lasted for a complete year, or a considerable portion of it, there was no ambiguity. However if he / she came to power a few short weeks before the beginning of an annual calendar cycle, it was apparently common practise *to make the first full year of the regal reign, the following year* and count from there. This practise caused the first regnal year to be longer than a normal year, such that the first regnal year then included *the small fraction of the previous year* which was noted as a separate period within the year..

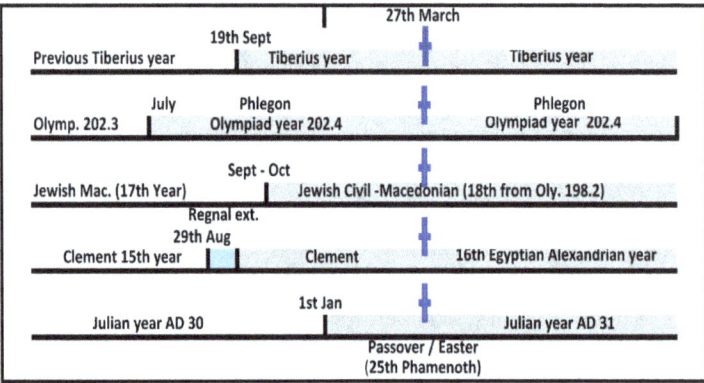

Calendar alignments for AD 31. For AD 30 and 32, subtract or add 1 year. Passover AD 30 was the 6th April and AD 32, the 14th April.

This meant that any calendar year beginning shortly after a Royal year, was treated the same as those immediately before, and therefore all the calendars were synchronised to reduce confusion. Whether an issue here or not, this Egyptian convention allowed their calendars to be locked to the same royal year.

Note: *This convention was dependent on the disposition of the following ruler and how many weeks were involved.*

Summing up: Regardless of which Egyptian calendars are used, the 1st year of the Tiberius co-regnal reign aligns with the AD 12-13 (Olympiad 197.4) and his official inauguration with AD 14-15 (Olympiad 198.2) and this has reconciled the *18th year of Eusebius* and the *16th of Clement*. It also seems that *the 16th year of Tiberius* is AD 29-30 (Olympiad 202.1) – or is it?

This depends on the conventions used, counting from the Tiberius official ascension date, and I have come to suspect that they were different in Egypt and elsewhere in Asia in the 1st

Ancient time cycles applied to the life of Christ

	Callippic Cycle 4	Callippic Cycle 5 (from 330 BC)			Callippic Cycle 6 (fr 330 BC)		
	⇐	Tri-Metonic Cycle 1			Tri-Metonic Cycle 2		
Metonic Cycles		1 / 40-41	2 / 41-42	3 / 42-43	4 / 43-44	5 / 44-45	(from 26 BC) (from 776 BC)
	BC	BC	BC-AD	AD	AD	AD	AD
O L Y M P I C	45-44	26-25	7-6 1	13-14 1	32-33 4	51-52 7	70-71
	44-43	25-24	6-5 9	14-15 9	33-34 2	52-53	71-72
	43-42	24-23	5-4 3	15-16 8	34-35 0	53-54 2	72-73
	42-41	23-22	4-3 1	16-17	35-36 3	54-55 0	73-74
	41-40	22-21	3-2 9	17-18 1	36-37	55-56 8	74-75
	40-39	21-20	2-1 4	18-19 9	37-38 2	56-57	75-76
	39-38	20-19	BC 1-1	19-20 9	38-39 0	57-58 2	76-77
	38-37	19-18	AD 1-2 1	20-21	39-40 4	58-59 0	77-78
	37-36	18-17	2-3 9	21-22 2	40-41	59-60 9	78-79
	36-35	17-16	3-4 5	22-23 0	41-42 1	60-61	79-80
	35-34 40	16-15 41	4-5 42	23-24 0	42-43 44	61-62 45	80-81 46
Y E A R S	34-33	15-14 1	5-6 1	24-25	43-44 3	62-63	81-82
	33-32	14-13 9	6-7 9	25-26 0	44-45 4	63-64	82-83
	32-31	13-12 1	7-8 6	26-27 0	45-46 2	64-65	83-84
	31-30	12-11	8-9	27-28 1	46-47 0	65-66	84-85
	30-29	11-10 1	9-10 1	28-29	47-48 6	66-67	85-86
	29-28	10-9 9	10-11 9	29-30 1	48-49	67-68	86-87
	28-27	9-8 2	11-12 7	30-31 2	49-50 2	68-69	87-88
	27-26	8-7	12-13	31-32 3	50-51 0	69-70	88-89

	Callippic Cycle 30	Callippic Cycle 31 (from 330 BC)			Callippic Cycle 32 (330 BC)		
	⇐	Tri-Metonic Cycle 35			Tri-Metonic Cycle 37		⇒
Metonic Cycles		105	106	107	108	109	(from 26 BC)
	AD	AD	AD	AD	AD	AD	AD
O L Y M P I C	1932-33	1951-52	1970-71 6	1989-90	2008-09	2027-28 3	2046-47
	33-34	52-53	71-72 8	1990-91 6	09-10 6	2028-29 4	47-48
	34-35	53-54	72-73 7	91-92 9	2010-11 9	2029-30 7	48-49
	35-36	54-55	73-74	92-93 2	11-12 7	2030-31 0	49-50
	36-37	55-56	74-75 6	93-94	12-13	2031-32 2	2050-51
	37-38	56-57	75-76 8	94-95 6	13-14 6	32-33	51-52
	38-39	57-58	76-77 8	95-96 9	14-15 9	33-34 7	52-53
	39-40	58-59	77-78	96-97 3	15-16 8	34-35 0	53-54
	1940-41	59-60	78-79 6	97-98	16-17	35-36 3	54-55
	41-42	1960-61	79-80 8	98-99 6	17-18 6	36-37	55-56
	42-43 46	61-62 47	1980-81 9	99-100 9	18-19 9	37-38 7	56-57 52
Y E A R S	43-44	62-63	81-82	2000-01 4	19-20 9	38-39 0	57-58
	44-45	63-64	82-83 6	01-02	2020-21	39-40 4	58-59
	45-46	64-65	83-84 9	02-03 6	21-22 7	2040-41	59-60
	1946-47	65-66	84-85 0	03-04 9	22-23 0	41-42 7	2060-61
	47-48	66-67 6	85-86	04-05 5	2023-24 0	42-43 0	61-62
	48-49	1967-68 8	86-87 6	05-06 6	2024-25	43-44 5	62-63
	49-50	68-69 6	87-88 9	06-07 9	2025-26 1	44-45	63-64
	1950-51	1969-70	1988-89 1	2007-08 6	2026-27 2	2045-46	2064-65

Metonic, Tri-Metonic and Callippic Cycles for the 1st and 21st centuries, with Ancient Olympiads numbered from 776 BC.

century. This is because the northern autumnal transition from one year to the next w.r.t to solar at the winter solstice, is similar to the transition of each day, ending before midnight at dusk. It is simply another prophetic 'day', solar system period.

If this is the case, it may leave an opening for the AD 30-31 Royal (regnal), Alexandrian and Olympic years to be treated differently; such that the last year - whether the 18th year of Eusebius or 16th year of Clement - *points to* the following year (and particularly the first six months of that year) which is analogous to the night time of a Jewish 24 hour day.

But what of the 25th Phamenoth and 25th of Pharmuthi?

Lunisolar calendars are unusual, in that users can start their year at the new moon, full moon or anywhere in between. Because alternate 29 and 30 day months are used, it means that if someone has a 'thing' for a certain number, they can make it the full moon for *every* lunar month of the year. However, although the Egyptian lunisolar calendar day will align with the 15th (or 14th) of Nisan, it sadly does not single out a particular year.

Conversely, the Alexandrian solar calendar in conjunction with the Jewish Nisan 15th and Gregorian calendar very much has the potential to do so, because it is locked to the Julian year and in turn the Julian is computer calculated to be just 2 days ahead of the proleptic Gregorian for the early 1st century.

For the Olympiad 202 years, only AD 31 is close to having an Alexandrian 25th day near Nisan 15, and that is the 27th March of the Gregorian calendar. This can be verified by counting the days from the 29th August (the Alexandrian new year's day) to March 27. It turns out that both the Alexandrian and Julian are 2 days ahead of the Gregorian (they align with the 31st August) and by adding in the days of the months through to the 27th March, we have 211 days landing on a Thursday - whether we use the 30 day (plus 5 day intercalary) Alexandrian months (and common year conventions) or the Gregorian.

Very significantly, the Alexandrian day corresponding to the Gregorian 27th and Julian 29th is **the 25th of Phamenoth!**

On the prophetic side, there is modest support for AD 30 or 31 from the Meton timeline and the end of the age period. This is because there is likely to be a match of the first Advent with the second Advent to come. This will splice the Lord's ministry periods together and likely complete the 70th Daniel 'week'..

One disappointing aspect in all this is that solar and lunar eclipses have not been as useful as one might have hoped.

A solar eclipse in late AD 29 is not one of interest and the partial lunar eclipse of 25th April AD 31 (at 23:02:48 Terrestrial Dynamic (TD) time and lasting for 123 minutes) was 56 degrees east over eastern Arabia, but did not coincide with the month of Nisan. Similarly, a total lunar eclipse on Passover April 14th, AD 32 was way over in the Pacific Ocean (along longitude 137 degrees west) and therefore well out of range in the Middle East. All three therefore, can be discarded.

In any case, a regular solar eclipse could not account for the three hours of darkness; unexplained blackness that would have made an unnerving and tumultuous day, with earthquake and chaos everywhere. And there was probably enough dust and smoke in the evening air to provide a 'blood moon' where normally it would not be expected. Rather foreboding in fact, after a day of heart-rending high drama, crisis and catastrophe.

That this occurred in both Bithynia and Jerusalem with earthquakes recorded at both sites, makes it all the more remarkable and points to Olympiad 202.2 and AD 31.

Returning to Daniel's 69 weeks and 483 years; if the last year fell in AD 26-27 and Jesus' ministry period began *after* that year, it means his ministry was conducted in the years 27-28, 28-29, 29-30 and also finished on Passover AD 31. However the Clement 16th Tiberius regnal year and the Eusebius 18th nominally support an AD 30 year, despite the earthquake and Alexandrian 25th Phamenoth date for AD 31. Also, an Egyptian lunisolar year would have dates tracking the 14-15th of Nisan for April AD 30 and March AD 31, but exactly how the October decameron (new year) was implemented, is not clear.

Significantly, Justin Martyr's description of the Passover ties with the evening of Thursday, 27th March, or (as others say) with Wednesday evening. It then continues through the Friday daylight hours, Unleavened Bread and Resurrection Sunday.

AD 32 and 33 lost their shine once the *'before the fourth year'* and *'second year'* versions of Phlegon emerged.

Appendix 5

Meton and Callippus

This Appendix follows on from the chronology based on the Macedonian and Egyptian calendars and how they were applied by Greek scholars such as Phlegon of Tralles, Clement of Alexandria and Luke in his Gospel. It has been included in order to establish whether the application of the Meton and Callippus cycles is consistent with the Seven Times timeline, the Phlegon Olympiads and the chronology of the 1,335th 'day' of Daniel.

I also hoped that some further clues might emerge to confirm which of the months Phamenoth and Pharmuthi (the seventh and eighth months of the Macedonian Greek calendar in Egypt) aligned with the Gregorian full moon from the NASA data. As previously noted, these months are called Artemisius and Daesius elsewhere by the Macedonians and are aligned with the months Nisan and Iyar of the Jewish Civil calendar.

They are also roughly aligned with March and April of the solar calendars, and to this end I started researching the 19 year lunisolar cycle which was named after Meton. The Tables in the previous Appendix help to illustrate the relationships between the cycles and the Julian dates we are interested in..

Meton lived in 5th century BC Greece and was the scholar who made the cycle famous by describing the structure of the intercalary months. It is a lunisolar calendar with a 235 lunar month sequence which very closely corresponds to 19 solar years. It was probably used centuries earlier in some form or other, and it is known that before about the 8th century BC, the extra lunar months were added fairly randomly. The king and his advisors would add a lunar month here and there, when their calendar became out of whack with the seasons.

Eventually the ancient scholars found that by adding seven (7) extra lunar months over 19 years, that the lunar and solar calendars would line up almost exactly at the end of the period. Furthermore if the insertions were spaced out as uniformly as possible, the seasons were always roughly aligned each year.

A little over a century later, a fellow Athenian called Callippus devised a 76 year cycle, by adding four Meton cycles together and this was used in Egypt and elsewhere. By subtracting a day every 76 years he could also correct the slight drift of the Meton cycles. Unfortunately this did not really solve the problem, but simply reversed the drift between the luni-solar calendar and a true solar one - but in the opposite direction.

Pondering on this issue, I began to wonder what advantage there was in bothering with the Callippus cycle at all, and why it became so popular. However a couple of things clicked; firstly it occurred to me that the gentlemen in Athens were the same guys who had used an Olympiad four year system since the year dot; meaning 776 BC. In addition both the Athenian version of the Macedonian year and the Olympiad year began around the northern summer solstice. It meant that by multiplying the Meton cycle by 4, the 76 year cycle automatically includes a common denominator of 19 Olympiad periods.

It began to look very much like the attraction of a 76 year cycle was not so much about further accuracy, ***but a way of aligning the lunisolar years with the Olympiads.***

From observations by ancient historians it is known that the first Callippic cycle began in the summer of 330-329 BC during Callippus' own lifetime. It is also known that in 26-25 BC when the Alexandrian solar year was first implemented in Egypt, that it was the first year of the 5th Callippic cycle.

The 5th Callippic cycle from 26 BC is also the specific period of interest for our study. One thing I did not pick up on originally, is that the first Callippus Cycle beginning in 330 BC is 9 years into the 24th Metonic Cycle counting from 776 BC.

It therefore does not splice into the existing Metonic sequence at the beginning of a Metonic cycle as one might expect.

This is somewhat surprising and suggests that Calippus had concerns on his mind other than timing logic, and one of those may well have been Alexander the Great. It is noteworthy that in Callippus' lifetime, Alexander the Great invaded Asia and the Callippic inauguration date was four years (or one Olympiad) after Alexander did so in 334 BC. Therefore it was probably a celebration of sorts in regard to Alexander's military success.

Interestingly, Alexander has often been considered a 'type' or 'antitype' for Jesus of Nazareth. Both men died quite young and both had a huge impact, even though their modus operandi was very different. In addition, Daniel's 2,300 'day' prophecy begins from 334 BC as well, and so it means that the Metonic and Callippic cycles could be prophetically significant.

In order to optimize the comparison between Jesus' first advent and his second, I have removed the 9 year offset in the chart by making the 401st Metonic cycle beginning in 766 BC align with the start of the 5th Callipic cycle (beginning in 330 BC). I have made the 1st tri-Metonic cycle begin here too.

This is not as artificial as it may seem, given that one of the reasons for presenting this chart is to compare both advents. However the overlap should be kept in mind from the point of view of interpreting the Eusebius-Phlegon Olympiad data.

Now the Meton cycles have lunar months added in the 3rd; 5th (or 6th) 8th; 11th; 13th (or 14th) 16th (or 17th) and 19th years.

It means that when the extra month is added in year 19, that the luni-solar calendar will align with the solar calendar almost exactly. It therefore follows (for example) that the previous year (the 18th in the Meton cycle) is about two thirds of a month early, and that these insertions have to be tracked to be able to determine what is happening in any particular solar year. It was sometimes done at the end of the year, but also half way through in some jurisdictions, e.g. the Jewish Civil calendar.

This becomes another trap when trying to convert an actual day from a luni-solar calendar to a Julian-Gregorian one.

Beyond these issues, there are potentially some clues for future prophetic timing. It was noted in the main text that each 'time, times and half a time' is 1,278 years in length, but that according to Daniel 12, the second period is extended 1,290 and then beyond to 1,335 years. In total this is an extra 57 years, which just happens to be equivalent to three (3) Metonic cycles.

We know that the 50 year jubilee cycle is useful and prophetically significant, and it may well be that a 57 year one would reveal more linkages as well. The secondary 1,335 year Muhammadan period outlined in the section 'Caliphates and crusades' - with its tie to 1967 and 1922 (the 1,290th year) - could be worth exploring on the same basis. It is offset 57 years from the main timeline ending in 2024, and therefore part of the tri-Metonic scheme.

In addition, Metonic cycles are based on real alignments of the solar year and the waxing and waning of the lunar months. They are therefore genuine candidates for biblical measuring periods, in the same way as 24 hour days, 7 day weeks, lunar months, solar years, zodiacal ages and 'times'. Thus, if we use 57 year periods (a tri-Metonic cycle) it may well be a prophetically viable way of measuring through the ages.

From a biblical point of view, it is also interesting to note that the year 2026 is the last year in the 31st Callippic cycle starting from 330-329 BC, but also the last year of the 36th tri-Metonic cycle measured from 26-25 BC.

This is bound to make comparisons between advents a little more difficult, but is still an improvement on a 9 year offset.

Note also, that over the 2,000 years, the drift between the Metonic cycles and the Gregorian calendar is approximately 9 days. This is about half the drift between the Julian and Gregorian calendars and will not materially affect the study here.

Now I have already suggested that war, earthquake and other signs that occurred in the first century may be a guide to the way things will unfold in the second advent, and that a careful and prayerful study might yield further insights.

A future Passover and anniversary Ascension date in the seven year 1,335 day milestone to come, could also be related to events around AD 30, given the Lord's ministry concluded at that time. It may even mean that some things such as war and earthquake will peak and subside as early as say, the Passover of 2026 or 2027 and that these lead-in events will then be followed seamlessly by those unique to the Second Coming.

So in summary: the Metonic Table compares both the relevant Callippic and tri-Metonic cycles for both Advents. The last cycle leading up to 2027 begins in 1951 for the Callippic cycle and 1970 for the tri-Metonic, and the similarities and differences between the 1st century BC-AD and 20th-21st first century cycles become clearer. The cycles have also been shaded to make 57 year comparisons easier, with the darker shading covering the first advent and likely second advent as well.

The Metonic patterns highlight the elapsed time between the 1967 (end of the Seven Times) and the 2024 '1,335th 'day' of our Daniel timeline - as well as all other combinations, 57 years apart. In short, the exercise illustrates the correspondence between the events at the beginning of the age and those at the end, and how they are likely to dovetail together.

Given the 36th tri-Metonic cycle since 26 BC ends in 2026-27, the next becomes the 37th beginning in 2027-28. In terms of the Lord as the 'Son of Man'; the 37th tri-Metonic cycle may directly follow the 1st tri-Metonic cycle; i.e. if the dual interpretation of Daniel's 70th week outlined earlier in the text is valid.

Thirty seven is also a special number associated with the Messiah, which is recognised by many followers of scripture including the author.

Although the start dates for the cycles are based on the original ancient Greek ones and in particular with the 26-27 BC start date of the Alexandrian calendar and the 5th Callippic cycle, it should be remembered (and I need to emphasise) that the Metonic cycles and the Olympiads are now offset 9 years in the charts from the 26-27 BC year for comparison's sake. Having said that, the Olympiad numbering in the text continues seamlessly from 776 BC, in order to be faithful to the Phlegon Olympiad data that is crucial to this study.

In addition, the interesting connection mentioned earlier between the lives of Alexander (and the Daniel 8 timeline) and Jesus himself, provide a basis for a prophetic connection and may be more significant than is first apparent. It means there is a tie-up here of sorts with the Callippic cycles, and although the tri-Metonic cycles do not start at the birth of Christ, but arbitrarily in 26-27 BC, we see they neatly encompasses the whole earthly life of Jesus; including his birth in the latter years BC.

The Metonic chart provided also shows the Olympiads numbered from 776 BC and interestingly, the 700th Olympiad ends immediately before the centre of Daniel's 1,335th 'day'.

The 50th tri-Metonic cycle numbered from 776 BC (rather than 26 BC) covers the 1.335th 'day' as well, with the preceding 49th cycle finishing in 2017-18.

I will leave it to the reader to explore these and other possibilities as desired. In any event, every indication is that the Kingdom of God will come into its fullness in the not too distant future. Somewhere in those years, we will experience the most humbling events that have ever likely been seen on earth.

Heavenly visitors

In the later chapters of this work, I suggested that when the timeline concludes, there is little further prophetic revelation.

Despite the truth of that statement from a Daniel point of view, there are texts which are worth review that I have not addressed; texts that could shed further light on events of the next few years and the end of our prophetic timeline. For example it would be nice to have further information on the events of the $1,335^{th}$ 'day' milestone, including the future areal extent of the Armageddon War and end-of-the-age earthquake calamity.

The nature of the Lord's return and exactly how and when the 'rapture' of believers occurs, is another question that has polarised many analysts of Bible prophecy. Will the rapture be exactly when the Lord descends on the Mount of Olives, or at some time in the remainder of the seven years - or indeed a longer period? How will it occur and will there be a consolidation of the 'stone kingdom' at this time; a time of peace on earth where love and prosperity prevail? Or will the end of the prophetic timeline also herald the abrupt end of the age and catapult humanity directly into the final judgement?

Certainly it makes absolute sense that the rapture will take place prior to the destruction of our earthly home and a chosen portion of humanity rescued at that time.

In recent years it has become obvious that there is great unrest in western democracies; including Israel itself, Britain and the United States. Europe is also undergoing signs of stress and disunity and trade manipulation and intellectual property theft are major concerns. This is all being playing out on a background, where global warming and pollution are threatening the planet like never before and population stresses seem impossible to manage. Global conflict is therefore a worrying possibility and all in all, we are seeing conditions which beg Divine intervention. Now these are all signs of the times and perhaps we do not need further warnings than what we now see before us. Nevertheless, many are interested in further signs and here I will address some of the 'signs in the sky' suggested by some folk in a little more detail.

Signs may be any manner of startling occurrence; including man initiated space ventures such as the Apollo missions to the moon in the late 1960s and early 70s, but also other such projects to come. It has also been suggested that our canopy of satellites is becoming increasingly vulnerable to collision and that a cascade effect of 'falling stars' is a real possibility.

But there is also the likelihood of alien or angelic activity and there certainly has been unidentified (UFO) activity seen in many parts, including over Jerusalem according to social media video and other reports. Whether otherworldly, military or natural (but unknown) phenomena is the question.

Some of the unidentified phenomena is likely to herald the second coming in a similar manner to the first.

From a personal perspective, I can think of two occasions where I have seen things at night that are in this category. Once was over an Australian state capital, an hour and a half or so after sunset. The object or objects were maneuvering erratically at high altitude apparently still in direct sunlight. The display was directly overhead and restricted to a relatively small angle of arc. It lasted about 20 minutes.

Beyond unidentified activity, the world of meteors, asteroids, comets and other natural phenomena are of interest, as is the embedded meaning of specific constellations and the passage of the planets and their conjunctions. In short, it has come to my attention that there are many looking for confirmation of the end of the age through various astronomical observations and heavenly interactions.

It seems a valid enough interest, given a currently popular interpretation of the Bethlehem 'star' which appeared one or more years after the birth of Jesus. Some believe the 'star' was an unusual conjunction of Jupiter (the king planet) Regulus (the king star) and Venus on a background of the Leo constellation.

I will not go into the details here, but interesting descriptions can be found in various articles.

In relation to Jupiter and the recent past, some rather extraordinary claims have been made concerning a planetary passage as recently as 23rd September 2017. It is being touted as significant and described as a 'virgin birth' sequence with the constellation Virgo in the background, the moon at the feet of the Virgin and stars (mainly from Leo) around about.

The scene presented is reminiscent of the picture in Revelation 12 of Israel. There is an assertion doing the rounds, that computer time analysis shows this is only the second time in 7,000 years that this phenomenon has occurred.

I have also seen commentary based on alternate data analysis, that suggests the event is much more common.

I am not sure what to make of all this, but I note it is usually described in the context of futurism and a future birth or arrival, with no acknowledgment that most of this image has already been fulfilled in the birth of Jesus over two thousand years ago. Despite this, the date is seven years before the second Jewish feast season of 2024 and therefore cannot be dismissed entirely. It could be a pointer to that year, with Leo standing for the 'Lion of Judah' and the 'moon under the feet' a final warning to the beast empire that its demise is at hand.

I have no conviction one way or another on this, but it is certainly interesting to some people.

It may well be there are also other astral conjunctions and images set against the zodiacs that have been missed. Reviewing the last several decades of planetary conjunctions and the positions of the moon from around 1967 and the end of the Seven Times onward (including the 2,000th anniversary of Jesus' birth) could also be illuminating.

Articles and videos can also be found on the subject of comets, asteroids, 'rogue planets' and lunar and solar eclipses.

In regard to comets as heralds of impending events, one particular comet seems to be of interest because its next perihelion (approach to the sun) is in 2024. Here is something that ties into our biblical timeline, but will be quite distant to the earth and on the other side of the sun on this occasion.

A fairly bright magnitude 5 transit is expected, and so if it flares into a display that can be easily seen with the naked eye, and say, sports a strong tail (or multiple tails) it could be acknowledged as a marker, perhaps approaching the significance of a Great Comet.

The comet in question is Comet 12P/Pons-Brooks, named after Jean-Louis Pons and William Robert Brooks who discovered it in the 19th century. It was also seen on earlier occasions and may have split into two or more comets at one point. If so, this makes it unlikely that extrapolating earlier visits is possible based on the 70.85 year cycle currently attributed to this comet. Notwithstanding this and other uncertainties, there are some claims that it may have appeared during the reign of Thutmose III and therefore marked the closing of the Hebrew Exodus, 1462 – 1456 BC, or coincide with the Thera eruption.

As interesting as this possibility will be to some, there are inconsistencies with some of the arguments being made by popular commentators about this comet. Unknowns in regard to its history in ancient times remain and the provenance of one of the supposedly Egyptian texts (the Tulli Papyrus) is very much in question. It is quite likely a forged text, that suggests a celestial event occurred in Thutmose III's 22nd regnal year.

Regardless of past appearances, it is nevertheless something to watch out for in 2024-2025, with the possibility that further historic evidence may be forthcoming in regard to earlier events. In fact, if a 'Great Comet' appearance was a marker, there is no compelling reason why it should be the same comet for both the Exodus and the end of the current age. 'Neat' one might say, but not essential.

On the negative side, there appears to be no definite mention of a comet in scripture.

Nevertheless, when the wise men visited at the Messiah's birth there may have been more than the conjunction of planets and a star. As distinctive as the multiple conjunctions over a claimed 9 months may have been; perhaps a comet, supernova or other unique spectacle also appeared at the time.

In support of this idea, there is a Chinese account of a comet appearance in March 5 BC (the second month of the second year) which was taken to signify the beginning of a new epoch. It apparently comes from Han Dynasty astronomical records. There is also a record ca. AD 31 (Emperor Guang Wu, 7th year), which potentially links Jesus' death and resurrection with an edict about sins being switched to one man.

Suffice to say, I have not had time to investigate these records, to determine whether the comments attributed are plausible, or whether some of the commentary may have been added subsequently. For example the story of Christ in China might have emerged in retrospect, via arriving travellers on the Silk Road. Chinese astronomers might then have checked their records, to see if there was anything significant at the time.

In regard to an impact event capable of initiating a massive earthquake sequence via an asteroid, comet, or other astronomical body, there seems to be no known object that is likely to crash into the earth in the 2024-2030 year period.

One large asteroid (Apophis) is calculated to pass within 18,600 miles (29,940 kilometres) of the earth in 2029, and this distance is much closer than the moon's orbit of 238,900 miles (384,570 kilometres). However astronomers consider it to be low risk on this pass compared to a random impact by the many postulated unknown objects in the solar system. So if the apocalyptic biblical earthquake is to be initiated by a bolide strike of some sort, it is almost certainly yet to be spotted and identified.

On the other hand, massive volcanic and earthquake events have occurred in the past, which have not required an extra jolt from space. Plate movement and collisions, rift formation, deep seated hotspots and huge basalt lava flows have all occurred before, and in some cases are likely to have been driven by inner convection currents alone.

The earth is very much alive and a moving beast inside.

In conclusion, this summary is an open ended introduction to the subject matter and unfortunately more questions than answers emerge. I expect the material will be mainly of interest to readers who are keen on undertaking some research of their own, or already have some useful insights.

Appendix 6

Exodus revisited

Given the fairly comprehensive nature of this prophetic study so far, I am somewhat reluctant to add to it now. In fact over 15 months since I learned of the identity of the Whore City of Revelation in late 2017. However there has been a need to improve the flow of the text and review the accuracy of the historic material. There have also been difficulties with publication and some loose ends to tidy up and clarify.

My hope is that in the end, it is all in line with the Lord's perfect timing in regard to the release of the material and the unfolding of future events. If it was released too early, some might be tempted to twist the information to justify their actions and thwart the prophecies.

This brings me to another issue. Since my earlier study of the contemporary dating of the Thera (Santorini) eruption, yet another study combining Carbon 14 data and tree ring analysis has been concluded. The original dating centred on 1627 BC, but was followed by another study suggesting a 1613 BC date based on a different tree ring dataset from North America. Now a third group, led by the University of Arizona in collaboration with University of Sheffield scientists in England, have conducted another research effort and their data analysis suggests that the eruption occurred between 1600 BC and 1525 BC.

This throws something of a spanner in the works in regard to my analysis in the chapter entitled The Exodus Cataclysm.

Perhaps my celebration over a schooner was a little premature. The University of Sheffield is one of the research organisations involved in fine tuning the C_{14} calibration curve and if their results prove to be a further improvement in accuracy, then their research obviously becomes difficult to ignore.

The span of seven and a half decades suggested by their work is surprisingly large, but may be a hint to the way things played out. If correct, it brings the eruption event some decades closer to the biblically computed date and it also ties in more closely with the reign of Ahmose I and the Ahmose Tempest Stele description of the disaster.

In the early to mid-16th century BC we also have the Flavius Josephus' dating of 612 or 592 years before the 4th year of Solomon's reign which comes in at 1582-1580 BC or 1562-1560 BC respectively.

So as unsettling as it is to have a new and unexpected dataset and interpretation, there could be a better alignment emerging between modern dating science and Egyptian archaeology; meaning the Amu / Hyksos invasion record, the Ahmose inscriptions and the written historical records.

At this point in time I have not seen the data, or the reasoning behind this new result.

In terms of the C_{14} analysis, I made the point earlier in the text that if the tested branches (house timbers or short lived carbonaceous material) were contaminated by any carbon produced from the initial eruption, that this could make any C_{14} determinations look older than they actually are.

So this would also make sense. Nevertheless I am now a little perplexed and concerned with the tree ring correlations of all three studies, given I expected that a sequence over one to several decades would be unique enough to be an unambiguous match. That now seems not the case, and the hope is that further analysis of existing and new data can clear up this mystery.

On the positive side, the results of the new analysis may reflect the protracted unfolding of all four stages outlined in the geological assessment. Perhaps the initial stage of the volcanic eruption occurred in the late 17th to early 16th centuries after all, but the later stages continued throughout the 16th century BC. The last massive explosion would then likely be mid century in the time of Ahmose, or a little later in the second half.

Despite the romance of the Thutmose III and Hatshepsut era, my preferred scenario is that the refugee aspect of the Exodus saga then continued until 1462 BC (430 years after Abraham left Harran) and concluded in the co-regnal period..

Appendix 7

Apocalypse vow

When we look up at the stars at night and see them in all their wonder, are we looking at a heavenly illusion? Is it a universe that once was, but now has nearly disappeared?

> I woke up in a dream like state after falling asleep early in the evening of March 10th 2012 (if the date on my computer was set correctly). I must have done a hard day's work and went off for a quick snooze.
>
> This was not to be an ordinary dream experience. In fact it was an 'in the Spirit one' usually described as the anointing or baptism of the Holy Spirit.
>
> I am confident of that, because in my late twenties and thirties I had this experience a number of times. The first time I sought it, but thereafter it happened when I was attending healing meetings to pray for others with critical health problems. Once it also occurred at a mid-week worship / prayer meeting.

My requests for healing were not always answered in the affirmative and disappointing in so far as that was concerned, but each time I had this incredible personal blessing.

On one occasion I remember after tumbling to the floor, being in a euphoric state for many minutes. Time doesn't seem to matter under these circumstances and you seem to be enveloped by the actual presence of the Lord.

Well the experience I am about to describe was like that, but it seemed to last considerably longer and it was accompanied by a message. I feel I have been inspired a number of times after wakening, including some of the numbers and timelines described in Daniel's prophecies in this volume. But this particular episode was quite different, in that the vision-like experience was not about racing to a calculator to process numbers in a new way, but rather out-of-the-blue propositions came out of a trance like state; such was the intensity and indelibility of the ideas at the time.

It is the nearest I have ever been to feeling like a conduit to some external source of new knowledge. I was still in this otherworldly 'state' when I went to the first word processor application at hand, which was to switch on Facebook and start tapping away. It lasted well over an hour.

After I'd gone back to bed and rose the following morning still more propositions seemed to tumble out, like the gushing of water out the end of a pipe.

There was some hilarity around several states of Australia and the United States as my (mostly not particularly religious) Facebook friends enquired as to my sanity. In part, the episode was addressing concepts I already knew about, but the 'propositions' were a sort of interpretation of the visionary experience.

I will not go into the details here, but summarise the experience with the main lesson.

Properly, I leave it to others and history to show whether this proves to be a true revelation from God (or his angelic host). Indeed I am recounting the experience with considerable trepidation and it is why I have left it to last. I do not want to detract from the firm findings of Daniel and John.

As an engineering and applied science focussed individual, I had always had an interest in extra-terrestrial and astronomical things and was aware of Einstein's Special and General Theories of Relativity since my youth. How these theories revolutionised our thinking and understanding alongside such things as 'quantum mechanics', nuclear theory and other associated esoteric matters.

What I saw and felt during this experience, was simply that I was being taken on a journey through those mainly 20th century findings at the very basic level I understood them.

In particular, that the correct interpretation of current cosmological observations such as dark energy (and perhaps some dark matter) and the high recessional velocities we see at the periphery of the known universe, may not be rightly understood. If our current model of the universe is incorrect, then the future could be unimaginably different to the expanding model we have been taught.

With that in mind, only a relatively simple vision was required to send a message.

> I remember being asked to look vertically up at the heavens and being told, that I was looking at the centre of the universe.
>
> As soon as I did this (albeit in my mind's eye) it was then suggested that no matter where I went on the earth; back to Egypt, Turkey, Israel, Britain, the Philippines, Antarctica or the U.S., that I would be looking at the centre of the universe there as well.

This of course was inexplicable, but I got a sense of the multi-dimensional nature of the universe and that what is observable is only a part of the story. It was also apparent that the current state of affairs was not going to last and that things will undergo a dramatic change in our region of space; a change totally inconsistent with the ever expanding mantra we have been lead to believe.

To set the scene in my own mind, I started to tap away with the following: 'The 'red shift' first observed by Hubble and studied by Lemaitre and others, shows that galaxies are receding ever faster with distance away in space. This is interpreted as an expanding universe. But is it? What if our visualisation is inadequate and our expanding model too poor / incomplete?'

'The Ptolemaic (earth centred) view of the solar system was accepted for over a thousand years and is mathematically correct. And yet the Copernican model of heliocentricity (sun-centeredness) is much easier to understand and is a much simpler model to demonstrate. Furthermore cosmology is saddled with so called 'dark energy / matter', much of which is yet to be identified. There are also extreme recessional velocities noted for galaxies on the fringe of the visible universe, which are yet to be explained. I smell a cosmological rat.'

I get a 'like' from a polite American friend. (The thing was that being in this otherworldly state I had virtually no self-consciousness about recounting what I felt and experienced).

I will not go through a blow by blow description of the next hour or so on Facebook, except to say that there were polite responses mingled with a 'What the…. are you talking about?' and considerable jocularity in others.

Basic, but important issues like "In a multidimensional universe the red shift shows only that galaxies are moving apart (or moving apart relative to one another)" were posited by 'yours truly' to emphasise a point. And central to the vision experience "No matter where you are on the earth, if you look directly upwards you are looking toward the centre of the universe".

This wasn't me thinking, there was no actual voice; but I *was* directed to look vertically upward nonetheless.

A South Australian friend added: *'Whoa.... you're messin' with my head!!'* (Ah, someone is alert).

A Newcastle friend (probably recovering from an episode of 'Home and Away'): *'You lost me at physics... Night, night xxx'.*

Newcastle friend still unable to completely disengage and probably still digesting the consequences of that night's 'H and A' episode': *'Lay OFF the weed!'*

My South Australian and American friends continue the banter to lighten the conversation...

The following morning I was still in a state of heightened awareness, but obviously no longer completely in the in-between state of mind. Eventually my nerdy side took over and I attempted to make sense of it all.

I came to the conclusion that what I had been shown, was that the multidimensional universe is collapsing outward under its own gravity, in an inherently non-intuitive manner.

It meant that the gravitational centre of the universe is way out there somewhere and that there are such enormous gravitational forces being exerted (and consequent distortions of space) that pointing to the sky anywhere, can be said to be pointing to the centre of the cosmos.

After further thought, I decided it was somewhat analogous to the Ptolemaic / Copernicus visualisation of the solar system. Our earth centric understanding of the solar system was valid, but not the best model for understanding the overall geometry.

In the same way, it seemed that our earth centric understanding of the cosmos is not the best way to appreciate the overall nature of the universe - especially as we cannot see it in its entirety.

So somehow as things began to gel with my layman's knowledge of cosmology, I understood from all this:

- The extraordinary away (recessional) velocities observable at the extremities of the universe are due to the (outward) collapse of the universe under its own gravity
- Dark energy / matter in the quantities sought to provide a satisfactory model of the observable universe, may be satisfied by the presence of an extraordinarily large black hole or antimatter equivalent
- This humungous central black hole which is swallowing all matter and energy, is at the same time now so close as to distort any sense of a 3D and perhaps a 4D reality
- Significantly, what we can see at the extremities of the known universe today, occurred so many billions of years ago, as to be meaningless as contemporary news
- This primeval black hole or antimatter system is also so large and nearby, as to effectively surround us and may have been created at the Big Bang

Einstein stated that the gravitational pull of the universe is such that if we sent a beam of light out in one direction, it would eventually return to us from the opposite direction. It effectively makes space curved. If that is true, this could account for the proposition that we can look outward from any point on the earth's surface and see the centre of the universe.

Now nothing like this had ever happened to me before, except for my 'anointing' experiences. I knew toward the end I was stumbling; trying to interpret something impossible to visualise. So what is the take home message here?

I may have over analysed the experience toward the end, but essentially what I understood was that the universe as it currently exists, is not what we can see in the sky.

This particularly applies the further we probe away from our earth observation post. In fact when we are talking about extraordinary away velocities at the edge of the universe, it is too easy to forget we are looking at, and talking about something that apparently happened in excess of 12 billion years ago.

Contemporary cosmology assumes that the density of matter throughout our universe is more or less uniform. It seems eminently plausible and the place to start when building a model.

However, if there has been a humungous, universal black hole out there from the beginning (let's call it the H-UB) which will one day swallow us and tear us apart - then we are in unimaginable trouble and we have no apparent warning. In fact the scenario painted here may be hinted at in scripture.

Isaiah 34:4 is an example:

'All the stars in the sky will be dissolved and the heavens rolled up like a scroll: all the starry host will fall like withered leaves from the vine, like shrivelled figs from the fig tree.' (NIV)

It could be that our universe was always going to collapse one day, and that if the most distant galaxies are billions of light years away (and we are looking so far back into history) that there is no possible way, we can know what is transpiring in those faraway corners of the cosmos. Those galaxies may have already been destroyed by the 'abyss' as described in scripture.

Not only will we have no warning, but the more distant parts of the visible universe (or even most of what we can see in the night sky) may not now exist. This collapse or 'fall' could be the physical reality of the moral 'fall' described in scripture.

Perhaps gravitational wave theory, supported by the most recent state-of-the-art astronomical observations and experimental data from large atom smashing experiments, could shed light on what is 'beyond the curtain'. Until then, and given the scenario described above, much of the universe may have already been swallowed up and 'rolled up like a scroll'.

PS. This rather bewildering experience just emphasises what Jesus and the prophets have already said; that the heavens would one day wind up, and most likely unexpectedly.

So in hindsight, it got me thinking that there was probably another reason I was blessed with this episode; it allowed me to empathise with the prophets that I was meant to write about in a much more personal way. It helped me to understand to a small degree what they went through, and how this transforming information arrived in the hands of humankind.

Bibliography

Bennet, Chris., *Egyptian Dates,* www.instonebrewer.com/TynedaleSites/Egypt/ptolemies/chron/egyptian/chron_eg_cal.htm 2011

Best, Myron G., *Igneous and Metamorphic Petrology,* W. H. Freeman and Company, USA 1982

Boggs, Sam Jr., *Principles of Sedimentology and Stratigraphy,* Merrill Publishing Company (Bell & Howell), Columbus, Ohio USA 1987

Brown, Francis, with Driver, S.R., Briggs, A., *The Brown-Driver-Briggs Hebrew and English Lexicon,* Hendrickson Publishers, Peabody, Massachusetts, USA 2012

Bullinger, E. W., *The Witness of the Stars*, Cosimo Inc New York, NY. Originally published in 1893

Carlisle, Richard, *Manual of Freemasonry,* William Reeves, London, Great Britain 1845

Collier, Mark, and Manley, Bill, *How to Read Egyptian Hieroglyphs,* University of California Press, 1998

Collins, Steven M., *The 'Lost' Ten Tribes of Israel...Found!,* CPA Books, Boring Oregon, USA 1992

De Sélincourt, Aubrey (translator), *Herodotus: The Histories,* Penguin Classics, Penguin Books Ltd., 80 Strand, London, England 1972

Dozeman, Thomas B., *Commentary on Exodus,* William B. Eerdmans Publishing Company, Grand Rapids, Michigan USA / Cambridge, UK 2009

George, Andrew (translator), *The Epic of Gilgamesh,* Penguin Group, Penguin Books Ltd., 80 Strand, London, England 1999.

Goard, Pascoe (Rev.), *The Documents of Daniel,* The Covenant Publishing Co., London Great Britain 1940

Guinness, H. Grattan (Dr. & Mrs.), *Light for the Last Days,* Morgan & Scott Ltd., London 1918

Haberman, Frederick, *Tracing our Ancestors,* The Covenant Publishing Co., London, UK 1962 (first edition 1934)

Heath, Alban (Rev.), *The Prophecies of Daniel in the Light of History,* The Covenant Publishing Co., London, Great Britain 1941

Hoffmeier, James K., *The Archaeology of the Bible,* Lion Hudson plc, Wilkinson House, Jordan Hill Road, Oxford, England 2008

Keller, Werner, *The Bible as History,* Hodder & Stoughton, Great Britain 1956

Knight, Christopher & Lomas, Robert, The *Second Messiah,* Century, The Random House Group Limited, London SW1V 2SA, UK 1997

Lennox, John C., *Against the Flow*, Monarch Books (Lion Hudson plc), Oxford, UK 2015

Magnusson, Magnus, *BC The Archaeology of the Bible Lands,* Book Club Associates by arrangement with The Bodley Head Ltd and the BBC 1977

McFall, Leslie, *A Translation Guide to the Chronological Data in Kings and Chronicles,* (Former Fellow, Tyndale House), mainframe.pdf, Cambridge, England 1991

National Aeronautics and Space Administration, *Five Millennium Catalogue of Lunar Eclipses,* https://eclipse.gsfc.nasa.gov/LEcat5/LE0001-0100.html

National Aeronautics and Space Administration, *Five Millennium Catalogue of Solar Eclipses,* https://eclipse.gsfc.nasa.gov/SEcat5/SE0001-0100.html

Phillips, Graham, *Act of God,* Pan Books, Pan Macmillan Ltd., London UK 2002

Phillips, Graham, *The Moses Legacy,* Pan Books, Pan Macmillan Ltd., London UK 2002

Rohl, David, *The Lords of Avaris,* Century, The Random House Group Limited, London SW1V 2SA, UK 2007

Ryan, William, & Pitman, Walter, *Noah's Flood,* Touchstone, Simon & Schuster, Inc. Rockefeller Center, New York, NY 2000

Saddler, Gilbert C., *Omens of the Age,* Destiny Publishers of South Africa, Johannesburg, SA 1948

Seibold, E. & Berger, W.H., *The Sea Floor; An Introduction to Marine Geology,* Springer-Verlag, Berlin, Germany 1993

Strong, James, *The Exhaustive Concordance of the Bible,* MacDonald Publishing Company, McLean, Virginia, USA, originally published 1890

Sullivan, Walter, *Continents in Motion,* McGraw-Hill Book Company, United States of America 1974

Telford, W.M., Geldart, L.P., Sheriff, R.E., Keys, D.A., *Applied Geophysics,* Cambridge University Press, Cambridge, UK 1976

Theodossiou, E., Danezis, E., Manimanis, V.N. and Grammenos, Th., *Ancient Macedonia and its Calendars,* http://nancybiska.com/ancient-macedonaia-and-its-calendars/

Thiele, Edwin R., *The Mysterious Numbers of the Hebrew Kings,* Kregel Publications, 1994, originally published 1951

Warner, Wayne E., *Kathryn Kuhlman, The Woman Behind the Miracles,* Vine Books, Servant Publications, Ann Arbor, Michigan, USA 1993

Whiston, William (translator), *The Works of Josephus,* Hendrickson Publishers, Inc., Peabody, MA, USA 1987

Whitcomb, John C., & Morris, Henry M., *The Genesis Flood,* The Presbyterian and Reformed Publishing Company, Nutley, NJ 1961

Wilson, Ian, *The Exodus Enigma,* Guild Publishing, George Weidenfeld and Nicolson Ltd., London, Great Britain 1986

Woolley, Sir Leonard, *Ur of the Chaldees,* Penguin Books Ltd., 80 Strand, London, England 1954.

As is common in this 21st century, a variety of online material from many web sites was consulted to expedite research into primary sources and to cross check information and opinion on ancient, and often scarce and ambiguous historical data. It includes encyclopaedia sites such as Wikipedia and Britannica and a plethora of other secular and Christian apologetic sources too numerous to itemise.

In regards to the dating of the Crucifixion of Jesus of Nazareth in Appendix 4, such sites as whatistruthbook.com, deadseaquake.info, Christian-apologist.com and bethlehemstar.com added valuable information. Other references are included in the bibliography above.

The usefulness of this online material cannot be overrated and is gratefully acknowledged by the author.

www.ingramcontent.com/pod-product-compliance
Lightning Source LLC
Chambersburg PA
CBHW070713020526
44107CB00078B/2357